Communications
in Computer and Information Science   **1732**

Rationale

The CCIS series is devoted to the publication of proceedings of computer science conferences. Its aim is to efficiently disseminate original research results in informatics in printed and electronic form. While the focus is on publication of peer-reviewed full papers presenting mature work, inclusion of reviewed short papers reporting on work in progress is welcome, too. Besides globally relevant meetings with internationally representative program committees guaranteeing a strict peer-reviewing and paper selection process, conferences run by societies or of high regional or national relevance are also considered for publication.

Topics

The topical scope of CCIS spans the entire spectrum of informatics ranging from foundational topics in the theory of computing to information and communications science and technology and a broad variety of interdisciplinary application fields.

Information for Volume Editors and Authors

Publication in CCIS is free of charge. No royalties are paid, however, we offer registered conference participants temporary free access to the online version of the conference proceedings on SpringerLink (http://link.springer.com) by means of an http referrer from the conference website and/or a number of complimentary printed copies, as specified in the official acceptance email of the event.

CCIS proceedings can be published in time for distribution at conferences or as post-proceedings, and delivered in the form of printed books and/or electronically as USBs and/or e-content licenses for accessing proceedings at SpringerLink. Furthermore, CCIS proceedings are included in the CCIS electronic book series hosted in the SpringerLink digital library at http://link.springer.com/bookseries/7899. Conferences publishing in CCIS are allowed to use Online Conference Service (OCS) for managing the whole proceedings lifecycle (from submission and reviewing to preparing for publication) free of charge.

Publication process

The language of publication is exclusively English. Authors publishing in CCIS have to sign the Springer CCIS copyright transfer form, however, they are free to use their material published in CCIS for substantially changed, more elaborate subsequent publications elsewhere. For the preparation of the camera-ready papers/files, authors have to strictly adhere to the Springer CCIS Authors' Instructions and are strongly encouraged to use the CCIS LaTeX style files or templates.

Abstracting/Indexing

CCIS is abstracted/indexed in DBLP, Google Scholar, EI-Compendex, Mathematical Reviews, SCImago, Scopus. CCIS volumes are also submitted for the inclusion in ISI Proceedings.

How to start

To start the evaluation of your proposal for inclusion in the CCIS series, please send an e-mail to ccis@springer.com.

Fuchun Sun · Jianmin Li · Huaping Liu ·
Zhongyi Chu
Editors

Cognitive Computation and Systems

First International Conference, ICCCS 2022
Beijing, China, December 17–18, 2022
Revised Selected Papers

Editors
Fuchun Sun
Tsinghua University
Beijing, China

Huaping Liu
Tsinghua University
Beijing, China

Jianmin Li
Tsinghua University
Beijing, China

Zhongyi Chu
Beihang University
Beijing, China

ISSN 1865-0929   ISSN 1865-0937 (electronic)
Communications in Computer and Information Science
ISBN 978-981-99-2788-3   ISBN 978-981-99-2789-0 (eBook)
https://doi.org/10.1007/978-981-99-2789-0

This Springer imprint is published by the registered company Springer Nature Singapore Pte Ltd.
The registered company address is: 152 Beach Road, #21-01/04 Gateway East, Singapore 189721, Singapore

Preface

This volume contains the papers from the First International Conference on Cognitive Computation and Systems (ICCCS 2022), which was held in Beijing, China, during December 17–18, 2022. ICCCS is a newly established international conference on the related fields of cognitive computing and systems, which has been initiated by the Technical Committee on Cognitive Computing and Systems of the Chinese Association of Automation.

ICCCS 2022 was hosted by the Chinese Association of Automation and the Chinese Association for Artificial Intelligence. It was organized by the Technical Committee on Cognitive Computing and Systems of the Chinese Association of Automation, the Cognitive Systems and Information Processing Society of the Chinese Association for Artificial Intelligence, and Tsinghua University.

Cognitive computing is an interdisciplinary field of research that focuses on developing intelligent systems that can process and analyze large amounts of complex data in a way that mimics human cognition. It is based on the idea that intelligent systems can be designed to learn, reason, and interact with humans in a more natural way, and that these abilities can be used to solve complex problems that would otherwise be difficult or impossible for humans to solve alone. By enabling agents to understand and process data in a more human-like way, cognitive computing can help organizations make better decisions, improve customer service, develop more effective treatments for diseases, and even predict and prevent natural disasters.

ICCCS aims to bring together experts from different expertise areas to discuss the state of the art in cognitive computing and intelligent systems, and to present new research results and perspectives on its future development. It is an opportunity to promote research and development in cognitive computing and systems, provide a face-to-face communication platform for scholars, engineers, teachers, and students engaged in cognitive computing and systems or related fields, and promote the application of artificial intelligence technology in industry.

ICCCS 2022 received 75 submissions, all of which were written in English. After a thorough peer reviewing process, 31 papers were selected for presentation as full papers, resulting in an approximate acceptance rate of 41%. The accepted papers addressed challenging issues in various aspects of cognitive systems, including not only theories and algorithms in machine learning, computer vision, decision making, etc., but also systems and applications in autonomous vehicles, computer games, intelligent robots etc.

We would like to thank the members of the Advisory Committee for their guidance, and the members of the Program Committee and additional reviewers for reviewing the papers. We would also like to thank all the speakers, authors, as well as the participants for their great contributions that made the first ICCCS successful. Finally, we thank Springer for their trust and for publishing the proceedings of ICCCS 2022.

December 2022

Fuchun Sun
Jianmin Li
Huaping Liu
Zhongyi Chu

Organization

Conference Committee

Honorary Chairs

Bo Zhang	Tsinghua University, China
Nanning Zheng	Xi'an Jiaotong University, China
Deyi Li	CAAI, China

Advisory Committee Chairs

Qionghai Dai	Tsinghua University, China
Lin Chen	Institute of Biophysics, CAS, China
Hong Qiao	Institute of Automation, CAS, China
Jifu Ren	Dedao University, China
Yaochu Jin	Bielefeld University, Germany & University of Surrey, UK

General Chairs

Fuchun Sun	Tsinghua University, China
Jianwei Zhang	University of Hamburg, Germany
Angelo Cangosi	University of Manchester, UK

Program Committee Chairs

Dewen Hu	NUDT, China
Stefan Wermter	University of Hamburg, Germany
Zengguang Hou	Institute of Automation, CAS, China

Organizing Committee Chairs

Huaping Liu	Tsinghua University, China
Guang-Bin Huang	NTU, Singapore
Hong Cheng	Chinese University of Hong Kong, China

Plenary Sessions Chairs

Wenqiang Zhang	Fudan University, China
Chenglin Wen	Guangdong University of Petrochemical Technology, China

Special Sessions Chairs

Fei Song	Science China, China
Yixu Song	Tsinghua University, China

Publications Chairs

Wei Li	Beihang University, China
Quanbo Ge	Tongji University, China

Publicity Chairs

Jianmin Li	Tsinghua University, China
Bin Fang	Tsinghua University, China

Finance Chair

Qianyi Sun	Tsinghua University, China

Registration Chair

Qingjie Zhao	Beijing Institute of Technology, China

Local Arrangements Chair

Zhongyi Chu	Beihang University, China

Electronic Review Chair

Xiaolin Hu	Tsinghua University, China

Contents

Decision Making and Cognitive Computation

Robot and Autonomous Vehicle

Computer Vision

U-YOLO: Improved YOLOv5 for Small Object Detection on UAV-Captured Images

Guowei Zhang, Xingyu Chen, Xun Tan, Jiahao Zhang, and Xuguang Lan[✉]

Institute of Artificial Intelligence and Robotics, Xi'an Jiaotong University,
Xi'an, China
{zgw1119,tanxun31,sakfzr}@stu.xjtu.edu.cn, xglan@mail.xjtu.edu.cn

Abstract. Small object detection on drone-captured images is a recently popular and challenging task. From the drone's perspective, the object scale varies significantly, and tiny objects lack distinguishable appearance information in complex backgrounds, which leads to undesired effects. To solve the issue mentioned above, we propose U-YOLO based on the original YOLOv5 model. We first extend the multi-scale feature fusion network and add a detection head for tiny objects. Secondly, we integrate the convolutional block attention model (CBAM) in the detection head to focus on the critical region of the feature map. Lastly, an attention feature fusion module based on contextual information is designed to combine local and global contextual details of small objects and attention mechanisms to enhance the multi-scale feature fusion capability. Experiments on dataset VisDrone2021-DET show that U-YOLO not only improves the detection performance on drone-captured scenarios but also has a real-time detection speed. Compared to baseline model (YOLOv5s), the mAP result of U-YOLO is increased from 19.07 to 24.1%, and the detection speed is 39FPS. It provides a good balance between detection accuracy and speed, promoting the progress of small object detection algorithms on UAV platforms.

Keywords: Object detection · Small object detection · Unmanned aerial vehicle · Visual attention

1 Introduction

Object detection algorithms need to retrieve a specific class of targets in the image and locate them accurately. Recent years have witnessed deep convolutional neural network based detectors [17,21,23,26,29] obtain encouraging detection results on some notable benchmark datasets such as MS COCO [18] and PASCAL VOC [8]. However, images or videos in these datasets are taken from a close-range horizontal perspective, where medium and large objects predominate. Directly applying previous models to deal with object detection tasks in drone-captured scenarios is not suitable and often fails to achieve the expected results.

F. Sun et al. (Eds.): ICCCS 2022, CCIS 1732, pp. 3–15, 2023.
https://doi.org/10.1007/978-981-99-2789-0_1

In Fig. 1, some images captured by drones intuitively illustrate the reason: First, different flight altitudes and observation perspectives result in significant target scale variations, and small objects lack of the appearance information to distinguish them from other instances or backgrounds. Second, drone-captured images contain many densely distributed targets, which causes overlap and severe occlusion between them. Third, the background information around the objects is complex and confusing, increasing the difficulty of classifying and locating. According to the three problems mentioned above, small object detection in complex drone-captured images is an indispensable and challenging task in the computer vision research field.

Fig. 1. Cases to illustrate the three problems in detection on drone-captured images. First row shows the scale variation. Second row shows the dense distribution of objects and confusion background on drone-captured images respectively

At present, most small object detection algorithms [3,4,6,9,28,35] are generally improved and optimized based on the conventional detection method designed for natural scene images. YOLO series [1,23–25] can well balance the detection accuracy and speed, and have been widely used in some low delay application scenarios like object detection in mobile phone or drone Embedded Edge Computing Platform. To meet the accuracy and efficiency requirements of different scenarios, the latest YOLOv5 [11] has five different network structures including YOLOv5n, YOLOv5s, YOLOv5m, YOLOv5l, YOLOv5x. In this paper, we choose YOLOv5s as the baseline model and propose a lightweight object detection model U-YOLO to promote the performance of the small object detection. YOLOv5s hardly meets the demand for drone-captured scenarios, and the reasons can be summarized as the following two aspects: (1) The original multi-scale feature fusion network only fuses the 8, 16, and 32 times down-sampling feature maps, which benefit medium and large object detection. However, the useful spatial, texture, and shape information is lost in down-sampling, making small object detection more difficult. (2) YOLOv5s concatenates different scale feature maps along the channel dimension. This simple linear fusion approach only increases the number of channels and cannot adaptively select the feature information with the most relevance to the current task.

In this paper, to further improve the precision of object detection on drone-captured images, we propose three improvement methods for the original YOLOv5s model. Our contributions in this paper are summarized as follows:

- We extend the original multi-scale feature fusion network and introduce a detection head for tiny objects.
- We integrate the Convolutional Block Attention Module (CBAM) [32] in detection heads to selectively focus on the object regions containing important information in the feature map and suppress irrelevant information.
- We design a context-based attention feature fusion module to utilize contextual information around the small object effectively, enhancing the multi-scale feature fusion capability and promoting object detection performance.

2   Related Work

2.1   Small Object Detection

Object occupying area less than or equal to 32×32 pixels will be defined as a small object in MS COCO [18] evaluation metric. and the targets In the aerial images may be much smaller than this criterion. In recent years, many methods have been devoted to improving the performance of small target detection, including data augmentation [5,15], multi-scale feature learning [2,16], and context-based object detection [12,27,36]. The purpose of data augmentation is to expand the amount and variability of the input samples so that the model has a higher generalization for unseen environments. Common data augmentation strategies include random scaling, rotating, flipping, CutOut [7], MixUp [33], Mosaic Augmentation [1], etc. Effective use of multi-scale features is of crucial importance for small object detection. Generally, shallow layers of CNN contain rich spatial and appearance features that facilitate target localization. Deep layers have semantic information that is helpful for the classification task. To combine the advantage of feature maps with different scales, Lin et al. [16] put forward a feature pyramid network (FPN). FPN constructed a top-down architecture with lateral connections to transfer spatial detail information from shallow to deep feature maps. Subsequently, many variants [19,20,29] of FPN are proposed to improve the detection performance of small objects. Contextual information plays an essential role in object detection. Targets usually appear in a specific environment and sometimes coexist with other related objects. Context-based methods improved the small object detection performance to a certain degree. Still, the coarse-grained global contextual information is not suitable for small target detection compared with fine-grained local contextual information.

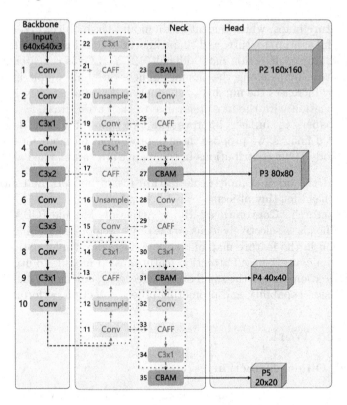

Fig. 2. The framework of the U-YOLO. The backbone uses the architecture CSPDark-net53 [1] with a SPPF layer, used for extracting image features. The multi-scale feature fusion network (Neck) adopts the architecture like PANet [19], and enhances feature information extracted by the backbone. Four detection heads with the CBAM module predict the category and location of the object at different scales.

2.2 Visual Attention Mechanism

Visual attention mechanism has been proved to improve the performance of various vision tasks, e.g., image classification, object detection, semantic segmentation, etc. category of attention strategy include channel attention [14,30], spatial attention [13,31], channel & spatial attention [22,32]. Channel attention selects important channels by generating attention masks across the channel domain, focusing on *what to pay attention to*. Spatial attention generates attention weight across s spatial domain to select the most relevant spatial regions, focusing on *where to pay attention to*. Channel & spatial attention generates channel and spatial attention masks separately to select important features. SENet [14] proposed a squeeze-and-excitation block to improve the representation ability of feature maps. The squeeze phase uses global average pooling to collect global spatial information. The excitation phase captures channel-wise relationships and generates an attention vector using fully-connected layers and non-linear activation layers. Convolutional Block Attention Module (CBAM) [32] is a simple but effective Channel & spatial attention mechanism, which stacks channel attention and spatial attention in series and enhances the importance of inter-channel as well as significant regions of the feature map.

3 U-YOLO

The framework of U-YOLO is demonstrated in Fig. 2. Three improvements make the original model more specialized in small object detection.

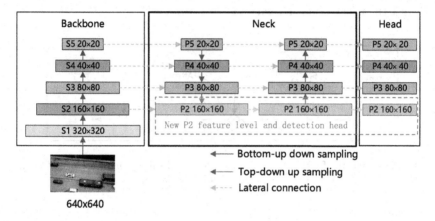

640x640

Fig. 3. The framework of the model after expanding the multi-scale feature fusion network. A new P2 feature level is used to aggregate the low-level, high-resolution S2 feature map. A detection head for extremely small object can improve the accuracy of tiny object detection.

3.1 Multi-scale Feature Fusion Network Extension

In the backbone, layers with the same scale of the output feature map are defined as a stage, as shown in Fig. 3, stages $S \in \{S1, S2, S3, S4, S5\}$. The original neck only fuses the $\{P3, p4, p5\}$ feature levels corresponding to $\{S3, S4, S5\}$ as the input of the three detection heads, these feature maps of deep layers lost the spatial and appearance information beneficial to small object detection in the down-sampling process. Besides, we investigate the VisDrone2021-DET train set and find that in 343201 ground-truth boxes, 29644 boxes' sizes are less than 3 pixels. So, we extend the original Neck and add a detection head for tiny objects. As shown in Fig. 3, the new P2 feature level fuses the low-level, high-resolution S2 feature map with a resolution of 160×160 pixels. Although expanding the multi-scale feature fusion network and adding one more detection head will increase the complexity and computation of the model, it is helpful for the model to improve the detection performance of tiny objects.

3.2   Detection Head with CBAM

On drone-captured images, objects usually appear in complex and confusing backgrounds. Attention mechanisms can select important regions and disregard irrelevant parts of the feature map. In this paper, Because the backbone and neck have powerful feature extraction and fusion capabilities, we integrate The Convolutional block attention module (CBAM) [32] into the four detection heads to refine the feature maps for the final classification and bounding box regression. The structure of CBAM is illustrated in Fig. 4. The Channel attention module and Spatial attention module are used to learn attention masks on the channel domain and spatial domain, adaptively select the important features and regions and resist the useless information.

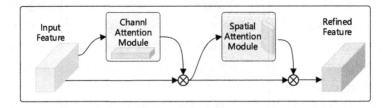

Fig. 4. The structure of CBAM. Two series-connected sub-modules are used to emphasize the important features in both channel and spatial dimensions.

3.3   Context-Based Attention Feature Fusion Module

Contextual information plays an important role in object detection. The relationship between objects and specific surrounding pixel areas can be used to supplement the feature information of a small target so as to improve the detection performance. Coarse-grained global contextual information is not specific to small target detection, to effectively utilize the contextual information, we design a Context-based Attention Feature Fusion Module (CAFF).

The CAFF module is shown in Fig. 5a. The original feature fusion method adopts simple and inefficient channel concatenation, only increasing the number of feature map channels. To explicitly explore contextual information around the small object and refine the feature map, We design the CAFF module to replace the original multi-scale feature fusion approach. CAFF is divided into a local context attention module (LCA), a global context attention module (GCA), and an identity connection. Local and global contextual information is collected and refined by LCA and GCA, respectively, taken F as input and the output F' can be formulated as:

$$\begin{aligned} \mathrm{F}' &= CAFF(\mathrm{F}) \oplus \mathrm{F} \\ &= \big(LCA(\mathrm{F}) \oplus GCA(\mathrm{F})\big) \oplus \mathrm{F} \end{aligned} \tag{1}$$

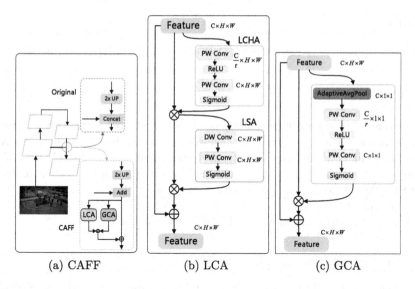

Fig. 5. The structure of CAFF. (a) The CAFF module used to replace the original multi-scale feature fusion approach. (b) Local context attention sub-module (LCA). (c) global context attention sub-module (GCA).

Figure 5b shows the structure of the local context attention module (LCA). To emphasize fine-grained local context information beneficial to small object detection, we stack the local channel attention sub-module (LCHA) and local spatial attention sub-module (LSA) in series. Taking an input feature map $F_l \in \mathbb{R}^{C \times H \times W}$, LCHA learns a 1D channel attention weight vector for each pixel, and LSA infers a 2D spatial attention weight mask. In detail, the LCA process can be summarized as follows:

$$
\begin{aligned}
F'_l &= LCA(F_l) \oplus F_l \\
&= \big(LSA\big(LCHA(F_l) \otimes F_l\big) \otimes \big(LCHA(F_l) \otimes F_l\big)\big) \oplus F_l
\end{aligned}
\tag{2}
$$

$$
LCHA(\cdot) = Sigmoid\,(PWconv2\,(ReLU\,(PWconv1(\cdot)))) \tag{3}
$$

$$
LSA(\cdot) = Sigmoid\,(PWconv\,(DWconv(\cdot))) \tag{4}
$$

In LCHA, the PWconv$(\cdot)$ denotes a point-wise convolution using 1×1 kernel, through a squeeze-and-excitation process like SENet [14], a channel-wise attention weight vector for every single point in the feature map is generated. In LSA, DWconv$(\cdot)$ means a depth-wise convolution using 3×3 kernel size, which is used to capture important local spatial regions for each channel, PWconv$(\cdot)$ can create a linear combination between independent channels.

The global context attention module (GCA) illustrated in Fig. 5c, squeezes the Global spatial information by adaptive average pooling (AAP) and captures the 1D channel-wise attention weight vector by point-wise convolution layers and non-linear activation function (ReLU and sigmoid). A GCA module can be formulated as:

$$\mathbf{F}'_g = (Sigmoid\,(PWconv2(ReLU(PWconv1(AAP(\mathbf{F}_g)))))) \otimes \mathbf{F}_g) \oplus \mathbf{F}_g \qquad (5)$$

4   Experiments

We introduce three improvements for the baseline algorithm YOLOv5s: Multi-scale Feature Fusion Network Extension (P2), detection head with CBAM (CBAM), and context-based attention feature fusion module (CAFF). The model will be evaluated on the VisDrone2021-DET dataset by five metrics: mAp@0.5:0.95 (mean average precision of 10 IoU thresholds, ranging from 0.5:0.95), mAP@0.5 (mean average precision which IoU threshold is 0.5), number of parameters, computation cost (GFLOPs) and inference speed.

4.1   Implementation Details

We implement U-YOLO on Pytorch 1.10.0. Models use an NVIDIA RTX2080Ti GPU for training and testing, and an NVIDIA GTX1050Ti GPU is used to test the inference speed. Before training, we use the KMeans clustering algorithm and genetic algorithm to redesign the anchor size for the VisDrone dataset. In the training phase, we use the pre-trained model of YOLOv5s to initialize the U-YOLO, the model is trained from scratch on VisDrone2021-DET trainset for 300 epochs, and the first three epochs are used for warm-up. We adopt the Adam optimizer for training and the initial learning rate is 0.001, a cosine annealing algorithm is used for learning rate decay. The size of the image is 640 pixels, and the mini-batch size is 32.

4.2   State-of-the-Art Comparison

As shown in Table 1, we compared our method with the state-of-the-art lightweight models on the VisDrone2021-DET dataset. Although the inference speed is lower than YOLOv3-tiny, YOLOv4-tiny, and YOLOv5s, the detection accuracy is significantly improved than the above methods. It is worth noting that our model surpasses the YOLOv5m model in terms of inference speed and mAP value. In terms of application, it can be seen that our model has better detection performance and also well-balanced accuracy and speed.

4.3   Ablation Studies

To study the effectiveness and importance of different components in the proposed U-YOLO, we conduct ablation experiments on the VisDrone2021-DET dataset and provide quantitative and qualitative analysis. The impact of each component is shown in Table 2 and Fig. 6.

The results of ablation experiments show that the three components enhance the detection performance of the original model to a certain degree. Extending

Table 1. The comparison of the performance in VisDrone2021-DET dataset

model	mAP@0.5:0.95 (%)	mAP@0.5 (%)	params (M)	GFLOPs	Speed (FPS)
YOLOv3-tiny [25]	9.2	22.9	8.86	**5.62**	47
YOLOv4-tiny [1]	11.1	25.7	**6.06**	6.96	**52**
SlimYOLOv3 [34]	16.9	34.6	8.0	22.4	34.3
YOLOv5s [11]	19.07	35.17	7.05	15.9	45
YOLOX-s [10]	20.1	36.29	9.0	26.8	29.6
YOLOv5m [11]	22.7	39.6	20.91	48.2	19.5
Ours	**24.1**	**42.9**	7.8	20.9	39

Table 2. Ablation study on VisDrone2021-DET dataset.

model	mAP@0.5:0.95 (%)	mAP@0.5 (%)	improve (%)
YOLOv5s	19.07	35.17	-
YOLOv5s+P2	22.67	39.93	3.6
YOLOv5s+P2+CBAM	23.21	41.07	0.54
YOLOv5s+p2+CBAM+CAFF	24.3	42.95	1.09

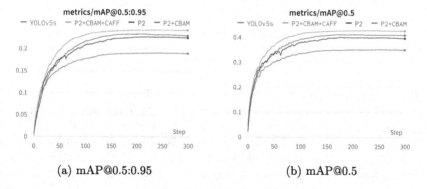

(a) mAP@0.5:0.95   (b) mAP@0.5

Fig. 6. Comparision of mAP@0.5:0.95/mAP@0.5 curves for different models. All models are improved on the basis of YOLO5s. P2: Multi-scale feature fusion network extension; CBAM: Detection head with CBAM; CAFF: Context-based Attention Feature Fusion Module.

multi-scale feature network and adding a detection head for tiny objects (P2) can obtain maximum gain, the mAp@0.5:0.95 value has increased from 19.07 to 22.67%, as it benefits from the shallow, high-resolution feature map that retains a wealth of spatial and appearance information for tiny objects. After integrating the CBAM module in the detection head (CBAM), the mAP@0.5:0.95 of the previous model increased by 0.54%. The CBAM module can adaptively select both important objects and regions from the inputs feature map of the detection head. Finally, adding the proposed CAFF module and integrating all components achieved the best result, compared with the original YOLOv5s, the mAP@0.5:0.95 of our method increased from 19.07 to 24.3%.

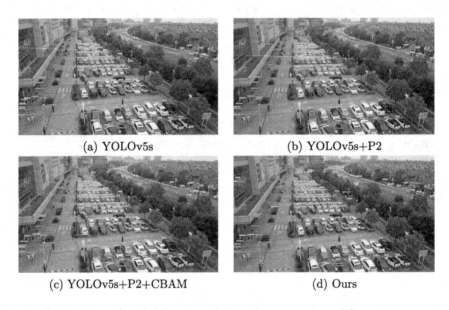

(a) YOLOv5s   (b) YOLOv5s+P2

(c) YOLOv5s+P2+CBAM   (d) Ours

Fig. 7. Detection results of different models in the same scene, different category use bounding boxes with different color.

The visualization results of the detection of different modules in the same scene are illustrated in Fig. 7. Compared with Fig. 7a of the original YOLOv5s, some long-distance vehicles or pedestrians with smaller physical sizes can be detected after adding the P2 module. Figure 7c shows the result of integrating CBAM in the detection head, and we can see that the problems of false detections and missed detections in the picture have been improved to a certain extent. Compared with the previous three models, the CAFF module has greatly enhanced the effect, some distant and heavily obscured targets can also be detected in Fig. 7d. It shows that the reasonable use of contextual information around the small targets can improve detection performance.

We have selected some representative images captured by drones to test the detection performance of our model intuitively. Figure 8 shows the detection

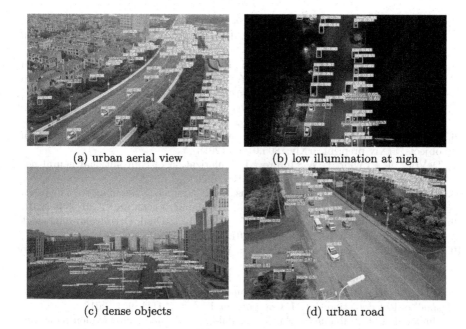

(a) urban aerial view   (b) low illumination at nigh

(c) dense objects   (d) urban road

Fig. 8. Some representative visualization results on drone-captured images.

results of the urban aerial view, low illumination at night, dense objects, and urban road, respectively.

5   Conclusion

In this paper, we proposed a lightweight small object detection algorithm U-YOLO on the basis of YOLOv5s. Three improvements can be summarized as follows:(1) To solve the problem of difficult detection of small targets on drone-captured images, we extend the multi-scale feature fusion network and add detection heads for tiny objects. (2) we integrate a lightweight but effective attention module CBAM in the detection head. (3) To better use of the contextual information around the small object, we design a context-based attention feature fusion module. The experimental results show that the detection accuracy of the U-YOLO model is significantly improved compared to the original algorithm, and the detection accuracy and speed can be well balanced.

References

1. Bochkovskiy, A., Wang, C.Y., Liao, H.Y.M.: Yolov4: Optimal speed and accuracy of object detection. arXiv preprint arXiv:2004.10934 (2020)
2. Cai, Z., Fan, Q., Feris, R.S., Vasconcelos, N.: A unified multi-scale deep convolutional neural network for fast object detection. In: Leibe, B., Matas, J., Sebe, N., Welling, M. (eds.) ECCV 2016. LNCS, vol. 9908, pp. 354–370. Springer, Cham (2016). https://doi.org/10.1007/978-3-319-46493-0_22

3. Cao, C., Wang, B., Zhang, W., Zeng, X., Yan, X., Feng, Z., Liu, Y., Wu, Z.: An improved faster r-cnn for small object detection. IEEE Access **7**, 106838–106846 (2019)

4. Chen, C., Liu, M.-Y., Tuzel, O., Xiao, J.: R-CNN for small object detection. In: Lai, S.-H., Lepetit, V., Nishino, K., Sato, Y. (eds.) ACCV 2016. LNCS, vol. 10115, pp. 214–230. Springer, Cham (2017). https://doi.org/10.1007/978-3-319-54193-8_14

5. Chen, Y., et al.: Stitcher: feedback-driven data provider for object detection. arXiv preprint arXiv:2004.12432 2(7) (2020)

6. Cui, L., et al.: Mdssd: multi-scale deconvolutional single shot detector for small objects. arXiv preprint arXiv:1805.07009 (2018)

7. DeVries, T., Taylor, G.W.: Improved regularization of convolutional neural networks with cutout. arXiv preprint arXiv:1708.04552 (2017)

8. Everingham, M., Van Gool, L., Williams, C.K., Winn, J., Zisserman, A.: The pascal visual object classes (voc) challenge. Int. J. Comput. Vision **88**(2), 303–338 (2010)

9. Fu, C.Y., Liu, W., Ranga, A., Tyagi, A., Berg, A.C.: Dssd: Deconvolutional single shot detector. arXiv preprint arXiv:1701.06659 (2017)

10. Ge, Z., Liu, S., Wang, F., Li, Z., Sun, J.: Yolox: exceeding yolo series in 2021. arXiv preprint arXiv:2107.08430 (2021)

11. Glenn-jocher: Ultralytics yolov5 (2020). https://github.com/ultralytics/yolov5

12. Guan, L., Wu, Y., Zhao, J.: Scan: Semantic context aware network for accurate small object detection. Int. J. Comput. Intell. Syst. **11**(1), 951 (2018)

13. Hu, J., Shen, L., Albanie, S., Sun, G., Vedaldi, A.: Gather-excite: exploiting feature context in convolutional neural networks. Advances in neural information processing systems 31 (2018)

14. Hu, J., Shen, L., Sun, G.: Squeeze-and-excitation networks. In: Proceedings of the IEEE Conference on Computer Vision and Pattern Recognition, pp. 7132–7141 (2018)

15. Kisantal, M., Wojna, Z., Murawski, J., Naruniec, J., Cho, K.: Augmentation for small object detection. arXiv preprint arXiv:1902.07296 (2019)

16. Lin, T.Y., Dollár, P., Girshick, R., He, K., Hariharan, B., Belongie, S.: Feature pyramid networks for object detection. In: Proceedings of the IEEE Conference on Computer Vision and Pattern Recognition, pp. 2117–2125 (2017)

17. Lin, T.Y., Goyal, P., Girshick, R., He, K., Dollár, P.: Focal loss for dense object detection. In: Proceedings of the IEEE International Conference on Computer Vision, pp. 2980–2988 (2017)

18. Lin, T.-Y., et al.: Microsoft COCO: common objects in context. In: Fleet, D., Pajdla, T., Schiele, B., Tuytelaars, T. (eds.) ECCV 2014. LNCS, vol. 8693, pp. 740–755. Springer, Cham (2014). https://doi.org/10.1007/978-3-319-10602-1_48

19. Liu, S., Qi, L., Qin, H., Shi, J., Jia, J.: Path aggregation network for instance segmentation. In: Proceedings of the IEEE Conference on Computer Vision and Pattern Recognition, pp. 8759–8768 (2018)

20. Liu, S., Huang, D., Wang, Y.: Learning spatial fusion for single-shot object detection. arXiv preprint arXiv:1911.09516 (2019)

21. Liu, W., et al.: SSD: single shot MultiBox detector. In: Leibe, B., Matas, J., Sebe, N., Welling, M. (eds.) ECCV 2016. LNCS, vol. 9905, pp. 21–37. Springer, Cham (2016). https://doi.org/10.1007/978-3-319-46448-0_2

22. Park, J., Woo, S., Lee, J.Y., Kweon, I.S.: Bam: Bottleneck attention module. arXiv preprint arXiv:1807.06514 (2018)

23. Redmon, J., Divvala, S., Girshick, R., Farhadi, A.: You only look once: unified, real-time object detection. In: Proceedings of the IEEE Conference on Computer Vision and Pattern Recognition, pp. 779–788 (2016)

24. Redmon, J., Farhadi, A.: Yolo9000: better, faster, stronger. In: Proceedings of the IEEE Conference on Computer Vision and Pattern Recognition, pp. 7263–7271 (2017)

25. Redmon, J., Farhadi, A.: Yolov3: an incremental improvement. arXiv preprint arXiv:1804.02767 (2018)

26. Ren, S., He, K., Girshick, R., Sun, J.: Faster r-cnn: towards real-time object detection with region proposal networks. Advances in neural information processing systems 28 (2015)

27. Shrivastava, A., Gupta, A.: Contextual priming and feedback for faster R-CNN. In: Leibe, B., Matas, J., Sebe, N., Welling, M. (eds.) ECCV 2016. LNCS, vol. 9905, pp. 330–348. Springer, Cham (2016). https://doi.org/10.1007/978-3-319-46448-0_20

28. Singh, B., Najibi, M., Davis, L.S.: Sniper: Efficient multi-scale training. Advances in neural information processing systems 31 (2018)

29. Tan, M., Pang, R., Le, Q.V.: Efficientdet: scalable and efficient object detection. In: Proceedings of the IEEE/CVF Conference on Computer Vision and Pattern Recognition, pp. 10781–10790 (2020)

30. Wang, Q., Wu, B., Zhu, P., Li, P., Hu, Q.: Eca-net: efficient channel attention for deep convolutional neural networks. In: 2020 IEEE/CVF Conference on Computer Vision and Pattern Recognition (CVPR) (2020)

31. Wang, X., Girshick, R., Gupta, A., He, K.: Non-local neural networks. In: Proceedings of the IEEE Conference on Computer Vision and Pattern Recognition, pp. 7794–7803 (2018)

32. Woo, S., Park, J., Lee, J.-Y., Kweon, I.S.: CBAM: convolutional block attention module. In: Ferrari, V., Hebert, M., Sminchisescu, C., Weiss, Y. (eds.) ECCV 2018. LNCS, vol. 11211, pp. 3–19. Springer, Cham (2018). https://doi.org/10.1007/978-3-030-01234-2_1

33. Zhang, H., Cisse, M., Dauphin, Y.N., Lopez-Paz, D.: mixup: beyond empirical risk minimization. arXiv preprint arXiv:1710.09412 (2017)

34. Zhang, P., Zhong, Y., Li, X.: Slimyolov3: narrower, faster and better for real-time uav applications. In: Proceedings of the IEEE/CVF International Conference on Computer Vision Workshops (2019)

35. Zhu, X., Lyu, S., Wang, X., Zhao, Q.: Tph-yolov5: improved yolov5 based on transformer prediction head for object detection on drone-captured scenarios. In: Proceedings of the IEEE/CVF International Conference on Computer Vision, pp. 2778–2788 (2021)

36. Zhu, Y., Zhao, C., Wang, J., Zhao, X., Wu, Y., Lu, H.: Couplenet: coupling global structure with local parts for object detection. In: Proceedings of the IEEE International Conference on Computer Vision, pp. 4126–4134 (2017)

Fruit Detection Based on Automatic Occlusion Prediction and Improved YOLOv5s

Yufeng Wang, Liang Ye$^{(\boxtimes)}$, Jing Zhao, and Huasong Min

Wuhan University of Science and Technology, Wuhan 430074, China
yeliang@wust.edu.cn

Abstract. The missed detection of orchard fruit recognition by robots is too high in the complex natural environment, because of different occlusion factors. Based on this practical problem, the apple fruit images collected from real orchard is taken as the research object, and an improved yolov5s fruit detection model is proposed based on automatic determination of occlusion information. On one hand, an algorithm for automatic judgment based on fruit occlusion information is added in the model training stage. In order to reduce the missed detection rate caused by non-maximum suppression (NMS), the unlabeled apple fruits are divided into three categories for training: no occlusion, branches and leaves occlusion, and fruit occlusion. On the other hand, improves the detection accuracy of occluded targets by introducing DIoU Loss and adding CBAM attention module in the original YOLOv5s network. Meanwhile, reduces false detection by adding false detection processing based on relative distance and intersection ratio after model NMS. Finally, the experimental results show that the proposed idea is feasible, the missed detection is significantly reduced, and the comprehensive evaluation index value (F1) reaches 97.8%. In compared with original yolov5s, the proposed method reduces missed detection rate from 6.5% to 2.0% without affecting the false detection rate, which is more suitable for robotic picking and yield estimation.

Keywords: Object detection · YOLOv5s · Fruit detection · Robot picking

1 Introduction

Smart agriculture has become a popular concept, with the rapid development of artificial intelligence technologies such as computer vision, robot control, and intelligent systems. Image information can be used to accurately judge crop growth, estimate crop yield, and help robots complete agricultural tasks such as picking [1]. Now, picking robots have been able to effectively solve the problems of insufficient labor, high labor intensity and low operation efficiency in manual operations. It ensures the timely harvesting and picking quality of crops at the same time [2]. However, fruits are sheltered in the natural environment to varying degrees. Because the orchard environment is complex, the light is changeable, the fruit is densely distributed, and the branches and leaves are everywhere. All these problems bring great challenges to the harvesting work of robots in orchards [3].

It is of great significance for robotic picking and yield estimation, which accurately identifies fruits in complex environments. At present, some progress has been made in fruit detection research at domestic and foreign. For example, Hongxing Peng et al. proposed an improved SSD fruit detection method with ResNet-101 as the backbone network [4], and Sun et al. used ResNet50 as the Faster RCNN feature extraction network to detect tomatoes [5]. Subsequently, Häni et al. carried out high-precision detection and counting of apples in the natural environment, which combined with deep learning methods and semi-supervised methods based on Gaussian mixture models [6]. The above detection methods all have good performance for the detection of uncovered fruits, but it is not suitable for the detection of occluded fruits in complex environments. The robustness and generalization of detection are poor when the fruits are densely distributed and severely occluded.

With the development of machine learning, the research of detection and recognition based on deep learning has become more and more popular. There has also been some progress in detection of occlusion problems. Such as, Chu et al. proposed a general detection model based on proposal to identify highly coincident objects under severe occlusion [7]. And Tian et al. effectively improved the detection accuracy of occluded objects by proposing a YOLO v3-dense model to detect overlapping and occluded apples [8]. The detection of occluded objects is mainly by improving the ability of model feature extraction at present [9]. Nonetheless, Gao et al. proposed a multi-classification apple detection method based on Faster R-CNN by classifying apples in natural state into four categories: unshaded, leaf shaded, branch or silk, and fruit shaded [10]. It effectively improved the average detection accuracy but the cost of manual annotation is too high.

In view of this, taking apples in orchard as the research and experimental object, a novel fruit detection method is proposed in this paper. Firstly, an automatic occlusion prediction algorithm based on image processing and analysis is proposed to pre-processing the training samples of training dataset. Training with the predicted occlusion labels of samples can significantly help the model reducing the missing detection caused by non-maximum suppression (NMS). Then the attention module and the DIoU loss function are added into the YOLOv5s network and improved it. Finally, a post processing for output of the network based on category unification is proposed, which eliminates false detection according to the relative distance and intersection ratio of the prediction bounding boxes. Several experimental results show that our method can achieve both lower missing rate and lower false rate than some state-of-the-art methods for fruit detection and be helpful to picking robots and yield estimation robots in complex environments.

2 Target Detection Based on Occlusion Information Automatic Judgment

2.1 Analysis of Occlusion Information for Target Detection

In the case of occluded objects, the missed detection and false detection are always higher than practical applications in the case of occluded objects. And the robustness of deep learning neural network-based classifiers is not as good as that of humans. Existing target recognition algorithms are not suitable for the detection of occluded targets under dense distribution. Because of they basically classify the detected targets under different conditions into one class for training and detection. When the target has no occlusion information, the occluded target and the uncovered target are regarded as the same type of target during the network training process. The learned features are not sensitive to occlusion, resulting in missed detections in the NMS stage. For the reason two or more objects that are close to each other and occlude each other are most likely to be suppressed as one object, when the predicted categories of the candidate boxes are the same.

At present, for the detection of occluded targets, the main problem is due to the lack of practical data to label occlusion information [11]. To address this problem, occluded objects and uncovered objects are regarded as different categories of objects. The occlusion targets are divided into target occlusion and non-target occlusion meanwhile. In this way, the target candidate frame predicted by the network will be obtained from candidate frames of different categories according to different features. The NMS stage still retains at least one candidate box for different classes as a result even if the objects are too close due to mutual occlusion. It effectively avoids missed detection caused by NMS. In order to verify this idea, two different training strategies were used to detect the apple images collected in the actual orchard environment, as in Table 1. Among them, strategy A trains all apples as one class. Strategy B adds a category label to each labeled apple detection frame based on strategy A, and divides apples into three categories for training: no occlusion, branch occlusion, and fruit occlusion.

Table 1. Sample distribution.

Training Strategy	Uncovered	Branches occlusion	Fruit occlusion	Total Sample
strategy A	-	-	-	3651
Strategy B	745	1171	1735	3651

For the above two strategies, the experimental comparison is carried out through the mainstream target detection network, and the detection results are in Table 2.

Table 2. Sample distribution.

Model	Training Strategy	Miss Rate (%)	False Rate (%)
Faster-RCNN	A	12.65	7.02
	B	3.61	10.74
YOLO v4	A	3.13	7.97
	B	1.90	16.11
YOLO v5	A	6.50	1.24
	B	1.99	2.49

The experimental results show that the missed detection rate of each detection network in strategy A is significantly higher than that in strategy B. In the detection model, two or more objects are no longer predicted to be the same class due to the addition of occlusion information, which are close to each other and occlude each other. Occluded objects will be retained in the NMS stage of the model because their different categories. The missed detection rate is significantly reduced. Although the false detection rate of strategy B increased slightly, it effectively improved the missed detection of model detection.

2.2 Automatically Generate Occlusion Information

Today, most of the existing detection datasets carry out target detection frame annotation according to the task or specific scene. Its workload increases with the variety and number of samples. If the occlusion information needs to be identified and marked during the labeling process, the labor cost and time cost will increase. Therefore, in order to reduce the cost of manual labeling, occlusion information of the detected target is automatically generated by the algorithm during the training process. In other words, the detected target is divided into no occlusion, branch occlusion, and fruit occlusion, in the case of only labeling the target frame. The algorithm firstly judges whether there is overlap between the target frames by the coordinate information of the target frame. Then it is judged whether there is target occlusion between the targets, after processing the overlapping area of overlapping callout boxes. For a single target whose annotation frame does not overlap, it is judged by segmentation whether the target is occluded by branches and leaves (see Fig. 1).

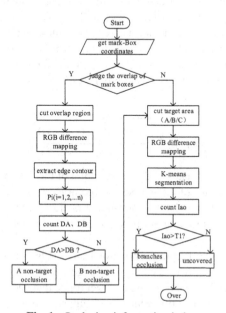

Fig. 1. Occlusion information judge.

For samples with overlapping annotation boxes, as in Fig. 2(a). It is necessary to judge whether there is fruit occlusion between the targets, and obtain the target occlusion relationship (see Fig. 2).

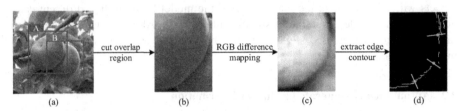

(a)   (b)   (c)   (d)

Fig. 2. Judgment of fruit occlusion. (a) target box overlap (b) clipping of overlapping areas (c) gray image (d) edge detection map.

Firstly, Fig. 2(b) is obtained by clipping the overlapping area between the target annotation boxes. Calculate the difference between R channel and G channel and B channel respectively. Then sum and take the average to obtain the pixel value I. Fig. 2(c) can be obtained after mapping I within 0 to 255, as in (1).

$$\begin{cases} I = \dfrac{(I_R - I_B) + (I_R - I_G)}{2} \\ I_{gray} = \dfrac{I \times 255}{\max(I) - \min(I)} \end{cases} \tag{1}$$

Figure 2(d) is obtained by extracting the edge contour of the target grayscale image by improving the canny operator 错误!未找到引用源. Take n pixels on the edge equally, and calculate the intersection of the normal vectors of these pixels in Fig. 2(a) (denoted as $P1, P2 \cdots Pn$). Finally, calculate the distances from all intersection points ($P1 \sim Pn$) to the center points of the A and B boxes. Take the mean value after summing the distances of all intersections of box A and box B respectively. Recorded as D_A and D_B, as in (2).

$$D_M = \dfrac{\sqrt{(x_m - x_i)^2 + (y_m - y_i)^2}}{n} \tag{2}$$

where, (x_m, y_m) is the coordinates of the center point of the desired target frame M, and (x_i, y_i) is the coordinates of the intersection point P_i ($i = 1, 2, \cdots n$).

If $D_A < D_B$, target A is judged as non-target occlusion, and target B is judged as fruit occlusion. On the contrary, target B is judged as non-target occlusion and target A is judged as fruit occlusion.

For samples without overlapping annotation frames (Fig. 3(a)) or targets determined to be non-target occlusions. It is necessary to judge whether the target is occluded by branches and leaves (see Fig. 3). Cut out the target frame that needs to be judged, and get Fig. 3(c) according to the RGB difference map. Figure 3(d) is obtained by performing K-means threshold segmentation on the Fig. 3(c). After that, taking the width, height

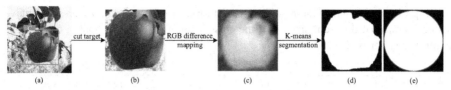

Fig. 3. Judgment of branches shade. (a) no overlap target boxes (b) cut-out of target (c) gray image (d) K-means split image (e) standard ellipse.

and center point of the segmented image as the major axis, minor axis, and center of the circle, as in Fig. 3(e). The calculation formula is shown in (3).

$$O = \frac{\left(x - \frac{w}{2}\right)^2}{\left(\frac{w}{2}\right)^2} + \frac{\left(y - \frac{h}{2}\right)^2}{\left(\frac{h}{2}\right)^2} \tag{3}$$

where, w and h are the width and height of Fig. 3(d), and $(\frac{w}{2}, \frac{h}{2})$ is the center point of Fig. 3(e).

Finally, calculate the intersection over union (IoU) of the pixels in the Fig. 3(d) and Fig. 3(e), and set the threshold $T1$. Among them, Pa represents the pixel point of the target area in Fig. 3(d), Po represents the pixel point of the standard ellipse area in Fig. 3(e). And IoU is recorded as I_{ao}, as in (4).

$$I_{ao} = \frac{|P_a \cap P_o|}{|P_a \cup P_o|} \tag{4}$$

If $I_{ao} > T1$, the target C is judged to be unobstructed. On the contrary, it is judged to be blocked by branches and leaves.

3 Model and Improvement

The YOLOv5 network is the latest product of the YOLO series. There are currently four versions: Yolov5s, Yolov5m, Yolov5l, and Yolov5x. We use YOLOv5s, which is the smallest network in the YOLOv5 series. This network model has few parameters, small size, high flexibility, high detection accuracy and high speed. It can effectively identify the detected target and meet the real-time nature of Apple detection.

The network structure of YOLOv5s is mainly divided into the following three parts: the backbone network Dacknet-53, the neck network PANet and the prediction network (see Fig. 4). In the model training phase, YOLOv5s uses technologies such as Mosaic data enhancement, adaptive anchor box calculation, and adaptive image scaling. It greatly improves the network training speed and reduces the model memory. The backbone network extracts features from the input image based on the Focus, CSP1_X and Spatial Pyramid Pool (SPP) module. After transferring the features to the neck network, the neck network enhances the detection ability of objects of different scales through the FPN_PAN structure. On account of the bidirectional fusion of low-level spatial features and high-level semantic features. The prediction network performs regression analysis on features of different sizes. Get the classification result, coordinate position and confidence after NMS.

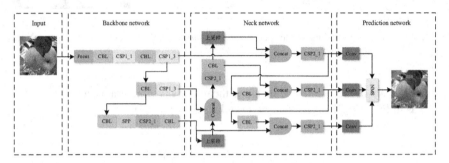

Fig. 4. YOLOv5s network framework.

3.1 Embedded Attention Module

A lightweight attention module is introduced in the design of the object recognition network to better extract object features, since shape and color of the detected objects are different from the background objects in the image. The detection accuracy of targets blocked by branches and leaves in the orchard is improved, and the missed detection rate of target detection is reduced. Convolutional Block Attention Module (CBAM) is a network structure proposed by Woo et al., which includes two independent submodules, Channel Attention Module (CAM) and Spatial Attention Module (SAM) [14]. It can realize the re-calibration of the original features by channel attention and obtain the attention features of the spatial dimension. At the same time, the connection of each feature in channel and space is improved. The calculated as follows.

$$M_c(F) = \sigma\left(W_1\left(W_0\left(F_{avg}^c\right)\right) + W_1\left(W_0\left(F_{max}^c\right)\right)\right) \qquad (5)$$

CAM calculation method as in (5). Where σ represents the sigmoid function, $W_0 \in R^{C/r \times C}$, $W_1 \in R^{C \times C/r}$, W_0 and W_1 are the input shared weights, F_{avg}^c and F_{max}^c represent feature maps generated by average pooling and max pooling in space.

$$M_s(F) = \sigma\left(f^{7 \times 7}\left(\left[F_{avg}^s; F_{avg}^s\right]\right)\right) \qquad (6)$$

SAM calculation method as in (6). Where $f^{7 \times 7}$ represents a convolution operation with a filter size of 7×7, F_{avg}^s and F_{max}^s represent the average pooling feature and max pooling feature of the channel.

The detection accuracy of the model is improved by embedding CBAM in the backbone network of the YOLOv5s architecture. Because of it can effectively improve the model's efficiency of extracting the features of occluded targets and locating target positions, and the amount of calculation is small (see Fig. 5).

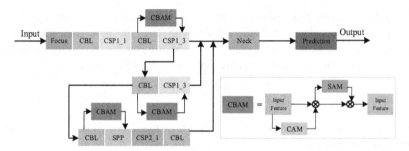

Fig. 5. Improved network structure.

3.2 Improvement of Loss Function

There are different degrees of shading between fruits since the dense distribution of fruits in orchards in the natural environment. In order to reduce the impact on the detection results when the fruits are occluded by each other, DIoU Loss is introduced to replace the bounding box loss function of the original YOLOv5s network model [15].

DIoU Loss considers the distance information of the center point of the bounding box based on the IoU and GIoU loss functions. Unable to learn and train When two boxes do not intersect, and IoU is 0 and IoU-loss is 1. For the case where IoU is 0, GIoU adds a penalty term to optimize the problem of non-intersect between two boxes. However, the relative position relationship cannot be distinguished. When the two boxes are in an inclusive relationship, the GIoU regression strategy will degenerate into the IoU regression strategy. DIoU Loss adds a penalty term to the IoU loss to directly minimize the normalized distance between the center points of the two bounding boxes, as in (7). It can achieve faster convergence than GIoU and make the regression more accurate and faster when it overlaps or even contains the target box.

$$L_{DIoU} = 1 - IOU + \frac{\rho^2(b, b^{gt})}{c^2} \tag{7}$$

where, b and b^{gt} represent the center points of the prediction box and the annotation box, ρ represents the Euclidean distance between the two center points, c represents the diagonal distance of the smallest closed area that can contain both the prediction box and the annotation box.

3.3 False Detection Handle Based on Category Unification

The targets are classified into three categories of no occlusion, branches occlusion, and fruit occlusion for training. The model may mistakenly believe that targets obscured by fruits or branches and leaves still exist behind the target when going through NMS, due to the dense target distribution and severe obscuration. Thus, the redundant frames will not be suppressed, resulting in the rise of detection false detections. To solve the above problem, a category-based uniform false detection process is added after detecting the network NMS.

The false detection processing operation is as follows: first, the model output label documents are traversed to find out the samples with false detection. Then, the left vertex distance S_{ij} between the predicted boxes (*box-i* and *box-j*) in the false detection samples is calculated according to the distance between two points formula, as in (8).

$$S_{ij} = \sqrt{\left(x_i - x_j\right)^2 + \left(y_i - y_j\right)^2} \tag{8}$$

where, (x_i, y_i) is the coordinates of the left vertex of *box-i*, (x_j, y_j) is the coordinates of the left vertex of *box-j*.

If S_{ij} is less than the distance threshold *T2*, calculate the IoU between box-i and box-j and denote as $IoU_{(box-i, box-j)}$, as in (9). Set the threshold *T3*, delete the prediction box with lower confidence if $IoU_{(box-i, box-j)} > T3$, and keep it vice versa.

$$\begin{cases} x_a = \max\left(x_i, x_j\right); y_a = \max\left(y_i, y_j\right); \\ x_b = \min\left(w_i, w_j\right); y_b = \min\left(h_i, h_j\right); \\ area_a = \left(w_i - x_i + 1\right) \times \left(h_i - y_i + 1\right); \\ area_b = \left(w_j - x_j + 1\right) \times \left(h_j - y_j + 1\right); \\ area_{ab} = \max(0, x_b - x_a + 1) \times \max(0, y_b - y_a + 1); \\ IOU_{(box-i,box-j)} = \dfrac{area_{ab}}{area_a + area_b + area_{ab}} \end{cases} \tag{9}$$

where, (x_i, y_i, w_i, h_i) and (x_j, y_j, w_j, h_j) indicate the position coordinates of box-i and box-j respectively.

4   Experiment

4.1   Test Environment

All experiments were conducted under Ubuntu 16.04 operating system, and the main configuration of the lab server was: Intel(R) Xeon(R) CPU E5–2678 v3 @ 2.50GHz × 48, 12GB GeForce RTX 2080Ti × 2GPU. We use CUDA version 10.2 parallel computing framework with CUDNN version 7.6 deep neural network acceleration library running up to 256 GB of memory, on which we build Pytorch 1.9 deep learning framework and program the network with Python 3.7 language for training and testing. The Pytorch 1.9 deep learning framework is built on this foundation, and the network is programmed in Python 3.7 for training and testing.

The images in the dataset were obtained through web crawlers and actual orchard shots. It contains 400 images of apples in natural environment. By augmenting the images, including cropping, zooming in and out, random rotation, brightness variation, and adding Gaussian noise, 3500 images were obtained with an image size of 448 × 448 pixels. The Labelimg tool is used to annotate the images in the dataset and generate a txt file. On which contains information about the image target category and the position of the apple in the image. The dataset was divided into training dataset, validation dataset, and test dataset according to the ratio of 7:2:1. The model training learning rate is set to 0.001, the batch size is set to 8, the decay coefficient is 0.9, the momentum is set to

0.98, and the number of training iterations is set to 2000. This algorithm is optimized using Stochastic Gradient Descent (SGD) algorithm. The thresholds $T1$, $T2$, and $T3$ are set to 0.7, 0.1, and 0.95, respectively.

4.2 Evaluation Indicators

In this study, the indicators of detection evaluation include precision (P), recall (R), comprehensive evaluation (F1), missed detection (Mr) and false detection (Fr), which are calculated as follows.

$$P = \frac{TP}{TP + FP} \tag{10}$$

$$R = \frac{TP}{TP + FN} \tag{11}$$

$$F1 = \frac{2PR}{P + R} \tag{12}$$

$$Mr = \frac{FN}{TP + FN} \tag{13}$$

$$Fr = \frac{FP}{TP + FN} \tag{14}$$

where, TP indicates correctly identified targets, FP indicates incorrectly identified targets, and FN indicates unidentified targets.

4.3 Results and Analysis

Apple Detection Effect of the Proposed Method

The detection effects of the proposed method were compared with those of several other commonly used target detection algorithms: Faster-RCNN, YOLO v4 [17], and YOLO v5s (see Fig. 6).

In order to quantitatively analyze our method and other comparison methods, the parameters set in Sect. 4.2 were used to conduct comparison tests for each method in the same data set, as in Table 3.

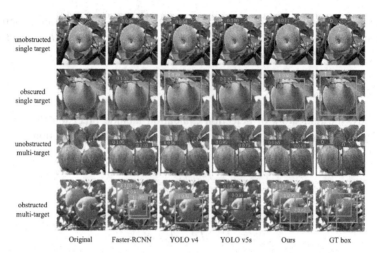

unobstructed single target

obscured single target

unobstructed multi-target

obstructed multi-target

Original   Faster-RCNN   YOLO v4   YOLO v5s   Ours   GT box

Fig. 6. Comparison results of different algorithms.

Table 3. Apple detection effect of different methods (%).

Model	Mr	Fr	R	R	F1
Faster-RCNN	12.65	7.02	92.64	87.36	89.96
YOLO v4	3.13	7.97	92.31	96.94	94.65
YOLO v5s	6.50	1.24	98.75	93.52	96.01
Ours	2.07	2.21	97.88	97.93	97.82

YOLO v5s algorithm has the highest accuracy rate, but it has too many missed detections and is not suitable for occlusion target detection. Our proposed method outperforms Faster-RCNN, YOLO v4 and YOLO v5s in terms of both detection accuracy and missed detection rate. The main reason is that there are more missed and false detections when the unclassified is trained on targets with dense distribution and heavy occlusion. In summary, our method effectively reduces the model miss detection rate and false detection rate. It is suitable for robotic harvesting and yield estimation in natural and complex environments.

Effect of Each Improvement Part
The following operations were performed to address the problems of missed detection and false detection. Firstly, the occlusion information of the detected target is automatically generated in the training phase. Next, the attention module of CBAM is embedded in the YOLOv5s network and employs DIoU Loss as its bounding box loss function. False detection processing was introduced at last. To verify the effectiveness of each improvement, different improvement strategies were used for experiments and quantitative statistics of the detection results were conducted, as in Table 4. As can be seen from the table, missed detection rate is significantly reduced by the preprocessing operation

Table 4. Experimental results under different Settings (%).

Baseline Network	Pre-processing	Attention Module	Loss Function	Post processing	Mr	Fr	P	R	F1
YOLO v5s					6.50	1.24	98.75	93.52	89.96
YOLO v5s	√				2.93	7.26	93.02	97.18	94.65
YOLO v5s	√			√	3.42	1.54	98.45	96.68	97.56
YOLO v5s	√	√			1.94	8.74	91.83	98.05	94.82
YOLO v5s	√	√		√	2.36	2.01	97.92	97.66	97.53
YOLO v5s	√		√		2.09	9.18	91.47	97.93	94.67
YOLO v5s	√		√	√	2.59	2.24	97.78	97.45	97.58
YOLO v5s	√	√	√		1.79	8.13	92.36	98.27	96.01
YOLO v5s	√	√	√	√	2.07	2.21	97.88	97.93	97.82

of automatic determination of occlusion information on the training set. Based on the above conditions, CBAM and DIoU Loss are introduced in YOLOv5s network, respectively. The table clearly indicates that both of them have a significant improvement on the recall. The false detection due to NMS in classification training is reduced by adding the false detection treatment based on the relative distance of prediction frames and the intersection ratio. Compared to baseline network, our method has fewer false and missed detections.

5 Conclusion

To address the problem of high leakage detection and false detection of fruits in natural complex environment. The scheme of automatic judgment algorithm of occlusion information to generate occlusion information, embedding CBAM attention module, using DIOU Loss as the bounding box loss function and adding false detection processing after non-extreme suppression is proposed.

The experimental results indicate that the training set with occlusion information is more suitable for the detection of densely distributed and heavily occluded targets, which effectively reduces the model detection leakage rate. In addition, the improved YOLO v5s model has a 4.4% reduction in missed detection rate and a 1.8% improvement in F1 compared to the original model. It is demonstrated that the improved YOLOv5s model detection based on automatic determination of occlusion information has a more obvious improvement in the detection performance of the occluded targets, which compared with Faster-RCNN, YOLO v4 and YOLO v5s convolutional neural networks. And availably improve the harvesting and estimation work of agricultural robots.

References

1. Wang, Q.I., Nuske, S., Bergerman, M., et al.: Automated crop yield estimation for apple orchards. In: Proceedings of International Symposium of Experimental Robotics, Springer International Publishing, pp. 745–758 (2013). https://doi.org/10.1007/978-3-319-00065-7_50
2. Ji, W., Zhao, D., Cheng, F., et al.: Automatic recognition vision system guided for apple harvesting robot. Comput. Electr. Eng. **38**(5), 1186–1195 (2012)
3. Gongal, A., Amatya, S., Karkee, M., et al.: Sensors and systems for fruit detection and localization: a review. Comput. Electron. Agric. 116(C), 8–19 (2015)
4. Peng, H.X., Huang, B., Shao, Y.Y.: General improved SSD model for picking object recognition of multiple fruits in natural environment. Trans. Chinese Soc. Agric. Eng.(Trans. CSAE) **34**(16), 155–162 (2018)
5. Sun, J., He, X.F., Ge, X., et al.: Detection of key organs in tomato based on deep migration learning in a complex background. Agriculture **8**(12), 8196 (2018)
6. Häni, N., Roy, P., Isler, V.: A comparative study of fruit detection and counting methods for yield mapping in apple orchards. J. Field Robot. **37**(2), 263–282 (2020)
7. Chu, X., Zheng, A., Zhang, X., et al.: Detection in crowded scenes: one proposal, multiple predictions. In: Proceedings of the IEEE/CVF Conference on Computer Vision and Pattern Recognition, pp. 12214–12223 (2020)
8. Tian, Y., Yang, G., Wang, Z., et al.: Apple detection during different growth stages in orchards using the improved YOLO-V3 model. Comput. Electron. Agric. 157, 417–426 (2019)
9. Saleh, K., Szénási, S., Vámossy, Z.: Occlusion handling in generic object detection: a review. In: 2021 IEEE 19th World Symposium on Applied Machine Intelligence and Informatics (SAMI), pp. 000477–000484 (IEEE, 2021)
10. Gao, F., Fu, L., Zhang, X., et al.: Multi-class fruit-on-plant detection for apple in SNAP system using Faster R-CNN. Comput. Electron. Agric. 176, 105634 (2020)
11. Wang, X., Xiao, T., Jiang, Y., et al.: Repulsion loss: detecting pedestrians in a crowd. In: Proceedings of the IEEE Conference on Computer Vision and Pattern Recognition, pp. 7774–7783 (2018)
12. Rong, W., Li, Z., Zhang, W., et al.: An improved CANNY edge detection algorithm. In: 2014 IEEE International Conference on Mechatronics and Automation, pp. 577–582. IEEE (2014)
13. Liu, Y., Lu, B., Peng, J., et al.: Research on the use of YOLOv5 object detection algorithm in mask wearing recognition. World Sci. Res. J. **6**(11), 276–284 (2020)
14. Woo, S., Park, J., Lee, J.-Y., Kweon, I.S.: CBAM: Convolutional Block Attention Module. In: Ferrari, V., Hebert, M., Sminchisescu, C., Weiss, Y. (eds.) ECCV 2018. LNCS, vol. 11211, pp. 3–19. Springer, Cham (2018). https://doi.org/10.1007/978-3-030-01234-2_1
15. Zheng, Z., Wang, P., Liu, W., et al.: Distance-IoU Loss: faster and better learning for bounding box regression. In: Proceedings of the AAAI Conference on Artificial Intelligence, 34(07), pp. 12993–13000 (2020)
16. Ren, S.Q., He, K.M., Girshick, R., et al.: Faster R-CNN: towards real-time object detection with region proposal networks. In: Proceedings of the 28th International Conference on Neural Information Processing Systems, Montreal, Canada, pp. 91–99. ACM (2015)
17. Bochkovskiy, A., Wang, C.Y., Liao, H.Y.: YOLOv4: optimal speed and accuracy of object detection [EB/OL]. [2020–10–23]. https://arxiv.org/abs/2004 10934

A Novel Autoencoder for Task-Driven Object Segmentation

Weijie Jiang[1] , Yuxiang Cai[2,3], Yuanlong Yu[1]([⊠]) , Rong Chen[1],
Jianglong Zhang[3], Weitao Zheng[3], Renjie Su[3], and Xi Wu[3]

[1] College of Computer and Data Science, Fuzhou University, Fuzhou 350108, China
{n180310005,yu.yuanlong}@fzu.edu.cn
[2] Shanghai Jiao Tong University, 800 Dongchuan Road, Shanghai 200240, China
[3] State Grid Fujian Information and Telecommunication Company,
Fuzhou 350001, China

Abstract. Salient object segmentation has attracted increasing attention in recent years. Though a lot of works based on autoencoder have been made, these methods can only recognize and segment one class of objects, but cannot segment the other classes of objects in the same image. To address this issue, this paper proposes a novel autoencoder to perform task-driven object segmentation, in which a control signal is added to the decoder to determine which class of objects need to be segmented. What's more, we establish a Bayesian rule in the decoder, in which the control signal is set as the prior, and the latent features learned in encoder is transferred to corresponding layer of decoder as observation, thus the posterior probability of each object with respect to the specific-class can be calculated, and the objects belonging to this class can be segmented. This proposed method is evaluated for task-driven salient object segmentation on a several benchmark datasets, including DUT-OMRON, ECSSD, DUTS-TE, etc. Experimental results show that our approach tends to segment accurate, detailed and complete objects, and improves the performance compared with the previous state-of-the-arts.

Keywords: Object Segmentation · Autoencoder · Bayesian Rule

1 Introduction

Inspired by the attention mechanism of the human visual system (HVS), salient target segmentation [19] has become an important image processing technique in the field of computer vision. Existing salient object segmentation methods are based either on bottom-up or on top-down visual attention mechanism, where the

This work supported by University-Industry Cooperation Project of Fujian Provincial Department of Science and Technology under grant 2020H6101 and Science and Technology Project of State Grid Fujian Electric Power Co., Ltd. under grant 52130M19000X.

bottom-up saliency is a primitive function of the human visual system and can be considered to deal with low-level attributes such as color, edge and texture, and the top-down guidance is used to deal with high-level cognitive factors.

Recently, researchers have paid attention to top-down attention mechanism, which has obtained better performance on salient object segmentation than the bottom-up mechanism. Most recent bottom-up based methods construct a autoencoder to perform object segmentation, in which two convolutional neural networks (CNNs) are constructed as the encoder and decoder. The encoder is a classification model used to learn high-level features, and the decoder is used to de-convolute the learned features to a saliency map. Though these methods can achieve somewhat satisfactory results, they suffer from some limitations. First, it is hard to segment accurate, fine and entire object due to uneven distribution of saliency and coarse range of high-level features. Second, if there are more than one object in the testing image, these methods will segment all objects or only the objects belonging to the class involved in training processing. Therefore, in testing processing, we cannot specify a certain class of objects to be segmented. This is unpractical in real applications.

To this end, this paper proposes a novel autoencoder for task-driven salient object segmentation in which encoder and decoder have the same number of layers. The same as existing methods, the encoder uses a classification model to learn high-level features. While differ from other methods, a control signal is imposed to the decoder to determine which class of objects need to be segmented. Our decoder module has two branches: a prior network and a deconvolutional network (deconvnet) for producing channel-wise attention and spatial attention separately. The prior network is used to calculate the posterior probability of each pixel with respect to the specific class, in which the control signal is set as the prior, and the latent features learned in encoder are set as observation. With the posterior probability, we can segment the objects belonging to the specific class accurately. Experimental results on several challenging salient datasets show that our approach can effectively deal with small-object problem and integrity problem and obtain more accurate location and more detailed edges when compared with other state-of-the-art methods.

The main contributions of this paper are summarized as follows:

- This paper focus on task-driven salient object segmentation. As our knowledge, it is the first time the problem has been proposed. In order to determine which class of objects should be segmented in testing processing, we impose a control signal in decoder.
- A Bayesian rule is introduced in each layer of decoder to calculate the posterior probability of each pixel with respect to the specific class, in which the control signal is set as the prior, and the latent features learned in encoder is transferred to corresponding layer of decoder as observation. With the Bayesian rule, the learned latent features that are highly related to the specific class can be detected, and the corresponding objects can be segmented accurately.

– Compared with other state-of-the-art methods, the proposed method can segment salient object more accurate and complete, and is more suitable for complex image segmentation.

2 Related Work

2.1 Salient Object Segmentation

The methods for salient object segmentation can be roughly divided into two categories: heuristic methods (non-deep learning) and deep learning methods. The heuristic methods often use low-level features for saliency segmentation. The low-level features are often effective in simple scenarios but they are not robust in some challenging cases. Many deep learning methods add low-level features to high-level features. Wang et al. [16] leverage local estimation and object proposals to do a global search for saliency maps. [26] aggregates multi-resolution features into a final fusion in a top-down manner, makes use of multi-scale representation. Li et al. [10] construct a multi-tasks model of a saliency segmentation branch and a contour detection branch. Liu et al. [11] embeds global and local information into the top-down pathway. From focusing on the latest deep learning methods in reviewing related work, we find a trend to develop global-local network.

2.2 Attention Mechanism

A brief review of recent work in attention mechanism within end-to-end trainable deep network architectures is given in this section.

Attention Without Extra Training Phase. These attention approaches [2,24] try to thoroughly select the most salient regions and locate objects for a specific target in a natural scene, which only use original architectures. Zhang et al. [24] proposed a new backpropagation scheme, called Excitation Backprop, to pass along top-down signals downwards in the network hierarchy via a probabilistic Winner-Take-All process (WTA) [14]. STNet [2] is another extension method of WTA by relaxing constraints, which establish a safe margin to be more easily selected as winners and group of selected winners according to some similarity measures to apply a more restrictive selection. However, experimental results shows that the performance of the above methods for localizing specific objects is not satisfactory enough.

Attention with Additional Training Phase. In general, attention can be considered as a reliable cue to tasks that select the most informative or distinctive regions and remove the noisy data which may belong to the other tasks. Attention is usually applied to a specific model part, represented by a trainable set of normalized weights. Mainstream approaches include assigning weights to different channels [5,27], spatial regions [12,13] or a mixture of both [3,15,25].

Our method proposes to locate the salient objects cue according to classification networks and then use channel-wise attention and spatial attention as a supplementation for accurate and complete salient object segmentation. The goal of our work is to locate and segment the task-driven object regions in images. Our method accept flexible-size RGB images as input, and it can generate soft contextual saliency for each pixel corresponding to the specific class.

3 Architecture

A generic architecture of the proposed method is shown in Fig. 1. Our method is based on the autoencoder. Differing from existing methods, the decoder is composed of a prior attention network and a deconvnet.

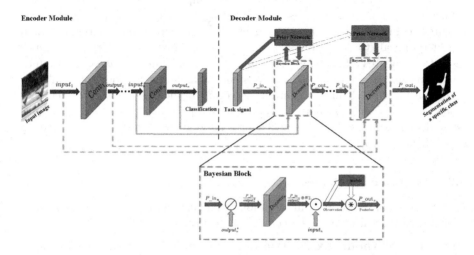

Fig. 1. Overview of the proposed method, which is based on autoencoder architecture. The decoder module is composed of a prior attention network and a deconvnet.

3.1 Network Architecture Overview

Our architecture is based on an autoencoder, inspired by a Bayesian approach [24] that combines task-driven signals with data-driven mappings. The encoder module is a classification network used to hierarchically learn multiple visual features from the input images. The decoder module is composed of a prior network and a deconvnet, which produces channel-wise attention and spatial attention, respectively. In our method, channel attention and spatial attention are probabilistically combined according to a Bayesian rule. We deliver the features learned from encoder module and the control signal to the decoder module as observation and prior, and combine observation with prior to compute posterior at each stage. Information at adjacent stages in the decoder module propagates according to corresponding information in the encoder module. The final output of the decoder module is a saliency map.

3.2   Encoder Module

The encoder uses a classifier network slightly modified to the traditional classifier network to cope with multi-target saliency detection. The last layer of our classification network replaces softmax layer with sigmoid layer $f(I;\theta)_i = \frac{1}{1+e^{-z_i^L}}$. We use cross entropy loss as loss function for our classification network, which can be written as

$$Loss = -\frac{1}{n}\sum_I [y \ln f(I;\theta) + (1-y)\ln(1 - f(I;\theta))] \tag{1}$$

3.3   Decoder Module

To mitigate the limitation that classification network only produce rough location, we introduce a decoder module as a multi-stage refinement process, which aims at extracting pure saliency maps by incorporating domain-specific features for the specific task. In Fig. 2, it has shown that channels in the classification network at higher layer have more task-specific characteristic but with lower spatial accuracy, and nodes at lower layer tend to localize more precise boundary with respect to the input image but less meaning. Motivated by this phenomenon, we use prior network to produce coarse-grained and meaningful channel-wise attention for a specific task. Our deconvnet first compute the current spatial attention, and then combine channel-wise attention and spatial attention to generate maps from coarse to fine.

Fig. 2. An intuitive description of task specificity in channels of feature maps. It can be divided into three parts due to specific tasks, including motorcycle, airplane and bicycle. The first row are original images, and the second row to the last row are the feature maps from the highest layer to the lowest layer. Outputs of each row are from the same layer. For each task, each row feature maps are in the same channel index.

We now discuss the two main branches: a prior network and a deconvnet, in our decoder module in details.

Prior Network. Our prior network contains three parts. The first part only concentrates on the relationship between task with feature maps in this layer. The second part employs global average pooling on the current feature maps to gain a global-feature descriptor. Then the final attention vector is generated by using a fully connected layer with a sigmoid function on outputs of the former two parts.

Suppose W_a^l, W_p^l and b_p denote the overall weights of the first part, the weights and bias of the third parts respectively. And q^l, g^l and p^l denote the outputs of the three parts respectively. Task signal and the feature maps in the lth layer are represent as t^c with respect to class c and f^l. The channel-wise attention is produced by

$$p^{l,c} = \frac{1}{1 + \exp{(W_p^l)^T[q^{l,c}; g^l] + b_p}}, \tag{2}$$

$$\tilde{f}_{cl}^{l,c} = p_{cl}^{l,c} \odot f_{cl}^l. \tag{3}$$

where $q^{l,c} = (W_a^l)^T t^c$, $g_{cl}^l = average_{i,j} f_{cl}^l$, and $\{i,j\}$ is the spatial location of a feature map, cl is the index of channels.

Note that the channel-wise attention vector is task-specific and further fed into each deconvolutional layer of decoder module regularly by applying dot product with the lth-layer feature maps, which select task-specific feature maps effectively.

Deconvnet. To effectively capture multi-scale features in the classification, channel-wise attention and spatial attention, we construct some Bayesian blocks by aggregating these features, as shown in Fig. 1. Each layer is required to calculate an approximate posterior, which is equivalent to estimating global and local features for each pixel.

We consider node A_s as a node activated by a specific task c, node B_n perhaps is related to this task. It can be factorized as

$$P(B_n) = \sum_{A_s \in parent(B_n)} P(B_n \mid A_s) P(A_s). \tag{4}$$

The conditional probability of node A_s toward node B_n is computed as

$$P(B_n \mid A_s) = \frac{z_{B_n}^l \cdot \overset{+}{w}_{A_s B_n}^{l+1}}{\sum_{B_i \in Son_{A_s}} z_{B_i}^l \cdot \overset{+}{w}_{A_s B_i}^{l+1}}, \tag{5}$$

where $z_{B_i}^l$ is the output of node B_i in the lth layer in encoder module, $w_{A_s B_i}^{l+1}$ is the weight connected node B_i to node A_s, and $\overset{+}{w} = max\{w, 0\}$.

In this paper, we insert channel-wise attention as prior into the above propagation process, as well as we consider the above mentioned probability as observation. We adopt Basysian rule to integrate this observation and prior. According to the Bayesian rule, the computation of all nodes' posterior for task c in the lth layer can be performed by matrix operations simply:

$$X^{l,c} = P^{l,c} \circledast Z^l \odot \left(W^{\overset{+}{l+1}} \otimes \frac{X^{l+1,c}}{W^{\overset{+}{l+1}} \otimes Z^l} \right), \tag{6}$$

where symbol X stands for posterior, symbol P stands for prior, and $\circledast$, $\odot$ and $\otimes$ denote multiplication operation across channels, multiplication operation at the pixel level and convolution operation respectively.

3.4 Training

During our training process, we need to train our classification network using image-level labels. In practice, we do not implement the training process of our classification network, and directly leverage the existing model weights. Parameters of our classification network are initialized by googlenet pretrained on ImageNet. We replace the fully-connected layer of googlenet with a convolutional layer. We mainly focus on the training process of the decoder module. The deconvnet branch is first initialized from the reverse version of classification network. For training this module, we need to obtain task-specific saliency maps and image-level labels associated with the input image, and updates deconvnet with fixed classification in an end-to-end mode at each training iteration. After that, we train the prior network with fixed deconvnet and classification network. The training process of prior network and deconvnet in our decoder module is carried out in a alternating-training way. In a word, our proposed architecture is pre-training with coarse labels and then fine-tuning with fine labels.

4   Experiments

In this section, experiments on eight datasets are implemented to verify the performance of the proposed method on task-driven salient object segmentation, and we compare the results with several state-of-the-art methods to validate the superiority of our approach.

4.1 Datasets

To show the performance of the proposed salient object segmentation architecture, we quantitatively and qualitatively evaluate it on seven generic salient datasets: DUT-OMRON [21], ECSSD [20], DUTS-TE [17], HKU-IS [8] and PASCAL-S [22], these datasets are widely used in most literatures for evaluation.

4.2   Implementation Details

We follow a common setting [18] to use DUTS-TR as our training set, and the rest of the above salient datasets as testing sets. In our experiments, any data augmentation is not used. The whole architecture is trained using stochastic gradient descent (SGD) in a end-to-end way. We train our decoder module with a learning rate of 0.05, 0.001 for channel attention network and spatial attention network respectively. We set momentum to 0.9 and weight decay to 0.0005. At each iteration, we use 224 as the input size for all datasets. The mini-batch size is set as 16. During testing phase, we use four average up-sampling layer instead of four reverses of max pooling between inception modules in googlenet to relieve sparsity of saliency maps.

4.3   Evaluation Metrics

We adopt six widely used metrics to evaluate the performance of all methods, i.e., Precision-Recall(PR) curves, $F_{measure}(F_\beta)$, Mean Absolute Error(MAE), Structure measure $(S_{measure})$ [4]. The precision and recall is computed by applying a threshold on the predicted saliency map to obtain a binary mask. $F_{measure}$ considers to balance precision and recall. Instead of using all $F_{measure}$ values, most methods usually use the maximum of F-measure. Here β^2 is set to be 0.5.

$$F_\beta = \frac{(1 + \beta^2)\, Precision \times Recall}{\beta^2 Precision + Recall}. \tag{7}$$

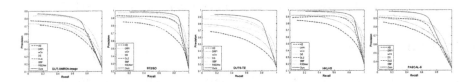

Fig. 3. Comparison of five saliency datasets in terms of the PR curve.

Table 1. Quantitative results of our method, including F_β, $S_{measure}$ and MAE, compared against other state-of-the art salient segmentation methods on five generic datasets. The best two results are shown in red and green respectively.

Dataset	DUT-OMRON			ECSSD			DUTS-TE			HKU-IS			PASCAL-S		
Methods/Metric	F_β ↑	S-measure ↑	MAE ↓	F_β ↑	S-measure ↑	MAE ↓	F_β ↑	S-measure ↑	MAE ↓	F_β ↑	S-measure ↑	MAE ↓	F_β ↑	S-measure ↑	MAE ↓
HS [20]	0.561	0.633	0.227	0.673	0.685	0.228	0.504	0.601	0.243	0.652	0.674	0.215	0.569	0.624	0.262
DRFI [7]	0.623	0.696	0.150	0.751	0.732	0.170	0.600	0.676	0.155	0.745	0.740	0.145	0.639	0.658	0.207
wCtr [28]	0.541	0.653	0.171	0.684	0.714	0.165	0.522	0.639	0.176	0.695	0.729	0.138	0.599	0.656	0.196
DS [9]		0.750	0.120	0.868	0.821	0.122	0.747			0.848	0.853	0.078	0.718		0.175
DLS [6]	0.644	0.725		0.826	0.806		-	-	-	0.807	0.799		0.712	0.723	
SBF [23]	0.649		0.110	0.833		0.091	0.657	0.743	0.109	0.821	0.829	0.078			0.133
RSDNet [1]	0.715	0.644	0.178		0.788	0.173	0.798	0.720	0.161			0.156	-	-	-
Ours	0.715		0.072	0.890	0.842	0.075	0.798	0.707	0.060	0.878		0.058	0.811	0.814	0.082

Table 2. Quantitative results of our iterative process, including F_β, $S_{measure}$ and MAE, on five generic datasets. The best three results are shown in red, green and blue respectively.

Dataset	DUT-OMRON			ECSSD			DUTS-TE			HKU-IS			PASCAL-S		
Iteration/Metric	$F_\beta\uparrow$	S-measure $\uparrow$	MAE $\downarrow$	$F_\beta\uparrow$	S-measure $\uparrow$	MAE $\downarrow$	$F_\beta\uparrow$	S-measure $\uparrow$	MAE $\downarrow$	$F_\beta\uparrow$	S-measure $\uparrow$	MAE $\downarrow$	$F_\beta\uparrow$	S-measure $\uparrow$	MAE $\downarrow$
prior_0	0.708	0.701	0.083	0.877	0.783	0.098	0.787	0.741	0.073	0.868	0.779	0.079	0.807	0.811	0.085
backnetwork_0	0.701	0.762	0.067	0.880	0.857	0.064	0.767	0.758	0.057	0.856	0.843	0.055	0.811	0.816	0.082
prior_1	0.707	0.693	0.081	0.877	0.777	0.097	0.782	0.730	0.073	0.861	0.768	0.078	0.813	0.812	0.082
backnetwork_1	0.701	0.701	0.085	0.874	0.788	0.100	0.792	0.761	0.070	0.870	0.800	0.073	0.812	0.813	0.083
prior_2	0.714	0.751	0.071	0.800	0.848	1.075	0.791	0.799	0.069	0.882	0.843	0.059	0.811	0.815	0.083
backnetwork_2	0.715	0.748	0.072	0.890	0.842	0.065	0.798	0.797	0.060	0.873	0.840	0.055	0.811	0.814	0.082

MAE measures the average pixel-wise absolute error between saliency map S and ground truth G,

$$MAE = \frac{1}{W \times H} \sum_{i=1}^{W} \sum_{j=1}^{H} \mid (G(i,j) - S(i,j)) \mid. \tag{8}$$

$S_{measure}$ takes object-aware similarity and region-aware similarity into consideration,

$$S_{measure} = \alpha \times S_o + (1-\alpha) \times S_r. \tag{9}$$

4.4 Performance Comparison on Saliency Datasets

We compare our method with other seven state-of-the-art methods, including 3 heuristic methods (HS [20], DRFI [7], wCtr [28]) and 6 deep learning methods (DS [9], DLS [6], SBF [23], RSDNet [1]).

Quantitative Evaluation. Figure 3 gives the PR curves of all methods on five salient datasets. From these results, we can observe that the curves of our proposed method are almost floating on the top part than that of other methods. And saliency maps produced by our method have balanced salient values rather than various levels of salient values, which have more steep slope. It is beneficial to segment the entire salient object directly without postprocessing.

The results of $F_{measure}$, $S_{measure}$ and MAE of all methods are shown in Table 1, the most results of Table 1 come from [19]. As it can be seen, our proposed method outperform other methods on most datasets in terms of all evaluation metrics, which do not use any post-processing techniques. In term of $F_{measure}$, our method improves 1.1%, 0.8%, and 9.6% than the second one on ECSSD, HKU-IS and PASCAL-S dataset, respectively. All $S_{measure}$ results of our method on these datasets almost rank first or rank second. In term of MAE, our method has an improvement of 20%, 12.8%, 33.3% and 15.9% on DUT-OMRON, ECSSD, DUTS-TE and HKU-IS dataset, respectively (Table 2).

Qualitative Evaluation. Figure 4 shows a visual comparison of our method with other methods. It can be seen that our method is capable of uniformly highlighting the entire salient object as well as suppressing the background noise. Further, our saliency maps are indeed finer than other methods, specifically in

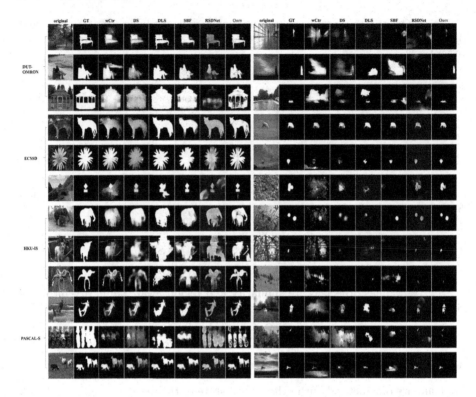

Fig. 4. Qualitative results of our method along with other state-of-the art methods.

the edge. Our method also does better in segmenting small salient objects in various challenging scenes. For example, desk object of the image in 1st row and 1st column lose leg information in wCtr, DS and SBF due to background factor, while our method can clearly differentiate it. For objects with complex boundaries (dog and flower of images in 4st row and 1st column, 5st row and 1st column), our method can draw irregular-shape objects. Objects of images in 2st column have small proportion. These results of other methods can be divided into three types. First, objects are not located. Second, saliency maps have a lot of noise. Third, saliency maps have coarse appearances. These three types can be solved by our method.

5   Conclusions and Future Works

Our architecture formulate channel-wise attention and spatial attention hierarchically via a Bayesian rule for segmenting more accurate and more complete objects progressively to finer and segment task-specific objects in a top-down manner. Experiments highlight the effectiveness and the value of our approach for salient object segmentation. It outperforms majority of existing methods and is far the closet method to human perception. Moreover, these ideas will provide insights for constructing novel and effective saliency segmentation architectures in future.

References

1. Amirul Islam, M., Kalash, M., Bruce, N.D.: Revisiting salient object detection: simultaneous detection, ranking, and subitizing of multiple salient objects. In: Proceedings of the IEEE Conference on Computer Vision and Pattern Recognition, pp. 7142–7150 (2018)
2. Biparva, M., Tsotsos, J.: Stnet: selective tuning of convolutional networks for object localization. In: Proceedings of the IEEE International Conference on Computer Vision, pp. 2715–2723 (2017)
3. Cao, C., et al.: Look and think twice: capturing top-down visual attention with feedback convolutional neural networks. In: Proceedings of the IEEE International Conference on Computer Vision, pp. 2956–2964 (2015)
4. Fan, D.P., Cheng, M.M., Liu, Y., Li, T., Borji, A.: Structure-measure: a new way to evaluate foreground maps. In: Proceedings of the IEEE International Conference on Computer Vision, pp. 4548–4557 (2017)
5. Hu, J., Shen, L., Sun, G.: Squeeze-and-excitation networks. In: Proceedings of the IEEE Conference on Computer Vision and Pattern Recognition, pp. 7132–7141 (2018)
6. Hu, P., Shuai, B., Liu, J., Wang, G.: Deep level sets for salient object detection. In: Proceedings of the IEEE Conference on Computer Vision and Pattern Recognition, pp. 2300–2309 (2017)
7. Jiang, H., Wang, J., Yuan, Z., Wu, Y., Zheng, N., Li, S.: Salient object detection: a discriminative regional feature integration approach. In: Proceedings of the IEEE Conference on Computer Vision and Pattern Recognition, pp. 2083–2090 (2013)
8. Li, G., Yu, Y.: Visual saliency based on multiscale deep features. In: Computer Vision and Pattern Recognition (2015)
9. Li, X., Zhao, L., Wei, L., Yang, M.H., Wu, F., Zhuang, Y., Ling, H., Wang, J.: Deepsaliency: multi-task deep neural network model for salient object detection. IEEE Trans. Image Process. 25(8), 3919–3930 (2016)
10. Li, X., Yang, F., Cheng, H., Liu, W., Shen, D.: Contour knowledge transfer for salient object detection. In: Ferrari, V., Hebert, M., Sminchisescu, C., Weiss, Y. (eds.) ECCV 2018. LNCS, vol. 11219, pp. 370–385. Springer, Cham (2018). https://doi.org/10.1007/978-3-030-01267-0_22
11. Liu, N., Han, J., Yang, M.H.: Picanet: learning pixel-wise contextual attention for saliency detection. In: Proceedings of the IEEE Conference on Computer Vision and Pattern Recognition, pp. 3089–3098 (2018)
12. Mnih, V., Heess, N., Graves, A., et al.: Recurrent models of visual attention. In: Advances in Neural Information Processing Systems, pp. 2204–2212 (2014)
13. Oktay, O., et al.: Attention u-net: learning where to look for the pancreas. arXiv preprint arXiv:1804.03999 (2018)
14. Tsotsos, J.K., Culhane, S.M., Wai, W.Y.K., Lai, Y., Davis, N., Nuflo, F.: Modeling visual attention via selective tuning. Artif. Intell. 78(1–2), 507–545 (1995)
15. Wang, F., et al.: Residual attention network for image classification. In: Proceedings of the IEEE Conference on Computer Vision and Pattern Recognition, pp. 3156–3164 (2017)

16. Wang, L., Lu, H., Ruan, X., Yang, M.H.: Deep networks for saliency detection via local estimation and global search. In: Proceedings of the IEEE Conference on Computer Vision and Pattern Recognition, pp. 3183–3192 (2015)
17. Wang, L., Lu, H., Wang, Y., Feng, M., Xiang, R.: Learning to detect salient objects with image-level supervision. In: IEEE Conference on Computer Vision and Pattern Recognition (2017)
18. Wang, T., Borji, A., Zhang, L., Zhang, P., Lu, H.: A stagewise refinement model for detecting salient objects in images. In: Proceedings of the IEEE International Conference on Computer Vision, pp. 4019–4028 (2017)
19. Wang, W., Lai, Q., Fu, H., Shen, J., Ling, H.: Salient object detection in the deep learning era: An in-depth survey. arXiv preprint arXiv:1904.09146 (2019)
20. Yan, Q., Xu, L., Shi, J., Jia, J.: Hierarchical saliency detection. In: Proceedings of the IEEE Conference on Computer Vision and Pattern Recognition, pp. 1155–1162 (2013)
21. Yang, C., Zhang, L., Lu, H., Xiang, R., Yang, M.H.: Saliency detection via graph-based manifold ranking (2013)
22. Yin, L., Hou, X., Koch, C., Rehg, J.M., Yuille, A.L.: The secrets of salient object segmentation (2014)
23. Zhang, D., Han, J., Zhang, Y.: Supervision by fusion: towards unsupervised learning of deep salient object detector. In: Proceedings of the IEEE International Conference on Computer Vision, pp. 4048–4056 (2017)
24. Zhang, J., Bargal, S.A., Lin, Z., Brandt, J., Shen, X., Sclaroff, S.: Top-down neural attention by excitation backprop. Int. J. Comput. Vision **126**(10), 1084–1102 (2018)
25. Zhang, M., Ma, K.T., Lim, J., Zhao, Q., Feng, J.: Anticipating where people will look using adversarial networks. IEEE Trans. Pattern Anal. Mach. Intell. (2018)
26. Zhang, P., Wang, D., Lu, H., Wang, H., Ruan, X.: Amulet: aggregating multi-level convolutional features for salient object detection. In: Proceedings of the IEEE International Conference on Computer Vision, pp. 202–211 (2017)
27. Zhou, B., Khosla, A., Lapedriza, A., Oliva, A., Torralba, A.: Learning deep features for discriminative localization. In: Proceedings of the IEEE Conference on Computer Vision and Pattern Recognition, pp. 2921–2929 (2016)
28. Zhu, W., Liang, S., Wei, Y., Sun, J.: Saliency optimization from robust background detection. In: Proceedings of the IEEE Conference on Computer Vision and Pattern Recognition, pp. 2814–2821 (2014)

Feedback Attention-Augmented Bilateral Network for Amodal Instance Segmentation

Junjie Dong[1], Huaping Liu[2(✉)], Jun Xie[1], XinYing Xu[3], and Zijun Lan[1]

[1] College of Information and Computer, Taiyuan University of Technology,
Taiyuan, China
[2] Department of Computer Science and Technology, Tsinghua University,
Beijing, China
hpliu@tsinghua.edu.cn
[3] College of electrical and power engineering, Taiyuan University of Technology,
Taiyuan, China

Abstract. Amodal instance segmentation is a recently novel research in the field of computer vision, which includes the prediction of the visible and occluded parts of each object instance. Many modern computer vision methods demonstrate excellent performance by using the mechanism of looking and thinking twice and the attention mechanism. In this paper, we propose a feedback attention-augmented bilateral network. Specifically, after the convolutional network extracts the first round features of the input image, a feedback connection for relearning is introduced to extract the second round features. Meanwhile, a spatial detail preservation network with multi-scale attention feature module is designed to preserve and capture the rich spatial details information. Finally, a simple but practical feature fusion module is used to effectively combine both shallow and deep features from the backbone branch and the spatial branch. We achieve better results on COCOA dataset and D2SA dataset. For COCOA dataset, our method outperforms different state-of-the-art models by 1.2% and 1.6% mAP under "All regions" and "Things only", and brings 9.4% improvement on D2SA dataset.

Keywords: Amodal instance segmentation · Bilateral network · Feedback attention-augmented · Spatial detail preservation

1 Introduction

Visual recognition tasks such as image classification [1–3], object detection [4–6], semantic segmentation [7,8] and instance segmentation [9,10] have witnessed dramatic progress in recent years. Instance segmentation, as an important image understanding method, aims to locate all object and classify them at the pixel level in the image. Nevertheless, instance segmentation is still not enough to fully understand the complex surrounding environment. For example, when objects have mutual occlusion relationships, instance segmentation can only identify and segment the visible parts, and can't predict the occluded parts of the object.

F. Sun et al. (Eds.): ICCCS 2022, CCIS 1732, pp. 41–55, 2023.
https://doi.org/10.1007/978-981-99-2789-0_4

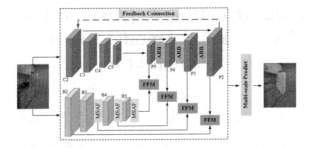

Fig. 1. Overview of the proposed feedback attention-augmented bilateral network for amodal instance segmentation. The top branch is the backbone network, the bottom branch is the spatial detail preservation network, the red dashed line denots the feedback connection, ARB: Attention Refinement Block, MSAF: Multi-scale Attention Feature, FFM: Feature Fusion Module. (Color figure online)

Humans, with their strong visual system, have the strong ability to see beyond the visible, i.e. to reason about the invisible, occluded parts of objects. In the field of computer vision, this ability is also called amodal perception [11]. Researchers drew inspiration from the human visual system and tried to develop a computer vision system with similar functions until most recently. Amodal instance segmentation [12] is an extension of the instance segmentation task. Its main task is to predict both the visible and the occluded part of each object, in order to help the computer vision system realize a wider range of perception of the object.

It is most challenging to reason about the complete physical structure of an object with only partial visibility. Amodal instance segmentation focuses on the reasonable extension of the invisible part at the pixel level. Therefore, it is not only necessary to identify and locate the object, but also to distinguish the interference caused by other surrounding object features. When predicting the occluded parts of the object, the same pixel may be assigned multiple labels. Existing researches [12,20–22] use the semantic segmentation to deal with amodal segmentation. However, the insufficient feature representation ability and the lack of spatial detail information makes the prediction results underfitting or even wrong.

Although the feedforward deep convolutional neural network [13] has achieved great success in computer vision, it should be noted that looking and thinking twice by using the feedback connection achieve better performance, because the feedback connection passing the high-level semantic information down to the low-level perception can help to better understand and extract the visual feature of expected objects. At the same time, the visual attention mechanism has been proven not only to tell us where to focus, but also to capture global context information and improve the feature representation ability. In addition, convolution and pooling operations with stride significantly reduce the spatial resolution of image feature maps, resulting in insufficient spatial detail information, which may cause the problem of feature misalignment. Thus, some researchers use skip connection to aggregate low-level spatial information and achieved good performance.

In order to obtain more advanced features and make full use of spatial information, we propose a feedback attention-augmented bilateral network for amodal instance segmentation [14]. The feedback connection can continuously learn image features; and the attention mechanism can establish long-term dependencies and capture global context information. Specifically, we add an attention refinement block to the backbone network. We introduce the feedback connection to relearn the output feature maps, and extract the image features twice. The reason is that a feature map convolved twice can correct previous states and regenerate more high-level representations. At the same time, we designed a spatial detail preservation branch with shallow convolutional layers parallel to the backbone network. Finally, we utilize feature fusion module to effectively combine both shallow and deep features from the backbone branch and the spatial branch, respectively. In this case, the network can aggregate advanced semantic information and abundant spatial detail information.

2   Related Work

2.1   Amodal 3D Object Detection

The topic of amodal perception has been widely studied in various fields of computer vision. One of the most important methods is the 3D object detection [15,16]. Shuran et al. [17] proposed the 3D Region Proposal Network to learn objectness from geometric shapes and the joint Object Recognition Network to extract geometric features in 3D and color features in 2D. Zhuo et al. [18] revisited the amodal 3D detection problem by sticking to the 2.5D representation framework, and directly relate 2.5D visual appearance to 3D objects. Hao et al. [19] introduced a 3D Convolutional Neural Network, and they treated detection and recognition as one regression problem in a single network. Compared with amodal 3D object detection perceiving the shape and semantics of the occluded parts by predicting the 3D bounding box, our work is more focused on object instances in 2D images.

2.2   Amodal Instance Segmentation

The concept of amodal instance segmentation has just emerged in the last few years. Li and Malik [20] were the first to provide a method for amodal instance segmentation. They first generate data by randomly overlaying one object instance onto another, thereby simulating the phenomenon of occlusion between objects. And they iteratively expand the modal bounding box into the direction of the high heatmap value and recalculate the heatmap to realize the amodal instance segmentation. Zhu et al. [12] provided a data set with amodal Ground Truth for amodal instance segmentation, denoted as COCO-amodal. Follmann et al. [21] proposed the ARCNN model, which is based on Mask RCNN, simultaneously predicted the visible mask, amodal mask and occlusion masks of each object by adding amodal mask head and occlusion mask head. Zhang et al. [22] introduced a semantics-aware distance map. The different regions of an object are placed into different levels on the map according to their visibility.

2.3   Attention Mechanism

Many current computer vision methods demonstrate excellent performance by using the attention mechanism. Hu et al. [23] proposed the SENet model, which explores the importance of the feature map each channel, and adaptively recalibrating the feature response of the channel. The work [24], which is related to self-attention module, mainly exploring effectiveness of non-local operation in spacetime dimension for videos and images. Sanghyun et al. [25] proposed convolutional block attention module (CBAM). This module sequentially infers attention maps along two separate dimensions, channel and spatial. Hiroki Tsuda et al. [26] focused on the feedback processing of the human brain, and solved the task of cell image segmentation through the feedback attention mechanism. In our work, we introduce the attention mechanism into the amodal instance segmentation task to adaptively focus on more relevant regions.

2.4   Feature Fusion in Deep Learning

In many recent works, the long skip connections are often used to aggregate features. Because long skip connections helps the network to obtain high-resolution semantic features by bridging the finer detailed features from the lower layers and the high-level semantic features from deeper layers. However, the fusion of features is usually implemented through addition or concatenation. Regardless of the variance of contents, features are assigned with fixed weights. Rupesh et al. [27] proposed Highway networks, which first introduced a selection mechanism in short skip connections. The work [28] proposed Parallel reverse attention network which is able to establish the relationship between areas and boundary cues. Yimian et al. [29] proposed an attentional feature fusion to combine features from different layers. In our work, we aggregate high-level semantic information and spatial detail information through soft selection or weighted averaging.

3   Proposed Method

3.1   Overview

The proposed feedback attention-augmented bilateral network is modified on the SLN method, and consists of three main components: the ResNet101 backbone network with feedback connection, the spatial detail preservation network (SPN) and feature fusion module (FFM). An overview of the network is depicted in Fig. 1. Firstly, the input images are fed into the ResNet101 backbone network to extract image features. Note that the feature map will be downsampled

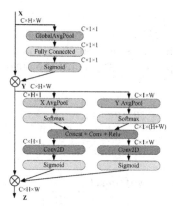

Fig. 2. Architecture of Attention Refinement Block(ARB).

to 1/32 of the original input image size. In order to generate larger size feature map, skip connection and attention refinement blocks are used to perform upsampling operations. Then we reconnect the output feature maps to the input through the feedback connection, forming a feedback attention-augmented structure for relearning and generating richer semantic details. Secondly, the input images are also fed into the spatial detail preservation network to further extract sufficient spatial information. This path can effectively aggregate spatial position detail information through the multi-scale attention feature module, which greatly improves efficiency. Thirdly, the outputs from the two paths are effectively combined by feature fusion module. Finally, we utilize multi-scale feature maps to obtain amodal instance segmentation prediction results.

3.2   Feedback Attention-Augmented Network

Our proposed method is based on ResNet101, which is used as a standard for image feature extraction. As illustrated in Fig. 1, the backbone network is ResNet101 with attention refinement block and feedback connection. Specifically, we denote the output of these last residual blocks as $\{C2, C3, C4, C5\}$ for conv2, conv3, conv4, and conv5 outputs, and they have strides of $\{4, 8, 16, 32\}$ pixels with respect to the input image [30]. Then we use skip connection and Attention Refinement Block(ARB) to upsample the output feature maps and denote these as $\{P2, P3, P4, P5\}$. Note that the P2 and C2 have the same channel and size. We do not include conv1 into the network due to its large memory footprint. The network architecture of ARB is shown in Fig. 2. The ARB is a dual attention mechanism, including channel and spatial attention. Given a input feature map $X \in R^{C \times H \times W}$, ARB firstly performs the global average pooling to capture global information. Then we use two fully-connected layers with Sigmoid function to obtain the channel attention vector. Finally, according to the channel attention vector, the different channels of the input feature map X are reweighted to obtain the feature map Y.

Given the channel weighted feature map $Y \in R^{C \times H \times W}$, we utilize two spatial extents of pooling kernels (H, 1) or (1, W) to encode each channel along the horizontal and vertical directions, and apply a softmax layer to calculate the spatial attention map respectively [31]. Thus, the output of the c-th channel at height h can be formulated as

$$z_c^h(h) = softmax(\frac{1}{W} \sum_{0 \leq i \leq W} y_c(h, i)) \tag{1}$$

Similarly, the output of the c-th channel at width w can be written as

$$z_c^w(w) = softmax(\frac{1}{H} \sum_{0 \leq j \leq H} y_c(j, w)) \tag{2}$$

Then we concatenate them and use 1×1 convolution followed by ReLU function to encode spatial information in both the horizontal and vertical directions, the output results can be formulated as

$$f = \delta(conv([z^h, z^w])) \tag{3}$$

where $[\cdot, \cdot]$ denotes the concatenation operation, δ denotes the ReLU function, and $f \in R^{C \times 1 \times (H+W)}$. We split f into two separate tensors $f^h \in R^{C \times H \times 1}$ and $f^w \in R^{C \times 1 \times W}$ along the horizontal and vertical directions, and utilize another two 1×1 convolutions followed by sigmoid function so that the f^h and f^w have the same channel as Y. Finally, we perform a multiplication between the channel weighted feature map Y and the horizontal and vertical output results f^h and f^w to obtain the output feature map Z.

The feedback mechanism is a loop iteration that allows the network to correct previous states and regenerate high-level representations. In our work, we utilize the feedback connection to transfer the P2 with high-level information to the input C2 with low-level information. Thus, a network with a recursive structure is constituted. The convolutional network extracts features of the input images, and introduces a feedback connection for relearning to extract the other features, and adaptively weights the features extracted at the two rounds. The output feature map O_m can be written as

$$O_m = P_m^1 + \alpha P_m^2 \tag{4}$$

where is α initialized as 0 and gradually learns to assign more weight, $P_m{}^1$ denotes the features extracted in the first round, $P_m{}^2$ denotes the features extracted in the second round, $m = 2, \ldots, 5$. In this case, we use the input feature map that was processed two times by convolution. The reason is that a feature map convolved twice can extract more advanced features than a feature map convolved once.

3.3   Spatial Detail Preservation Network

Amodal instance segmentation is to reasonably predict the shape and boundary of the occluded parts of the object at the pixel level, which requires not only sufficient semantic information, but also rich spatial detail information. Intuitively,

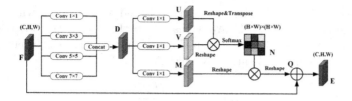

Fig. 3. Multi-scale Attention Feature Module(MSAF).

the feature maps of the deep layer contains rich high-level semantic information, while the features of the shallow layer are more helpful for describing the shape and boundary of the object. Many modern amodal instance segmentation methods have poor segmentation quality. One of the important reasons is that they ignore spatial information for predicting fine results for amodal instance segmentation tasks.

In our proposed method, the outputs of the backbone network with attention refinement block and feedback connection mainly obtain the rich semantic information while the spatial detail information is only partially preserved. Therefore, we propose a Spatial detail Preservation Network (SPN) to preserve the spatial size of the input image and encode affluent spatial information. Meanwhile, To actively capture the rich contextual relationships over global features, we introduce a Multi-scale Attention Feature Module(MSAF). The module encodes a wider range of contextual information through multi-scale feature aggregation and spatial attention mechanisms, thus obtaining abundant spatial detail information. Next, we elaborate the structure of spatial detail preservation network.

As illustrated in Fig. 1, the spatial detail preservation network is a shallow convolutional neural network that contains multi-scale attention feature module(MSAF). Note that the $\{B2, B3, B4, B5\}$ and $\{P2, P3, P4, P5\}$ have the same channel and size respectively. The B2 contains two 3×3 convolutional layers. However, from B3 to B5, each stage consists of one MSAF module and two 3×3 convolutional layers. All convolutional layers are followed by Batch Normalization [32]and ReLU. The only goal of the spatial detail preservation network is to encode the spatial details of the image. Figure 3 presents the structure of the MSAF.

Given the feature map $F \in R^{C \times H \times W}$, we first use a variety of different scale convolution kernels to extract features, such as 1×1, 3×3, 5×5, 7×7 convolution kernels. The number of output feature map channels of each convolution kernel scale is reduced by half, and then perform the concatenation operation in the channel dimension to obtain the output feature map D, where $D \in R^{2C \times H \times W}$. The different scale convolution kernels can effectively extract features of objects with large spatial changes and different scales, which enhances the adaptability to object scales. Then we feed the feature map D into two 1×1 convolution layers W_u and W_v to generate two new feature maps U and V, respectively, the output can be written as

$$U = W_u{}^T F \quad \text{and} \quad V = W_v{}^T F \tag{5}$$

where $U, V \in R^{C' \times H \times W}$. Then we reshape U and V to $R^{C' \times K}$, where K = H × W is the number of pixels. After that we perform a matrix multiplication between the transpose of feature map U and V, and we utilize softmax function to calculate the multi-scale attention map $N \in R^{K \times K}$:

$$n_{ij} = \frac{\exp(U_i^T V_j)}{\sum_{i=1}^{K} \exp(U_i^T V_j)} \tag{6}$$

where n_{ij} indicates the correlation between the i-th position and the j-th position.

Meanwhile, we transform the feature map F to another representation $M \in R^{C' \times H \times W}$ by using the 1 × 1 convolution and reshape M to $R^{C' \times K}$. Then perform a matrix multiplication between the feature map N and the transpose of M and reshape the result to $R^{C \times H \times W}$. Finally, an element-wise sum operation is performed with the input feature map F to get the final feature map E. We formulate the function as

$$E_i = \sum_{j=1}^{K} (n_{ij} M_j^T)^T + F_i \tag{7}$$

The output feature map E shows that the MSAF can aggregate multi-scale global context information. Therefore, our proposed spatial detail preservation network not only encodes rich spatial detail information, but also captures long-range contextual information in the spatial dimension, which makes the network more efficient in preserving spatial information.

3.4   Feature Fusion Module

In order to make full use of high-level semantic information and spatial detail information, we design a Feature Fusion Module to aggregate these two kinds of information. Specifically, Given two feature maps $X, Y \in \mathbb{R}^{C \times H \times W}$, we assume that Y is the feature map from the backbone network with attention refinement block and feedback connection, and X comes from the spatial detail preservation network. We first perform a element-wise sum operation on the feature maps X and Y, then accomplish feature fusion through soft selection. Figure 4 presents the structure of the Feature Fusion Module (FFM), where the blue dashed line denotes $1 - G(X + Y)$. Thus the FFM can be expressed as

$$Z = \mu G(X + Y) \otimes X + \lambda(1 - G(X + Y)) \otimes Y \tag{8}$$

where $Z \in R^{C \times H \times W}$ is the fused feature, $G(\cdot)$ denotes the convolution, sigmoid and multiplication operation, μ, λ are learnable parameter. The FFM enable the network to conduct a soft selection or weighted averaging between X and Y.

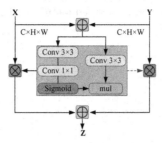

Fig. 4. Feature Fusion Module (FFM).

4   Experiments

4.1   Implementation Details

We implement our method based on Pytorch. The backbone network is ResNet101, and use the parameters pretrained on the COCO2014 dataset to initialize the network. We employ stochastic gradient descent algorithm as the optimization function in the training process. The initial learning rate is set to 0.001. Momentum and weight decay coefficients are set to 0.9 and 0.0001 respectively. The size of the input image is 1024 × 800, and Batchsize are set to 1. The loss function and other hyperparameters of our model are set according to the SLN method. We set training time to 18 epochs for COCOA dataset and 24 epochs for D2SA dataset. For data augmentation, we apply random left-right and different lightings during training for D2SA dataset.

4.2   Datasets and Evaluation Metrics

COCOA Dataset. The COCOA dataset consists of 5073 images, of which 2500, 1323 and 1250 images are used for training, validation and testing, respectively. In the COCOA dataset, most objects and background stuff regions are annotated with amodal masks. The annotations contain the category, amodal mask, and additional visible or invisible masks for occluded objects. All of the objects are classified into two categories:'things' and 'stuff', where the'things' represents an object of practical significance(e.g. person, horse, car, bicycle), the 'stuff' usually indicates background(e.g. sky, rode, grass).

D2SA Dataset. The D2SA dataset is a recently proposed amodal dataset for supermarket products. There is only minor occlusions in the D2SA dataset. Similarly, occluded objects are additionally annotated with visible and invisible masks. In our experiment, we adjusted the dataset distribution to include 2500 training, 719 testing images.

Table 1. Comparisons between the proposed method and other state-of-the-art methods on the COCOA test dataset

Method	All regions					Things only					Stuff only				
	mAP	AR^{10}	AR^{100}	AR^P	AR^H	mAP	AR^{10}	AR^{100}	AR^P	AR^H	mAP	AR^{10}	AR^{100}	AR^P	AR^H
AmodalMask	5.74	13.5	29.23	31.0	21.3	6.12	16.5	33.1	37.0	23.6	0.78	5.4	18.1	16.1	18.0
ARCNN	4.1	10.2	21.3	22.0	13.3	4.4	12.0	23.9	34.7	15.2	0.3	4.8	13.8	15.1	10.1
ARCNN++	6.6	15.3	32.4	34.8	17.1	7.8	19.5	37.6	40.8	19.9	0.5	3.3	17.1	19.9	12.5
SLN	8.4	16.6	36.5	40.1	22.5	9.6	20.5	40.5	43.6	24.9	0.8	5.3	25	31.3	18.6
OURS	9.6	17.1	35.6	40.0	21.9	11.2	20.8	39.3	43.4	24.3	0.9	6.4	24.9	31.4	18.0

Table 2. Ablation study on COCOA test dataset. FA represents Feedback Attention-augmented Network, SPN represents Spatial detail Preservation Network

Method	BaseNet	FA	SPN	All/mAP	Things/mAP
SLN	Res101			8.4	9.6
ours	Res101	✓		8.9	10.3
ours	Res101		✓	9.1	10.5
ours	Res101	✓	✓	9.6	11.2

Table 3. The results comparison between our method and ARCNN, SLN on the D2SA dataset

Method	All			Partial occlusion			Heavy occlusion		
	mAP	AR^{10}	AR^{100}	mAP	AR^{10}	AR^{100}	mAP	AR^{10}	AR^{100}
ARCNN	23.4	47.3	73.1	6.2	38.0	72.6	0.6	17.8	45.0
SLN	33.7	56.0	80.4	10.4	47.9	79.7	1.5	25.8	58.9
OURS	43.1	60.7	81.3	19.5	54.1	81.1	1.6	29.8	61.1

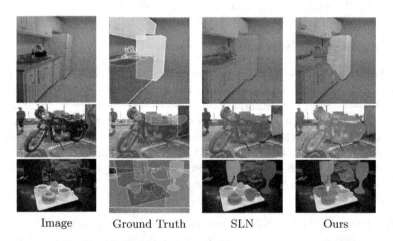

Image   Ground Truth   SLN   Ours

Fig. 5. Example results on the COCOA test dataset. Our method obtains better performance.

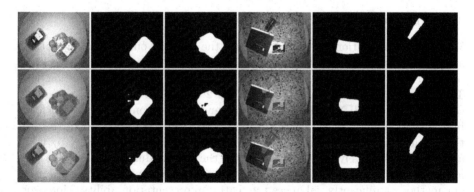

Fig. 6. Visualization results on the D2SA test dataset. From top to bottom respectively input images and Ground-Truth, SLN and our predicted results.

Evaluation Metrics. We use the mean Average Precision (mAP) and the mean Average Recall (AR) to measure the proposed method. As is common practice, the mAP are averaged over ten equally spaced Intersection over Union (IoU) thresholds from 0.5 to 0.95 to highlight more precise results. The AR is computed by averaging the segmentation recall over the same set of thresholds. For the COCOA dataset, we denote the AR for both 10 and 100 segments per image as AR^{10} and AR^{100}. We calculate the mAP, AR^{10} and AR^{100} in the three categories of All regions, things and stuff. Meanwhile, we also report the AR^{100} under partial and heavy occlusion, which we demote as AR^P and AR^H. For the D2SA dataset, we also report the mAP, AR^{10} and AR^{100} for all objects and occluded objects.

4.3   Comparing with Other Methods on COCOA Dataset

We compare our method with AmodalMask, ARCNN, ARCNN++ and SLN on the COCOA dataset. Note that the ARCNN ++ denotes ARCNN with visible mask. As shown in Table 1, we can see that the proposed method improve the performance remarkably. Specifically, the mAP obtaines by the proposed method under "All regions" and "Things only" is about 1.2% and 1.6% higher than SLN on COCOA dataset, respectively. Similarly, our method obtains the consistent performance improvements in AR^{10} metrics. But, the AR^{100} decrease slightly. And AmodalMask and ARCNN show poor segmentation accuracy (they obtain more than 3% mAP drop than the proposed method). Different from the previous methods, we introduce the two-path framework, where one branch is used to extract high-level semantic information, and the other branch captures spatial detail information. And the proposed method can achieve better performance, which demonstrates the effectiveness and efficiency of our amodal semantic segmentation method.

4.4 Ablation Study on COCOA Dataset

We conduct an ablation study to investigate the effects of our feedback attention-augmented network(FA) and spatial detail preservation network(SPN). We introduce our modules gradually and test the mAP values of each combined models. The comparison results are reported in Table 2. Using the ResNet101 network to compare with the SLN under "All regions"and "Things only", employing feedback attention-augmented network brings 0.5% and 0.7% improvement, respectively. Meanwhile, employing SPN individually outperforms the SLN by 0.7% and 0.9%, respectively. When we integrate the two modules together, the performance further improves to 9.6% and 11.2%. These results demonstrate that our method significantly enhances the feature representation ability. Moreover, the multi-scale spatial detail preservation network with shallow convolution layers encodes and capture global spatial details information. In this case, more advanced semantic information and abundant details of spatial information complement each other, which solves the problem of feature misalignment to a certain extent and realizes feature alignment.

4.5 Visualization Results on COCOA Dataset

To further verify the effectiveness of our method, we pose the visual results in Fig. 5. Some spatial details and the boundaries of occluded objects are clearer, such as the "refrigerator" in the first row, the "car" in the second row and the 'cup' in the third row. Compared to ground-truth annotations, our prediction results are more reasonable and accurate whether it is the visible parts or the occluded parts of the object. These visualizations results demonstrate that the proposed method can better predict the semantics and shape of the occluded objects.

4.6 Results on D2SA Dataset

We also compare our method with ARCNN and SLN on the D2SA dataset. We reimplemented the SLN method. Quantitative results are shown in Table 3. We can observe that our method achieves better performance regardless of the mAP or AR. The proposed method boosts the accuracy to 43.1%, and brings 9.4% improvement. Moreover, the AR^{10} outperforms the SLN by 4.7%. From the perspective of segmentation accuracy, our proposed method enhances the feature expression ability, and effectively captures and preserve the spatial detail information, and obviously improves the quality of amodal segmentation. The qualitative results of our method on D2SA dataset in Fig. 6 are very promising. We found that our method predicts the occluded parts more reasonable and closer to the Ground-Truth. The experimental results prove that the feedback attention-augmented network improves the feature representation ability, meanwhile, the spatial detail preservation network captures multi-scale spatial position information.

5 Conclusion

In this paper, we propose a feedback attention-augmented bilateral network for amodal instance segmentation, which is able to learn more advanced semantic features and preserve spatial details information. We introduce the attention mechanism and feedback connection into backbone network so that image features are extracted twice by convolution. The reason is that a feature map convolved twice can extract more advanced features than a feature map convolved once. Moreover, the SPN with multi-scale attention feature module is designed to preserve the abundant spatial details information and capture global context information in the spatial dimension. The ablation experiments show that our two-path framework obtains more precise segmentation results. Meanwhile, both qualitative and quantitative results on the COCOA and D2SA datasets demonstrate the effectiveness and efficiency of our proposed method. In addition, it is also important to improve the computational efficiency and achieve real-time processing as much as possible, which will be studied in future work.

References

1. Krizhevsky, A., Sutskever, I., Hinton, G.E.: Imagenet classification with deep convolutional neural networks. In: Advances in Neural Information Processing Systems, pp. 1097–1105 (2012)
2. He, K., Zhang, X., Ren, S., Sun, J.: Deep residual learning for image recognition. In: Proceedings of the IEEE Conference on Computer Vision and Pattern Recognition (CVPR) (2016)
3. Xu, J., Pan, Y., Pan, X., Hoi, S.C., Yi, Z., Xu, Z.: RegNet: self-regulated network for image classification. IEEE Trans. Neural Networks Learn. Syst. (2022)
4. Liu, W., Anguelov, D., Erhan, D., Szegedy, C., Reed, S., Fu, C.-Y., Berg, A.C.: SSD: single shot MultiBox detector. In: Leibe, B., Matas, J., Sebe, N., Welling, M. (eds.) ECCV 2016. LNCS, vol. 9905, pp. 21–37. Springer, Cham (2016). https://doi.org/10.1007/978-3-319-46448-0_2
5. Ren, S., He, K., Girshick, R., Sun, J.: Faster R-CNN: towards real-time object detection with region proposal networks. In: IEEE Transactions on Pattern Analysis and Machine Intelligence (PAMI) (2016)
6. Lin, T.Y., Goyal, P., Girshick, R., He, K., Dollár, P.: Focal loss for dense object detection. In: Proceedings of the IEEE International Conference on Computer Vision (ICCV) (2017)
7. Long, J., Shelhamer, E., Darrell, T.: Fully convolutional networks for semantic segmentation. In: Proceedings of the IEEE Conference on Computer Vision and Pattern Recognition (CVPR) (2015)
8. Chen, L.C., Papandreou, G., Schroff, F., Adam, H.: Rethinking atrous convolution for semantic image segmentation. arXiv preprint arXiv:1706.05587 (2017)
9. He, K., Gkioxari, G., Dollár, P., Girshick, R.: Mask R-CNN. In: Proceedings of the IEEE International Conference on Computer Vision (ICCV) (2017)
10. Bolya, D., Zhou, C., Xiao, F., Lee, Y.J.: Yolact: real-time instance segmentation. In: Proceedings of the IEEE International Conference on Computer Vision (ICCV) (2019)

11. Wagemans, J., et al.: A century of Gestalt psychology in visual perception: II. Conceptual and theoretical foundations. Psychol. Bull. 1218–1252 (2012)

12. Zhu, Y., Tian, Y., Metaxas, D., Dollár, P.: Semantic amodal segmentation. In: Proceedings of the IEEE Conference on Computer Vision and Pattern Recognition(CVPR) (2017)

13. LeCun, Y., Bottou, L., Bengio, Y., Haffner, P.: Gradient-based learning applied to document recognition. Proc. IEEE **86**(11), 2278–2324 (1998)

14. Yu, C., Wang, J., Peng, C., Gao, C., Yu, G., Sang, N.: BiSeNet: bilateral segmentation network for real-time semantic segmentation. In: Ferrari, V., Hebert, M., Sminchisescu, C., Weiss, Y. (eds.) ECCV 2018. LNCS, vol. 11217, pp. 334–349. Springer, Cham (2018). https://doi.org/10.1007/978-3-030-01261-8_20

15. Agnew, W., et al.: Amodal 3D reconstruction for robotic manipulation via stability and connectivity. In: Conference on Robot Learning (CoRL) (2020)

16. Xiang, Y., Xie, C., Mousavian, A., Fox, D.: Learning RGB-D feature embeddings for unseen object instance segmentation. In: Conference on Robot Learning (CoRL) (2020)

17. Song, S., Xiao, J.: Deep sliding shapes for amodal 3d object detection in RGB-D images. In IEEE Conference on Computer Vision and Pattern Recognition(CVPR) (2016)

18. Deng, Z., Jan Latecki, L.: Amodal detection of 3d objects: inferring 3D bounding boxes from 2D ones in RGB-depth images. In: IEEE Conference on Computer Vision and Pattern Recognition (CVPR) (2017)

19. Sun, H., Meng, Z., Du, X., Ang, M.H.: A 3D convolutional neural network towards real-time amodal 3D object detection. In: IEEE/RSJ International Conference on Intelligent Robots and Systems (IROS) (2018)

20. Li, K., Malik, J.: Amodal instance segmentation. In: Leibe, B., Matas, J., Sebe, N., Welling, M. (eds.) ECCV 2016. LNCS, vol. 9906, pp. 677–693. Springer, Cham (2016). https://doi.org/10.1007/978-3-319-46475-6_42

21. Patrick Follmann, R.K., Härtinger, P., Klostermann, M.: Learning to see the invisible: end-to-end trainable amodal instance segmentation. In: IEEE Winter Conference on Applications of Computer Vision (WACV) (2019)

22. Zhang, Z., Chen, A., Xie, L., Yu, J., Gao, S.: Learning semantics-aware distance map with semantics layering network for amodal instance segmentation. In: ACM International Conference on Multimedia (MM) (2019)

23. Hu, J., Shen, L., Sun, G.: Squeeze-and-excitation networks. In: IEEE Conference on Computer Vision and Pattern Recognition (CVPR) (2018)

24. Wang, X., Girshick, R., Gupta, A., He, K.: Non-local neural networks. In: IEEE Conference on Computer Vision and Pattern Recognition (CVPR) (2018)

25. Woo, S., Park, J., Lee, J.-Y., Kweon, I.S.: CBAM: convolutional block attention module. In: Ferrari, V., Hebert, M., Sminchisescu, C., Weiss, Y. (eds.) ECCV 2018. LNCS, vol. 11211, pp. 3–19. Springer, Cham (2018). https://doi.org/10.1007/978-3-030-01234-2_1

26. Tsuda, H., Shibuya, E., Hotta, K.: Feedback attention for cell image segmentation. In: Bartoli, A., Fusiello, A. (eds.) ECCV 2020. LNCS, vol. 12535, pp. 365–379. Springer, Cham (2020). https://doi.org/10.1007/978-3-030-66415-2_24

27. Srivastava, R.K., Greff, K., Schmidhuber, J.: Training very deep networks. In: 29th Annual Conference on Neural Information Processing Systems (NIPS) (2015)

28. Fan, D.-P., et al.: PraNet: parallel reverse attention network for polyp segmentation. In: Martel, A.L., et al. (eds.) MICCAI 2020. LNCS, vol. 12266, pp. 263–273. Springer, Cham (2020). https://doi.org/10.1007/978-3-030-59725-2_26

29. Dai, Y., Gieseke, F., Oehmcke, S., Wu, Y., Barnard, K.: Attentional feature fusion. In: IEEE/CVF Winter Conference on Applications of Computer Vision (WACV) (2021)
30. Lin, T.Y., Dollár, P., Girshick, R., He, K., Hariharan, B., Belongie, S.: Feature pyramid networks for object detection. In: IEEE Conference on Computer Vision and Pattern Recognition (CVPR) (2017)
31. Hou, Q., Zhou, D., Feng, J.: Coordinate attention for efficient mobile network design. arXiv preprint arXiv:2103.02907 (2021)
32. Ioffe, S., Szegedy, C.: Batch normalization: accelerating deep network training by reducing internal covariate shift. In: International Conference on Machine Learning (ICML) (2015)

Squeeze-and-Excitation Block Based Mask R-CNN for Object Instance Segmentation

Kouta Nagasawa, Shin Ishiyama, Huimin Lu[✉], Tohru Kamiya, Yoshihisa Nakatoh, Seiichi Serikawa, and Yujie Li

Kyushu Institute of Technology, Fukuoka 804-0015, Japan
{nagasawa.kouta119,lu.huimin945}@mail.kyutech.jp

Abstract. Deep learning-based methods have taken center stage in image recognition, such as AlexNet and deep learning-based method. At present, Image recognition based on deep learning has been widely used in agriculture, factory automation, automated driving, medical fields and so on. In the fields of automated driving and medical care, the accuracy of the image recognition directly affects human lives. For these reasons, the importance of improving the accuracy of image recognition is clear. In this paper, we focus on instance segmentation tasks. The method used is Mask R-CNN, which is the basis of current state-of-the-art methods. The network structure based on ResNet, and we tried to improve the accuracy by adding Squeeze-and-Excitation Block (SE Block). According to the result of experiments, it is proved that this method has certain advantages for object instance segmentation.

Keywords: Instance Segmentation · Mask R-CNN · SE Block

1 Introduction

In recent years, methods based on deep learning [1–4] have become the mainstream in image recognition. Image recognition based on deep learning theory is widely used in the fields of agriculture, factory automation, automated driving, medical fields and so on. In the fields of automated driving and medical field, the accuracy of the image recognition is directly related to human life. According to the reasons, we can acquire the importance of improving the accuracy for the task of image recognition.

As part of the automation for factory operations, this task is required in which multiple identical parts are detected, and then detecting and picking up parts that are easy to pick up. In this project, we need to identify multiple objects of the same class. In such a situation, object recognition based on instance segmentation is effective. For this reason, we have tried to improve the accuracy of existing instance segmentation methods by adding a Squeeze-and-Excitation Block (SE Block). Since the Mask R-CNN used in this research has a large number of parameters, real-time performance is often compromised by network improvements, but SE Block is expected to improve accuracy with less processing. Considering this, we tried to improve the accuracy of instance segmentation by incorporating SE Block into Mask R-CNN.

F. Sun et al. (Eds.): ICCCS 2022, CCIS 1732, pp. 56–64, 2023.
https://doi.org/10.1007/978-981-99-2789-0_5

2   Related Research

Mask R-CNN is state-of-the-art method for object detection and instance segmentation that was selected as the Best Paper at the International Conference on Computer Vision (ICCV), the premier conference in the world of computer vision [5]. The method is based on Faster R-CNN [6] and has been improved in several methods in addition to adding segmentation capabilities.

The network structure of the Mask R-CNN is shown in Fig. 1. First, features are extracted from the image using a convolution layer in the backbone section. We adopted ResNet for the feature extraction networks. Second, the region proposal network (RPN) is used to learn and select candidate object regions. Finally, bounding box recognition and object classes for each RoI are obtained as output in the head section. RoI Align is replace of RoI Pooling as the feature fusion module for Faster R-CNN, which add a mask branch for object instance segmentation.

The mask branch predicts a segmentation mask for each region of interest proposals. This enables instance segmentation in addition to object detection and class prediction in Faster R-CNN.

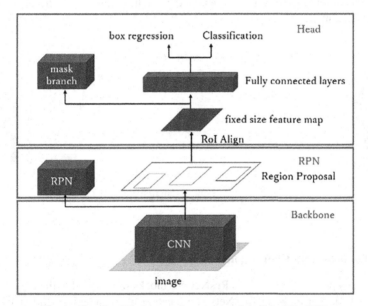

Fig. 1. The simplified network structure of the Mask R-CNN.

3   Proposed Methodology

3.1   Outline

Our method proposes the module of Squeeze-and-Excitation Block (SE Block) based on Mask R-CNN for instance segmentation. The reason for this is that ResNet is used in the backbone of the Mask R-CNN network, and the SE Block can be easily incorporated into the ResNet. Due to the short execution time due to the inclusion of Se blocks, the overall program is real-time and effective.

3.2 Squeeze-and-Excitation Block

The central component of a convolutional neural network (CNN) is the convolution operator, which fuses to construct informative features both spatial and channel information in the local receptive field at each layer. Squeeze-and-Excitation Networks (SENet) focuses on the relationship between channels.

Specifically, the system adaptively focuses on each relationship between channels and outputs weighted features. Figure 2 shows the calculation process. First, pixel averages per channel are taken by Global Average Pooling. This part is called Squeeze, which means to narrow down the information per channel. Figure 3(a) shows a normal Residual Block and Fig. 3(b) shows a Residual Block that SE Block added. Here, to extract the relationship between channels, each element is passed through two fully connected layers and then normalized to a value between 0 and 1 by applying a sigmoid function to each element. Finally, the output is obtained by multiplying the weights characterized so far with the input.

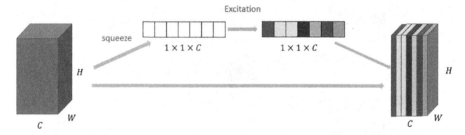

Fig. 2. Squeeze-and-Excitation Networks.

3.3 Differences from Conventional Methods

When actually adding SE blocks to ResNet, SE blocks were added to the identity block and conv block that make up ResNet. The location of the addition is after the batch normalization at the end of each block, as shown in Fig. 4. In the excitation portion of the SE Block, $r = 8$ was used to reduce the number of channels in fully connected layers.

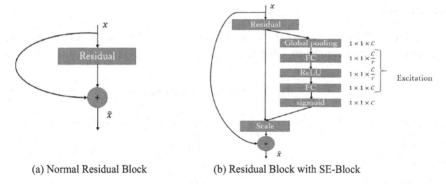

(a) Normal Residual Block   (b) Residual Block with SE-Block

Fig. 3. The comparison between Normal Residual Block and Residual SE-Block.

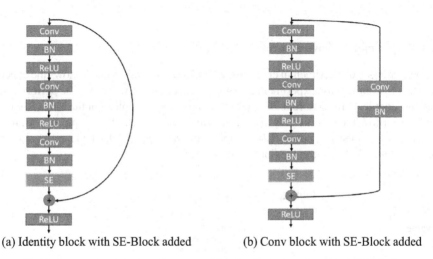

(a) Identity block with SE-Block added   (b) Conv block with SE-Block added

Fig. 4. The comparison between Normal Residual Block and Residual SE-Block.

4 Experiments

4.1 Experimental Methods

In this paper, we conducted controlled experiments use the SE Block added to the ResNet-FPN [7] of the backbone for feature extraction network based on Mask R-CNN.

The following experiment compares ResNet50-FPN with 50 layers of ResNet-FPN (conventional method) and SE-ResNet50-FPN with SE Block (proposed method) for the backbone part of ResNet in Experiment 1. Experiment 2 compares ResNet101-FPN with 101 layers (conventional method) and SE-ResNet101-FPN with SE Block (proposed method).

4.2 Datasets

COCO dataset [8] with annotated data on 80 object classes was used as datasets. The number of datasets is shown in Table 1.

Table 1. The number of datasets.

Classification	Number
train data	118,287
test data	40,775
validation data	5000

4.3 Experimental Environment

Table 2 shows the experimental environment for Experiment 1. Table 3 shows the experimental environment for Experiment 2. For the evaluation, we will prove the performance for our module on the different PC. Python3.6 was used to implement the Mask R-CNN, and Tensorflow 1.10 was used as the deep learning framework. We complete the training strategy by batch size 2, epoch 640, optimizer momentum of 0.9. The learning rate was initially set to 0.001, but in the final stage was set to 0.0001.

Table 2. The environment of Experiment 1.

Name	Parameters
OS	Ubuntu20.04[LTS]
CPU	Intel(R) Core(TM) i9-10900 K@3.70 GHz × 20
memory(CPU)	33.6 [GB]
GPU	Nvidia GeForce RTX 2080 SUPER
memory(GPU)	8 [GB]
development environment	Visual Studio Code

Table 3. The environment of Experiment 2.

Name	Parameters
OS	Ubuntu18.04[LTS]
CPU	Intel(R) Core(TM) i9-10900X@3.70 GHz × 20
memory(CPU)	67.2 [GB]
GPU	Quadro RTX 8000
memory(GPU)	48 [GB]
development environment	Visual Studio Code

4.4 Experimental Results

With the weight files obtained from the training in Experiments 1 and 2, test set were made on 500 test images. The inference results for Experiment 1 are shown in Table 4 and those for Experiment 2 in Table 5.

Mean Average Precision (mAP)[9] were used as experimental indices.The values of mAP varies with the IoU threshold. IoU in the table represents the threshold of IoU, and IoU = 0.50:0.95 represents the average mAP of inference while increasing the threshold of IoU by 0.05 from 0.50 to 0.95. As for area, it means that the inference is limited to the following object sizes: small (smaller than 32[pixel] × 32[pixel]), medium (larger than 32[pixel] × 32[pixel] and smaller than 96[pixel] × 96[pixel]), large (larger than 96[pixel] × 96[pixel]).

Table 4. Inference Results for Experiment 1.

Evaluation index	Conventional Method	Proposed Method
mAP(IoU = 0.50:0.95)	0.299	0.310(+0.011)
mAP(IoU = 0.50)	0.478	0.499(+0.021)
mAP(IoU = 0.75)	0.316	0.323(+0.007)
mAP(IoU = 0.50:0.95) (area = small)	0.160	0.167(+0.007)
mAP(IoU = 0.50:0.95) (area = medium)	0.331	0.358(+0.027)
mAP(IoU = 0.50:0.95) (area = large)	0.458	0.454(−0.004)
Execution time[s/image]	0.2087	0.2045(−0.0042)

Table 5. Inference Results for Experiment 2.

Evaluation index	Conventional Method	Proposed Method
mAP(IoU = 0.50:0.95)	0.364	0.376(+0.012)
mAP(IoU = 0.50)	0.541	0.560(+0.019)
mAP(IoU = 0.75)	0.396	0.406(+0.010)
mAP(IoU = 0.50:0.95) (area = small)	0.192	0.192(+0)
mAP(IoU = 0.50:0.95) (area = medium)	0.412	0.424(+0.012)
mAP(IoU = 0.50:0.95) (area = large)	0.539	0.565(+0.026)
Execution time[s/image]	0.1960	0.2026(+0.0066)

Figure 5 shows the results of inference on a test image using the weights obtained in Experiment 1. In this result, the proposed method detects the rightmost bowl, which is not detected by the conventional method, and eliminates the false detection of the bottom-right tie. Figure 6 shows the results of inference on a test image using the weights obtained in Experiment 2. The conventional method detects the two suitcases in the foreground as a single object, while the proposed method is able to detect the two objects separately.

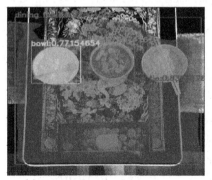

(a) Instance segmentation by conventional method	(b) Instance segmentation by proposed method

Fig. 5. Instance Segmentation Results (1).

4.5 Consideration

As for the 6 items evaluated of accuracy, 5 items show improvement in accuracy in both of experiments. Execution time decreased in Experiment 1 and increased in Experiment 2. However, since both are within the margin of error, it is expected that the execution time is almost the same between the conventional and the proposed methods. These results indicate that the addition of the SE Block can achieve improved accuracy with a small increase in time computation.

| (a)Instance segmentation by conventional method | (b) Instance segmentation by proposed method |

Fig. 6. Instance Segmentation Results (2).

In Experiment 1, mAP showed little improvement in accuracy for small and large objects compared to medium objects. In Experiment 2, we improve the detection accuracy for the small object. Both of the Experiments 1 and 2, small objects had coarser pixels in the image, which may have made it difficult for the SE Block weighting that consider between each channel to work correctly.

The reason for the lack of improvement in accuracy for large objects in Experiment 1 is thought to be that ResNet50-FPN was used for the backbone portion, but because the layer was shallow, it was not able to extract advanced features for large objects, and the benefits of the SE Block were not utilized. On the other hand, in Experiment 2, the deeper ResNet101-FPN was used for the backbone portion, which is thought to have improved accuracy for large objects. From this, it can be inferred that it is important to increase the layer depth in image processing by deep learning.

5 Conclusion

In this paper, we changed the original method of Mask R-CNN to improve the accuracy of image-based instance segmentation. Our method proposes the SE Block to fuse ResNet in the feature extraction network. According to the experimental result, it proves that the accuracy was improved with almost no increase in processing time. For future prospects, we would like to further improve the accuracy by using a deeper layer of ResNet or changing the activation function.

Acknowledgements. This paper was partially supported by NSFC No.62206237, JSPS No.22K12093 and JSPS No.22K12094.

References

1. Krizhevsky, A., Sutskever, I., Hinton, G.E.: ImageNet classification with deep convolutional neural networks. In: Proceedings of the 25th International Conference of Neural Information Processing Systems–Vol. 1, 12, pp.1097–1105 (2012)
2. Huimin, L., Rui, Y., Zhenrong, D.: Chinese image captioning via fuzzy attention-based DenseNet-BiLSTM. In: ACM Transactions on Multimedia Computing Communications and Applications (2020)
3. He, K., Gkioxari, G., Dollar, P., Girshick, R.: Mask R-CNN. arXiv 1703.06870, vol. 3 (2018)
4. Hu, J., Shen, L., Albanie, S., Sun, G., Wu, E.: Squeeze-and excitation networks. arXiv 1709.01507, vol. 4 (2019).
5. IEEE, Inside Signal Processing Newsletter. https://signalprocessingsociety.org/newsletter/2018/01/iccv-2017-best-paper-award-mask-r-cnn (Accessed 10 Jan 2022)
6. Ren, S., He, K., Girshick, R., Sun, J.: Faster R-CNN: towards real-time object detection with region proposal networks. arXiv 1506.01497, vol. 3 (2016)
7. Lin, T.-Y., Dollar, P., Girshick, R., He, K., Hariharan, B., Belongie, S.: Feature pyramid networks for object detection. arXiv 1612.03144, vol. 2 (2017)
8. COCO dataset. https://cocodataset.org/ (Accessed 15 Jan 2022)
9. Padilla, R., Netto, S.L., da Silva, E.A.B.: A Survey on Performance Metrics for
10. Object-Detection Algorithms 2020 International Conference on Systems, Signal and Image Processing,1–3 July 2020.

PointNetX: Part Segmentation Based on PointNet Promotion

Kai Zhao[1]([✉]), Huimin Lu[2], and Yujie Li[3]

[1] Qingdao University, Qingdao, Shandong, China
1171479689@qq.com
[2] Kyushu Institute of Technology, Kitakyushu, Japan
[3] Yangzhou University, Yangzhou, Jiangsu, China

Abstract. Recently, point cloud learning has become widely utilized in a variety of domains, including autonomous driving, robotics, and computer vision. PointNet, a pioneer in point cloud processing, uses max pooling to address the disorder of point clouds. However, PointNet's method of mapping points to high-dimensional space, and then obtaining global features through maximum pooling still leads to a large loss of feature information. To this end, we suggest a new PointNet-based segmentation and classification network called PointNetX. PointNetX expands the network's depth and the number of neurons. Simultaneously, we extracted the features of different layers to compensate for the loss caused by pooling and used a better rotation matrix to adjust the Angle. On the other hand, we add label smoothing to the loss function, and use Ranger optimizer in training to realize various tasks of point clouds. Experiments show that PointNetX has better performance in part segmentation as well as object classification than PointNet.

Keywords: Deep Learning · Point Cloud · Segmentation

1 Introduction

Recently, the progress of point cloud research has aroused intense interest. This paper primarily studies the improvement of part segmentation by PointNet [1] on the 3D point cloud. The convolution processing of 2D images has been very mature in the past few years. Nevertheless, owing to the disordered and irregular characteristics of the point cloud, convolution fails to acquire the spatial information of points and adapt to the N! order of the point set. In earlier research, researchers were only able to translate point cloud data to voxel space or multi-view, and then do feature learning and shape classification. However, due to the sparsity of the point cloud, such a project-based network will cause serious space waste, computing time loss, and other problems.

By using symmetric functions, PointNet, a pioneer in dealing directly with point cloud information, directly provides an effective and straightforward structure to address the issue of point cloud disorder. PointNet firstly maps point clouds to high-dimensional space by using several MLP layers for point-by-point learning for each point and then uses Max pooling to extract global features. The resulting global features can be directly scored or transferred to other networks for different tasks. Nevertheless, the direct use of symmetric functions would certainly lead to a loss of information. Also, with the deepening of MLP layers and the increase of neurons, the dimension of feature mapping becomes higher and higher. The model may suffer from overfitting and poor generalization ability.

How to reduce the loss of feature information by Max pooling is our research's primary direction. In our new model, feature fusion is the key to problem solving. We extract the features of point clouds in multiple feature Spaces of different dimensions in order to collect detailed information on different layers. Different features obtained by Max pooling are combined to achieve feature fusion. In this way, details can be added to the features mapped to higher dimensions. Experiments demonstrate that the global features extracted by our model could retain more information, which is also effective for classification tasks. In the segmented network, we replaced a local point feature in the original model with a more informative local point feature to increase the accuracy by replacing useless features.

Based on PointNet, we utilize feature fusion techniques for acquiring global features with higher accuracy and apply them to classification as well as segmentation networks. We create a rotation matrix that is more favorable for network classification and segmentation based on the point cloud's rotation invariance. In order to avoid overfitting the model, we utilize the loss function of label smoothing in order to enhance generalization. Ranger optimizer replaces Adam optimizer in the training network. These improvements could ensure that our model would find better results in a faster time. Experiments demonstrate that our model is clearly superior to the original model. In summary, the main contributions of this paper can be summarized as follows:

(a) Using feature fusion technology applied to classification and segmentation networks, we improve the depth of the network layer and the number of neurons based on PointNet to provide high-precision global features. A better rotation matrix has been designed. Local point features with more information are used in the segmentation network for enhancing segmentation accuracy.

(b) Ranger optimizer has been used in classification and segmentation training. The loss function with label smoothing was used by the network model, instead of the cross entropy loss function.

We will introduce the previous work and methods in Sect. 2. Simultaneously, the shortcomings of the previous methods have been highlighted. Section 3 would introduce our model and specific improvement methods in detail. The findings of our comparison with various approaches to the data set are presented in Sect. 4, along with the results of the ablation experiment and visualization. In Sect. 5, we will summarize our work and discuss future research directions.

2   Related Work

In this section, we first review the relevant parts of the methods applied in our model.

2.1   Projection-Biassed Approach

In 2D images, CNN has been widely employed. Pixels in 2D regular grids [2] show why multi-layer neural networks are better. The researchers want to apply this technique to 3D data, but deep learning on point clouds is still being explored. As a result, different learning methods have been generated based on different representations of 3D data. Volume CNN: [3–5] is a classic work of applying a 3D convolutional neural network to voxelized shapes. These methods solve the problem of disorder through voxels. Nevertheless, owing to the sparsity of voxel data, voxel convolution brings huge space consumption as well as computational cost. The volume representation is limited by its resolution. To this end, hierarchical structures and compact graph structures (such as octree [6, 7], Kd tree [8], and hash [9]) have been introduced for reducing the computation and storage costs of these methods. The general idea is to skip the convolution step in Spaces where there is no point information to save computation and storage space [10, 11]. The point cloud has been divided into grids, each grid is represented by grid average point and Fisher vector respectively, and convolved with a 3D kernel in order to execute effective point cloud processing.

Multi-view CNN: [5, 12] Attempts to render 3D point clouds or shapes into 2D images, and then apply a 2D CONV network to classify them. With the well-designed image CNN, this series of methods have achieved excellent performance in shape classification and retrieval tasks [13]. However, this method of converting 2d images has obvious disadvantages. Although 3d shape information can be retained to a certain extent through projection changes. However, in the process of projection, some structural information of the 3D model will be lost, resulting in the loss of features. This projection method requires 3d models to be aligned vertically, which also limits the application scenarios of this method.

2.2 Networks that Deal Directly with Point Clouds

So is there an easier way to input as a point cloud and do deep learning without preprocessing? PointNet is skilled in this, and its core is to use the symmetric function Max for solving the point clouds' disorder. The author maps point cloud information to high-dimensional space through a multi-layer MLP layer in order to avoid excessive information loss caused by direct Max pooling. Some characteristics of the original cloud are preserved by adding two layers of T-Net, based on the rotation invariance of the point cloud. T-net is great for classification, but not so great for segmentation, particularly semantic segmentation.

Most of the present methods focus on the network structure of extracting local point information. Point cloud analysis using convolution [14], graphs [14], or attention mechanisms [15] can yield good results. [16] also proposed a method of supplementing local features with long range point relation. Although the results of these methods are satisfactory, they mainly rely on a well-designed local extractor, which leads to fierce competition to explore local point information. However, complex extractors are not without their drawbacks. Due to the prohibitive overhead of computation and memory access, these complex extractors hinder the efficiency of applications in natural scenarios. For example, most 3D point cloud applications so far are still based on simple methods such as PointNet(and PointNet + +). However, the application of the above advanced methods is rarely seen in the literature.

[17] proposed kernel correlation and graph pool to improve the similarity network. [18] RNN is used to process features merged from ordered point cloud slices. Neighborhood structures are recommended for point Spaces and element Spaces. DGCNN [19] adds the coordinates of points to the network architecture of PointNet and the distance from the domain points. These methods are based on symmetric pools and those in [20–22] guarantee order invariability, but they also work on a PointNet basis. So we designed our PointNetX network, which has the same simple architecture as PointNet but is much more competitive. It can easily replace network architectures that use PointNet as a feature extractor and provide performance gains without much modification.

3   Proposed Method

PointNet has room for improvement in point cloud feature extraction and segmentation networks. We will initially introduce our improved model in this section. Then we'll discuss our modified loss function and how the Ranger optimizer helps us do the rest.

3.1   Network Structure

In the model, we mainly use the technique of feature fusion. Low-level features are less abstract and contain more location and detailed information. However, due to less

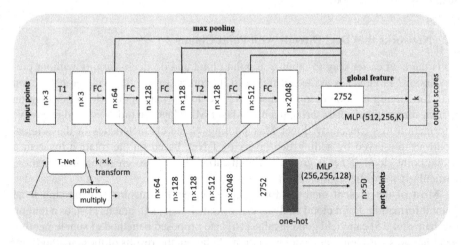

Fig. 1. PointNetX **network architecture.** Take N points as input, transform the position and pose of input and feature levels, extract point features of multiple features of different sizes by max pooling, and finally merge them into global features. The global feature outputs the scores of k categories through several fully connected layers. The segmented network is realized on the basis of a classified network. It splices global features and local points to obtain segmentation fractions. T1 and T2 are the aligned transformation networks of input points and features. FC is a mapping of the fully connected layer to higher dimensionality. The MLP stands for multi-layer perceptron, and the number represents the number of neurons. One-hot is a vector of size 16 that represents the class of input shapes.

convolution, low-level features have lower semantics and more noise. Stronger semantic information is present in high-level features, but the perception of details is poor and feature abstraction levels are high. The key to improving the segmentation model is how to combine them efficiently, take advantage of their advantages as well as discard their disadvantages. Here, we employ an early fusion approach. As illustrated in Fig. 1, the point cloud we input needs to contain location information (x,y,z). Normals, colors, and other features can be added depending on the task. Similarly, to solve the feature loss caused by the symmetric function as much as possible, we first map the input point cloud information to multiple MLP layers. It is worth mentioning that we increased the number of neurons and the depth of the network in some layers of the model so that our features would be mapped to higher dimensions. For the expression of the point cloud, the high-dimensional space must be information redundancy.

However, when the information is redundant, the symmetric function is used to extract features, which can reduce the loss of information and retain enough point cloud information. To eliminate the influence of translation and rotation on point cloud classification and segmentation, we utilize two spatial transforms in order to achieve alignment in space. The first input transform is to adjust the spatial midpoint cloud. Intuitively, rotating an object in a direction that is more conducive to classification or segmentation, such as turning it toward the front, makes sense. Aligning the extracted 128-dimensional features is the second feature transform, which modifies the point cloud at the feature level. We add a Max pooling layer after these Spaces of different dimensions in order to extract the current features. This makes up for PointNet's lack of local features to some extent. Note that we did not extract global features for each mapping, because too many similar features can affect the model. Finally, the features obtained from several times of pooling are spliced to form a 2752 dimensional global feature. The global feature acquired by fusion is more discriminant than the single feature. The post-fusion characteristics not only maximize the integrity of the original point cloud information as much as possible but also add more low-dimensional details. The new global features greatly decrease the impact of pooling and are still useful for other tasks with point clouds. To calculate the probability of each category in the classification task, we just need to map the global characteristics to the k categories using multiple MLP layers.

In our segmentation network, the local point features after each convolution are spliced to form point features for part segmentation. However, in1, the author only spliced the local point features of the first few layers. During the experiment, we found that the performance of the model would be significantly improved when we added n × 2048 dimensional local point information. We removed the n × 128 after the second transform and replaced it with local point information of N × 2048. It plays a considerable role in supplementing information.

Simultaneously, we noticed that our test results frequently failed to reach the accuracy of training, and there was a phenomenon of over-fitting. The hypothesis stated in [23] states that the outcomes are worse when the batch-normalization and Dropout layers are combined. In a segmented network, Dropout is positioned behind every BN layer with p set to 0.3. Do not add dropout modules in segmented networks. Thus, that solves the variance deviation issue. Then the generalization ability of the model has been enhanced and the over-fitting situation is reduced.

3.2 Alignment Network

Based on the rotation invariance of the point cloud, the same point cloud should remain unchanged after certain rigid changes in space. Inspired by PointNet, we predict affine transformation matrices through a small network and apply this transformation directly to the coordinates of input points. Small networks are structured like PointNetX and consist of basic modules for cross-layer feature extraction, feature fusion, maximum pooling, and fully connected layers. This method can also be used for subsequent alignment of feature Spaces. We can insert another small network at the feature level and predict a feature transformation matrix to align the point clouds at the feature level. The difference, however, is that the two aligned networks are different in size to ensure relatively fast computation time. But after adding feature fusion and modifying the model size, they both perform better than PointNet on rotation results and are portable.

Nevertheless, the dimension of the transformation matrix in the feature space is much higher, compared to that of the input transformation matrix, which will have the opposite effect on the accuracy of the network. We adopt the method given by PointNet to add a regularization term to the training loss, and constrain the edge transformation matrix to a nearly orthogonal matrix:

$$L_{reg} = ||I - AA^T||_F^2 \tag{1}$$

where A represents the feature alignment matrix of alignment network prediction. (1) Will be added into the total loss function as part of optimization.

3.3 Ranger Optimizer

Here we replace the SGD optimizer and Adam [24] optimizer with the Ranger optimizer. Ranger optimizer combines two very good optimizers Radam [25] and LookAhead [26].

Radam is advanced in that it can dynamically turn adaptive learning rates on or off according to variance dispersion, providing a method that does not require tunable learning rate preheating. This is because the Adam optimizer tends to fall into poor local optimization at the start of training when it is not warmed up. Radam has both the advantages of Adam [24] and SGD, which could ensure fast convergence speed and is not easy to fall into the local optimal solution. Its accuracy is superior to SGD when learning rates are high. However, once the variance stabilized, RAdam was equal to Adam and SGD during the rest of the training period. In other words, RAdam improved mainly at the beginning of training.

LookAhead was inspired by a recent understanding of the loss of surfaces of deep neural networks and provides breakthroughs for robust and stable exploration throughout the training process. Lookahead, however, is a new development that maintains two sets of weights and then interpolates between them. In effect, it enables a set of faster weights to "look ahead" or explore, while slower weights have been left behind to provide longer term stability. Results the variance in the training process is reduced, the sensitivity to suboptimal hyperparameters is greatly reduced, and the need for tuning a large number of hyperparameters is reduced. This is to achieve faster convergence when completing various deep learning tasks. An impressive breakthrough, LookAhead can be easily understood using the following comparison. Imagine you are at the top of a mountain, surrounded by peaks of different heights. There is only one route that descends to the bottom of the hill; the others do not. It's hard to explore on your own because you have to pick a path down, assume it's a dead end, and find your way out again. But if you have a friend who stays at or near the top, help you back up if the path you pick is good. You'll probably have a better chance of finding the optimal path because exploring the terrain will be quicker and you'll be less likely to get stuck in an unfortunate crevasse. LookAhead decreases the number of hyperparameters, which need to be adjusted and enables faster convergence of different deep learning tasks with minimal computational overhead.

The combination of the two optimizers, which is highly synergistic, enables the Ranger optimizer to perform better and more quickly in various facets of deep learning optimization.

3.4 Loss

In multi-classification, one-hot tags are often used to calculate the cross entropy loss. However, in the case of simple cross entropy loss, only the loss of correct label position is considered, while the loss of wrong label position is ignored. As a result, the model may fit very well on the training set, but because the loss of the wrong label position is not calculated, the probability of prediction error is relatively high, which is often called overfitting.

Thus, we choose the loss function with label smoothing. It could perturb the target variable, decrease the certainty of its prediction by the model, and obtain the effect of reducing over-fitting. The formula is given below:

$$Loss = e(i) + \epsilon \sum \frac{ce(j)}{N}(1 - \epsilon)c \qquad (2)$$

where $ce(x)$ represents the standard cross-entropy loss of x (e.g. $-\log(p(x))$), and ϵ is a small positive number. In this article, we set ϵ to 0.2. i is the correct class and N is the number of classes.

The final loss function is the sum of (1) and (2).

$$TotalLoss = Loss + L_{reg} \qquad (3)$$

4 Experiments

In this section, we would demonstrate the improvement of our work on 3D object classification and object partial segmentation. Although we concentrate on part segmentation in our research, our method can also be applied to other tasks quite successfully.

4.1 Object Classification

The same networks that we use to capture global features in segmentation can also be used in classification. We evaluated our model against the ModelNet40 [3] shape classification benchmark. There are 12,311 CAD models from 40 human-made object categories, split into 9,843 for training and 2,468 for testing. We use our proposed feature extraction network in the experiment. We are consistent with PointNet in activating functions, and ReLU activating functions. We also compared several other good approaches, as shown in Table 1.

In Table 1, we compare our model with voxel-based and multi-view approaches. Compared with these methods, the overall accuracy of our network is significantly improved. In the ModelNet40 classification, we also see that our classification is significantly stronger than PointNet with the same input data size (1024 points) and features (coordinates only).

Table 1. ModelNet40 shape classification Our PointNetX has competitive performance compared to these approaches.

Method	input	accuracy	accuracy
		avg.class	overall
Subvolume [5]	vox	86.0	89.2
MVCNN[6]	img	**90.1**	-
PointNet[1]	point	86.0	89.2
PointNet + + [29]	Point		90.7
SO-Net[30]	Point	87.3	90.9
PointNetX	point	88.7	**91.1**

4.2 Object Part Segmentation

Part segmentation is a difficult task, which needs accurate segmentation of parts with complex shapes and fine structures. Two areas dominate the challenges it faces.

First of all, the shape parts with the same semantic label have greater geometric variation and fuzziness. Secondly, the method needs to be robust to noise and sampling. Given a 3D scan or mesh model, the task is to determine which category each point or surface belongs to (for instance, an airplane wing, or a car tire).We used the Shapenet [27] dataset given in Pointnet + + [29]. Each point in the dataset is equipped with a normal direction in order to better describe the underlying shape. During the training process, our data adopted the method in DGCNN [19], which randomly scaled and shifted the point cloud. We don't process the data while the testing phase is underway.

We compare our framework to deep learning-based techniques in Table 2 [1, 28] and conventional learning-based techniques [27]. The point of intersection (IoU) on the union is used as an evaluation indicator and is averaged across all parts classes. The cross-entropy loss is minimized during training. In most parts of the segmentation, our method has no small advantage.

In contrast to these methods, our model could achieve the best results in mean IoU. Excellent results can also be achieved on most parts. In contrast to Pointnet, our advantages in Instance Average mIoU and class Average mIoU are improved by 1.9% and 2.7% respectively. Compared to PointNet + + [29] and SO-Net [30], we also have a good advantage in Instance Average mIoU and multiple specific parts. This proves that our network is competitive.

Table 2. ShapeNet **part segmentation results.** Average mIoU over instances (Ins.) and are reported.

	mean	aero	bug	cap	car	chair	ear phone	guitar	knife	lamp	laptop	motor	mug	pistol	rocket	skate board	table
Yi[28]	81.4	81.0	78.4	77.7	75.7	87.6	61.9	92.0	85.4	82.5	95.7	70.6	91.9	85.9	53.1	69.8	75.3
PN[1]	83.7	83.4	78.7	82.5	74.9	89.6	73.0	91.5	85.9	80.8	95.3	65.2	93.0	81.2	57.9	72.8	80.6
SSCNN[29]	84.7	81.6	81.7	81.9	75.2	90.2	74.9	93.0	86.1	84.7	95.6	66.7	92.7	81.6	60.6	82.9	82.1
PN++[30]	85.1	82.4	79.0	87.7	77.7	90.8	71.8	91.0	85.9	83.7	95.3	71.6	94.1	81.3	58.7	76.4	82.6
PointNetX	85.1	83.4	82.6	92.4	77.5	90.8	76.0	91.3	85.9	82.2	96.0	69.1	93.1	81.1	58.4	75.1	82.9

4.3 Ablation Study

In this section, we conduct ablative studies of the modules proposed in this work and demonstrate the advantages of our PointNetX in order to confirm the functionality and efficacy to the many components we provide. The 0 model uses PointNetX's network architecture, common cross entropy loss functions, and the Adam optimizer in training. Table 3 provides an overview of the experimental findings.

Table 3. Ablation tests on ShapeNet

Model	Ranger	Label Smoothing	MIoU
0	✗	✗	84.4
1	✓	✗	84.9
2	✗	✓	84.5
3	✓	✓	85.1

4.4 Visualizations

Figure 2 illustrates some examples of predicted versus true results on the ShapeNet part test data set. As shown in the figure, the result of our segmentation is visually satisfactory.

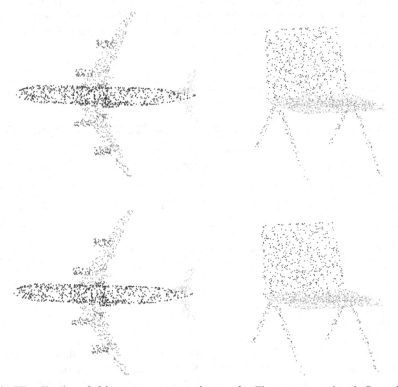

Fig. 2. Visualization of object part segmentation results. First row: ground truth. Second row: predicted segmentation. From left to right: airplane, chair.

5   Conclusion

In this work, we came up with a segmented network called PointNetX. PointNetX is an improved point-cloud segmentation network based on PointNet. It can easily replace PointNet in other tasks without excessive modification. We adopt the feature fusion method in the global feature extraction network. The point clouds mapped to different dimensions are maximized to obtain several different global features. Finally, we concatenate several global characteristics to compensate for the information lost during maximum pooling. As a result, our network has improved for other tasks, such as object classification. We use the Ranger optimizer in our training, which allows us to better approximate or achieve optimal values. In the network model, we use the loss function of label smoothing to achieve better accuracy and generalization effect. Experimental results verify the effectiveness and accuracy of the improved method. However, our improvement has not yet involved the local characteristics of the point cloud, and we will take the local characteristics of the point cloud as the main research direction in future work.

References

1. Qi, C.R., Su, H., Mo, K., et al.: Pointnet: deep learning on point sets for 3D classification and segmentation. In: Proceedings of the IEEE Conference on Computer Vision and Pattern Recognition, pp. 652–660 (2017)
2. LeCun, Y., Bottou, L., Bengio, Y., et al.: Gradient-based learning applied to document recognition. Proc. IEEE **86**(11), 2278–2324 (1998)
3. Wu, Z., Song, S., Khosla, A., et al.: 3D shapenets: a deep representation for volumetric shapes. In: Proceedings of the IEEE Conference on Computer Vision and Pattern Recognition, pp.1912–1920 (2015)
4. Maturana, D., Scherer, S.: Voxnet: a 3D convolutional neural network for real-time object recognition. In: 2015 IEEE/RSJ International Conference on Intelligent Robots and Systems (IROS). IEEE, pp. 922–928 (2015)
5. Qi, C.R., Su, H., Nießner, M., et al.: Volumetric and multi-view CNN for object classification on 3D data. In: Proceedings of the IEEE Conference on Computer Vision and Pattern Recognition, pp. 5648–5656 (2016)
6. Riegler, G., Osman Ulusoy, A., Geiger, A.: Octnet: Learning deep 3D representations at high resolutions. In: Proceedings of the IEEE Conference on Computer Vision and Pattern Recognition, pp. 3577–3586 (2017)
7. Wang, P.S., Liu, Y., Guo, Y.X., et al.: O-CNN: octree-based convolutional neural networks for 3D shape analysis. ACM Trans. Graph. (TOG) **36**(4), 1–11 (2017)
8. Klokov, R., Lempitsky, V.: Escape from cells: deep KD-networks for the recognition of 3D point cloud models. In: Proceedings of the IEEE International Conference on Computer Vision, pp. 863–872 (2017)
9. Shao, T., Yang, Y., Weng, Y., et al.: H-CNN: spatial hashing based CNN for 3D shape analysis. IEEE Trans. Visual Comput. Graphics **26**(7), 2403–2416 (2018)
10. Hua, B.S., Tran, M.K., Yeung, S.K.: Pointwise convolutional neural networks. In: Proceedings of the IEEE Conference on Computer Vision and Pattern Recognition, pp. 984–993 (2018)
11. Ben-Shabat, Y., Lindenbaum, M., Fischer, A.: 3D point cloud classification and segmentation using 3D modified fisher vector representation for convolutional neural networks. arXiv preprint arXiv:1711.08241 (2017)
12. Su, H., Maji, S., Kalogerakis, E., et al.: Multi-view convolutional neural networks for 3d shape recognition. In: Proceedings of the IEEE International Conference on Computer Vision, pp. 945–953 (2015)
13. Savva, M., Yu, F., Su, H., et al.: Shrec16 track: largescale 3D shape retrieval from shapenet core55. In: Proceedings of the Eurographics Workshop on 3D Object Retrieval, p. 10 (2016)
14. Li, G., Müller, M., Qian, G., et al.: DeepGCNs: making GCNs go as deep as CNNs. IEEE Trans. Pattern Anal. Mach. Intell. (2021)
15. Guo, M.H., Cai, J.X., Liu, Z.N., et al.: PCT: point cloud transformer. Computat. Vis. Media **7**(2), 187–199 (2021)
16. Xiang, T., Zhang, C., Song, Y., et al.: Walk in the cloud: learning curves for point clouds shape analysis. In: Proceedings of the IEEE/CVF International Conference on Computer Vision, pp. 915–924 (2021)

17. Shen, Y., Feng, C., Yang, Y., et al.: Mining point cloud local structures by kernel correlation and graph pooling. In: Proceedings of the IEEE Conference on Computer Vision and Pattern Recognition, 4548–4557 (2018)

18. Huang, Q., Wang, W., Neumann, U.: Recurrent slice networks for 3D segmentation of point clouds. In: Proceedings of the IEEE Conference on Computer Vision and Pattern Recognition, pp. 2626–2635 (2018)

19. Wang, Y., Sun, Y., Liu, Z., et al.: Dynamic graph CNN for learning on point clouds. ACM Trans. Graph. (TOG) 38(5), 1–12 (2019)

20. Dieleman, S., De Fauw, J., Kavukcuoglu, K.: Exploiting cyclic symmetry in convolutional neural networks. In: International Conference on Machine Learning, PMLR, pp. 1889–1898 (2016)

21. Zaheer, M., Kottur, S., Ravanbakhsh, S., et al.: Deep sets. Adv. Neural Inf. Process. Syst. 30 (2017)

22. Ravanbakhsh, S., Schneider, J., Poczos, B.: Deep learning with sets and point clouds. arXiv preprint arXiv:1611.04500 (2016)

23. Li, X., Chen, S., Hu, X., et al.: Understanding the disharmony between dropout and batch normalization by variance shift. In: Proceedings of the IEEE/CVF Conference on Computer Vision and Pattern Recognition, 2682–2690 (2019)

24. Kingma, D.P., Ba, J.: Adam: a method for stochastic optimization. arXiv preprint arXiv:1412. 6980 (2014)

25. Liu, L., Jiang, H., He, P., et al.: On the variance of the adaptive learning rate and beyond. arXiv preprint arXiv:1908.03265 (2019)

26. Zhang, M., Lucas, J., Ba, J., et al.: Lookahead optimizer: k steps forward, 1 step back. Adv. Neural Inf. Process. Syst., 32 (2019)

27. Yi, L., Kim, V.G., Ceylan, D., et al.: A scalable active framework for region annotation in 3D shape collections. ACM Trans. Graph. (ToG) 35(6), 1–12 (2016)

28. Yi, L., Su, H., Guo, X., et al.: Syncspeccnn: synchronized spectral CNN for 3D shape segmentation. In: Proceedings of the IEEE Conference on Computer Vision and Pattern Recognition, pp. 2282–2290 (2017)

29. Qi, C.R., Yi, L., Su, H., et al.: Pointnet++: deep hierarchical feature learning on point sets in a metric space. Adv. Neural Inf. Process. Syst., 30 (2017)

30. Li, J., Chen, B.M., Lee, G.H.: So-net: self-organizing network for point cloud analysis. In: Proceedings of the IEEE Conference on Computer Vision and Pattern Recognition, pp. 9397–9406 (2018)

Enhancement for Low-Contrast Images with Dynamical Saturating Nonlinearity and Adaptive Stochastic Resonance

Guodong Wang[1], Xi Wang[2], Yumei Ma[1(✉)], Zhenkuan Pan[1], Xuqun Zhang[1], and Jinpeng Yu[3]

[1] College of Computer Science, Technology Qingdao University, No. 308, Ningxia Road, Qingdao 266071, People's Republic of China
mayumei@qdu.edu.cn
[2] Communication Dispatching Section Qingdao Emergency Center, No. 120, Jinsong 3rd Road, Qingdao 266035, People's Republic of China
[3] College of Automation, Qingdao University, No. 308, Ningxia Road, Qingdao 266071, People's Republic of China

Abstract. In recent years, image processing based on stochastic resonance has received more and more attention. In this paper, a dynamic saturated nonlinear stochastic pooling network model based on adaptive stochastic resonance is proposed. And it combines the dynamical saturating nonlinearity with the stochastic resonance, which can realize adaptive iteration and enhance low-contrast images. At the same time, a new method to solve the optimal parameters of nonlinear systems is proposed, which is used to solve the optimal parameter values of dynamical saturating nonlinearity in this paper. This new image enhancement method is tested on low-contrast images in the LOL dataset. By comparing the new model and some other image enhancement algorithms, it demonstrates that the results not only have good visual perception, but also obtain more excellent evaluation indicators value.

Keywords: stochastic resonance · dynamical saturating nonlinearity · adaptive algorithm · image enhancement

1 Introduction

Now, in some cases, there are many images with low dynamic range of intensity values due to lack of sufficient brightness and exposure. Image enhancement is required for improving the visual effect of these low-light and low-contrast images and obtaining more visual information from them. The enhancement of low illumination images is widely used in night vision detection [1], underwater images [2], medical imaging [3] and many other fields. Many image enhancement algorithms are based on the spatial domain, including traditional histogram equalization (HE) [4], contrast-limited adaptive histogram equalization, (CLAHE) [5], gamma correction [6] and so on. However, artifacts is produced when there is a high gradient in the image [7]. In gamma correction, it

F. Sun et al. (Eds.): ICCCS 2022, CCIS 1732, pp. 77–88, 2023.
https://doi.org/10.1007/978-981-99-2789-0_7

is proposed in [8] that smooth curve is used, and then automatically enhance the brightness of the image through a weighted distribution. Although artifacts are avoided in this method, the enhancement effect is not ideal. The single-scale Retinex (SSR) [9] and the multi-scale Retinex algorithm (MSR) [10] were proposed in the 20th century. However, in the SSR method, the brightness of the picture has not been significantly improved, and MSR produces halos where there is a large difference in brightness. The Retinex-Net [11] significantly improved the contrast of the image, but the algorithm would blur the edges of the image due to noise interference. The brightness value of the image can be effectively improved by self-supervised image enhancement network (SSIEN) [12], but some of the original color information will lose because the brightness value is too high.

It is well known that noise is inevitable in image processing. In the process of image transmission, noise is often generated due to the interference of some external factors, which will affect the quality of the image. However, the discovery of stochastic resonance (SR) has changed this phenomenon. The stochastic resonance mechanism was discovered for the first time in 1981 [13], and it was explained that in the process of signal transmission, the signal can be enhanced by adding an appropriate amount of noise, making the original noise a favorable factor. In 1998, adaptive stochastic resonance (ASR) theory was proposed [14] and developed rapidly [15]. It is widely used in fault detection [16] and ultraviolet absorption spectroscopy [17] and other fields.

In this paper, a dynamic saturated nonlinear stochastic pooling network model based on adaptive stochastic resonance and a new method to solve the optimal parameters of nonlinear systems are proposed. The model can perform adaptive iterative processing on low-contrast images and enhance the contrast. When the mean square error of the network output reaches the minimum value, it stops. At the same time, the output image after eliminating noise disturbance is fitted with the original image by the least squares method (LSE) to obtain the optimal dynamical saturating nonlinearity (DSN) parameter.

2 Model

2.1 Dynamical Saturating Nonlinearity

In this paper, a dynamically saturated nonlinear system is used to conduct related experiments. Its one-dimensional expression is modelled as Eq. (1),

$$dx/dt = -x + \alpha tanh(\beta x) + s(t) + \xi(t) \qquad (1)$$

where x is the neuron membrane potential, α is the self-coupling coefficient, β is the slope of the saturated system, s(t) is the input signal, and $\xi(t)$ is the external white Gaussian noise.

After the image is iteratively processed in the model, the contrast can be improved gradually. The model obtained by solving Eq. (1) by the Euler-Maruyamas iteration method [18] is shown in Eq. (2),

$$x(n + 1) = x(n) + \Delta t(-x(n) + \alpha tanh(\beta x(n)) + s(t) + \xi(t)) \qquad (2)$$

where n represents the data of the nth iteration, and Δt represents the step size ($\Delta t = 0.01$).

2.2 Adaptive Stochastic Pooling Network

The discovery of SR makes noise a beneficial part, and this advantage will be take to add a moderate amount of noise to an image to enhance image contrast. The stochastic pooling network was proposed by Zozor [19] in 2007, which can use a moderate amount of noise to generate stochastic resonance to enhance the signal. In this model, a dynamically saturated nonlinear system will be used to build the basic unit of the stochastic pooling network. The model is shown in Fig. 1.

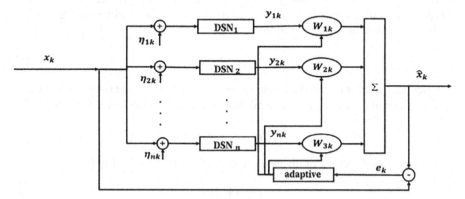

Fig. 1. Dynamic saturated nonlinear stochastic pooling network based on ASR

In this paper, the parameters α and β of the dynamic saturated nonlinear system when the noise intensity is minimum are obtained by using the pooling network. The appropriate parameters will appear stochastic resonance phenomenon, and then the corresponding parameters and the dynamic saturated nonlinear system are substituted into them. In addition, Kalman and LMS adaptation algorithms will be experimented with in combination with Kalman gain matrices. The specific steps of the adaptive stochastic resonance parameter optimization are as follows,

Step 1: Quantization processing. Random noise is added to the input signal x, and the output y_k of each subsystem node can be got by process it through the nonlinear node function g(x) of formula (3).

$$g(x) = \begin{cases} 1, x + \eta_k > \theta \\ 0, x + \eta_k \leq \theta \end{cases} \tag{3}$$

where η_k is Gaussian white noise, $\theta = 0$ is the threshold of network nodes. $\mu_\varepsilon = 0$ and $\sigma_\varepsilon^2 = 0.01$ are their mean and variance.

Step 2: Calculate and update the weight vector and weight error covariance matrix. After each iteration, the network output and optimal learning gain will be calculated, and the mean squared error of the network output and network input will be judged whether it reaches the minimum value. If not, the weight vector and weight error covariance matrix will be update.

The solution method of the network output $\hat{x}_k$ for each iteration is shown in Eq. (4):

$$\hat{x}_k = W_k^T y_k + W_0 \tag{4}$$

where W_k is the weight vector after each iteration, and the weight W_0 is initialized as a zero matrix of size N × N.

The calculation method of the optimal learning gain g_k[18] after each iteration is shown in Eq. (5):

$$g_k = G_k \tilde{y}_k = \frac{\Phi(k)\tilde{y}_k}{\tilde{y}_k^T \Phi_L(k) y_k + \sigma_\delta^2} \tag{5}$$

where G_k is the Kalman gain matrix, $\Phi(k)$ is the weight error covariance matrix, $\Phi(0) = \sigma^2 I$, σ^2 is a positive value, and I is the identity matrix. $\tilde{y}_k$ is the observed value of y_k, which is calculated as shown in Eq. (6):

$$\tilde{y}_k = y_k - E[y_k] \tag{6}$$

Calculate the error e_k and the mean squared error J between the network output and the network input in this iteration. The calculation method of error e_k is shown in Eq. (7):

$$e_k = \hat{x}_k - x \tag{7}$$

The calculation method of the mean square error J is shown in Eq. (8):

$$J = E[e_k^2] \tag{8}$$

The value of J is be calculated after each iteration and judged whether it reaches the minimum value. If the minimum value is not reached, the weight vector and weight error covariance matrix are updated. The weight vector is updated each time as shown in Eq. (9):

$$W_{k+1} = W_k + g_k\left(\tilde{x}_k - w_k^T \tilde{y}_k\right) \tag{9}$$

where $\tilde{x}_k$ is the observed value of x_k. The weight error covariance matrix update method is shown in Eq. (10):

$$\Phi(k+1) = \Phi(k) - g_k \tilde{y}_k \Phi_\kappa \tag{10}$$

The outputs of all iterative networks are accumulated to obtain the estimated value $\hat{x}$ of the input signal after the disturbance term is removed, as shown in Eq. (11):

$$\hat{x} = \sum_{k=0}^{N} \hat{x}_k \tag{11}$$

Step 3: Calculate the dynamic saturated nonlinear system parameters. The optimal system parameters $\alpha = 2$ and $\beta = 1$ are obtained by fitting LSE between the network output $\hat{x}$ after eliminating the disturbance term and the input signal x.

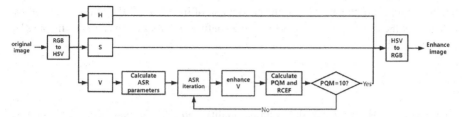

Fig. 2. Flow chart of contrast enhancement algorithm based on ASR

2.3 Image Enhancement Process

The model for enhancing image contrast has been shown in the previous chapter. In this chapter, the overall processing flow of the experiment will be introduced. The experimental flow chart is shown in Fig. 2.

The main steps of the image enhancement process are as follows:

Step 1: The original image is converted from RGB (Red-Green-Blue) color space to HSV (Hue-Saturation-Value) color space by using MATLAB functions. Because the algorithm only operates on the V component, the H component and the S component are not changed, so some color-related information of the picture is preserved.

Step 2: Determine the optimal parameters using an adaptive algorithm. The Kalman-LMS adaptive algorithm is used to remove the disturbance of the image, and the parameters α and β in the dynamic saturated nonlinear system are obtained by LSE fitting between the de-disturbed output and the original high-contrast image.

Step 3: Adjust image contrast using dynamic saturated nonlinear system adaptive iterative model. The V component is adjusted by the adaptive stochastic resonance dynamic saturation nonlinear model, and the perceptual quality measurement (PQM) is continuously calculated in the iterative process, and stops when the PQM is close to 10 and the maximum relative contrast enhancement factor (RCEF) index value is obtained.

Step 4: Convert the contrast-enhanced image from HSV space to RGB space and obtain the final contrast-enhanced image.

3 Analysis of Results

3.1 Image Performance Evaluation Indicator

In this paper, the relative contrast enhancement factor (RCEF), distribution separation measure (DSM), perceptual quality measure (PQM), target-to -background enhancement based on entropy (TBEE) and target-to-background enhancement based on standard deviations (TBES) five performance metrics were selected to measure the image enhancement effect.

RCEF is judged by measuring the mean and global variance of the original and enhanced images. The amount of contrast enhancement can be expressed by the Michelson contrast index [20], which can be summarized as the contrast quality index M, as shown in Eq. (12):

$$M(a) = \frac{s_a^2}{\mu_a} \tag{12}$$

where s^2 and μ are the global variance and mean of image a. RCEF can be obtained by calculating the ratio of contrast quality values before and after image enhancement [21]. As shown in Eq. (13):

$$RCEF = \frac{M_0}{M_1}$$

(13)

where M_0 and M_1 represent the contrast quality value of the enhanced high-contrast image and the contrast quality value of the original low-contrast image before enhancement.

PQM is a reference-free image evaluation metric that considers the block and blur effects of the image, as shown in Eq. (14) [22]:

$$PQM = \alpha + \beta B^{\gamma_1} A^{\gamma_2} Z^{\gamma_3}$$

(14)

where $\alpha = 245.9$, $\beta = 261.9$, $\gamma_1 - 0.0240$, $\gamma_2 = 0.0160$, $\gamma_3 = 0.0064$.

B is the block degree estimated as the average difference between block boundaries, A and Z are the mean absolute difference between image samples obtained by estimating the image signal and zero-crossing rate. According to Mitra, the value of PQM should be as close to 10 as possible to obtain the best perceptual quality [23].

The change of the background area between the enhanced image and the original low-contrast image will be presented by DSM [24], and DSM is used to evaluate the enhancement degree of the background area, as shown in Eq. (15):

$$DSM = \left(\left(\left|\mu_u^T - \mu_u^B\right|\right) - \left(\left|\mu_f^T - \mu_f^B\right|\right)\right)$$

(15)

where μ_f^B, μ_f^T, μ_u^B, μ_u^T represent the mean values of the background area (B) and the target area (T) of the original low-contrast image (f) and the enhanced image (u).

Further calculation of DSM, TBE$_E$ and TBE$_S$ two indicators will be got [24]. TBE$_E$ is used to evaluate index of entropy. The calculation method of entropy is shown in Eq. (16):

$$E(e) = \sum_{k=0}^{255} p(k)\log_2\left[p(k)\right]$$

(16)

where k represents the gray level of the image, and $P(k)$ represents the probability of each gray level appearing.

The calculation of TBE$_E$ and TBE$_S$ are shown in Eq. (17) and (18):

$$TBE_E = \left(\frac{\mu_u^T}{\mu_u^B} - \frac{\mu_f^T}{\mu_f^B}\right)\frac{e_f^T}{e_u^T}$$

(17)

$$TBE_s = \left(\frac{\mu_u^T}{\mu_u^B} - \frac{\mu_f^T}{\mu_f^B}\right)\frac{\sigma_f^T}{\sigma_u^T}$$

(18)

where e_f^T, e_u^T, σ_f^T, σ_u^T represent the entropy and standard deviation of the target area before and after image enhancement processing.

Generally, the larger the DSM and RCEF index values, the better the visual perception quality after image enhancement. Therefore, the iteration should be stopped when the PQM value is closest to 10 because we need to obtain the best perceptual quality. At the same time, the RCEF value of the enhanced image should be calculated. For DSM, TBEE and TBES indicators, the higher their values, the better the enhancement of the image.

3.2 Data Comparison and Analysis

3.2.1 Method Comparison

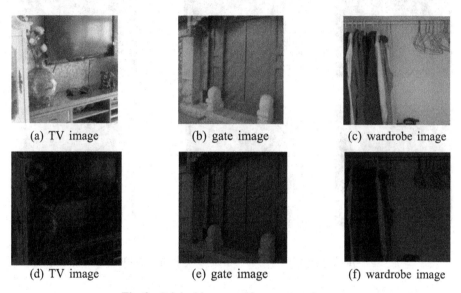

(a) TV image   (b) gate image   (c) wardrobe image

(d) TV image   (e) gate image   (f) wardrobe image

Fig. 3. Original image and low-contrast image

In order to demonstrate the superiority of this method (the variance of external noise is 0.00001), three experimental images will be used and compared with five image enhancement algorithms of CLAHE, Gamma, SSR, Retinex-Net, and SSIEN. The parameters used in the method of this experiment are the values measured in the experiment, and the number of iterations is related to the image. Figure 3(a)–(c) are three original images, (d)–(f) are the corresponding low-contrast images.

3.2.2 TV Image

Figure 2 is a comparison diagram of the enhancement effect between the DSN method in this paper and the five classic methods (TV image 128 * 128). Figure 4(b)–(f) are the processed results of CLAHE, Gamma, SSR, Retinex-Net and SSIEN methods, respectively. Obviously, the contrast enhancement effect of the image processed by the traditional method is not significant, and the brightness improvement of the image restored by the Gamma method is not obvious. Although the brightness value of the image is improved significantly by Retinex-Net and SSIEN, the image brightness value is too high. As a result, part of the color information is lost, resulting in color deviation and unnatural visual perception. The results show that the contrast enhancement of the ASR iterative method proposed in this chapter is better than other methods in terms of visual effect, and the images recovered by the ASR iterative method are clear without artifacts and moderate in color.

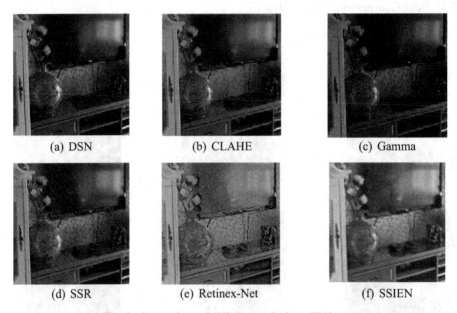

Fig. 4. Comparison of different methods on TV image

Table 1. Performance indicators of TV images under different enhancement methods

Method	TV image (128 × 128)				
	PQM	RCEF	DSM	TBE$_E$	TBE$_S$
CLAHE	9.4745	2.0268	57.9286	−0.1319	−0.0890
Gamma	10.1979	3.5896	42.2250	−0.2050	−0.0273
SSR	9.6202	3.9114	63.1710	−0.3250	−0.0305
Retinex-Net	9.3537	3.7267	63.0074	−0.1109	−0.0194
SSIEN	9.4878	4.2214	68.5000	−0.1524	−0.0045
DSN	9.9985	**5.2708**	**75.4779**	**−0.0813**	**−0.0025**

Table 1 shows the comparison of five performance indicators between the DSN method in this paper and the five classical methods for enhanced pictures (TV image

128 * 128). When the PQM values of the six methods are all close to 10, the larger the RCEF index value, the better the image enhancement effect. According to the data in Table 1, the RCEF index value of the image enhanced by Gamma is lowest, and it is 2.0268. The RCEF index value of the image enhanced by SSR and Retinex-Net are relatively close, the RCEF index value of DSN is the best, and the value of SSIEN method is sub-optimal. In terms of DSM, TBEE and TBEs, the values of the new method of DSN are also the best. According to Fig. 2 and Table 1, the enhanced images of the new method are better in both visual quality and index values.

3.2.3 Gate Image

The comparison chart of the enhancement effect of the DSN method in this paper and the five classic methods (door map 256 * 256) is shown in Fig. 5. The brightness of the image enhanced by the Gamma method is still somewhat low, and although the Retinex-Net and SSIEN methods are better in noise suppression, they also have the problem of too high image brightness, and the color is biased, which affects the visual perception.

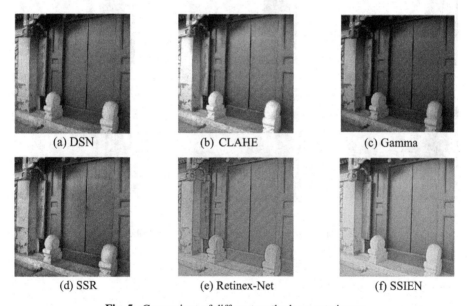

| (a) DSN | (b) CLAHE | (c) Gamma |
| (d) SSR | (e) Retinex-Net | (f) SSIEN |

Fig. 5. Comparison of different methods on gate image

The comparison of the five performance indicators of the gate image between the DSN method and the five classical methods are as shown in Table 2. According to Table, when the PQM is close to 10, the new method is still the best in the four indicators of RCEF, DSM, TBEE and TBEs. Among the five comparison methods, the image enhanced by Retinex-Net is relatively good in the DSM index, but its RCEF value is not ideal. At the same time, in the image comparison of the gate, the enhancement effect of the new method is still good, and the color is true.

Table 2. Performance indicators of gate images under different enhancement methods

Method	Gate image (256 × 256)				
	PQM	RCEF	DSM	TBE$_E$	TBE$_S$
CLAHE	9.2726	4.8245	66.5835	−0.0943	−0.0141
Gamma	9.8657	4.6561	54.4177	−0.1106	−0.0072
SSR	9.3611	4.1970	56.0020	−0.1831	−0.0158
Retinex-Net	9.5484	3.2111	68.1048	−0.4478	−0.1790
SSIEN	10.5063	4.2429	67.2523	−0.2743	−0.0667
DSN	9.9998	**5.7440**	**87.2321**	**−0.0575**	**−0.0046**

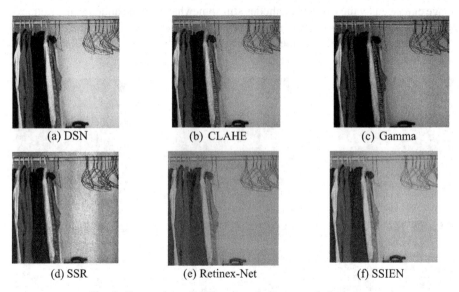

(a) DSN   (b) CLAHE   (c) Gamma

(d) SSR   (e) Retinex-Net   (f) SSIEN

Fig. 6. Comparison of different methods on wardrobe image

3.2.4 Wardrobe Image

The enhancement effect of six methods on the wardrobe image are shown in Fig. 6. The brightness of the image enhanced by the Gamma method is still somewhat low, and the image enhanced by the SSR method has a lot of artifacts, and the restoration effect is not good. At the same time, we can see that compared with the previous TV images and door images, with the increase of the picture resolution, the restoration effect of the new method and the traditional method is improved, and it is getting closer and closer to the original high-contrast image. It can also be seen in Table 3 that the value of the DSM indicator has improved significantly.

Table 3. Performance indicators of wardrobe images under different enhancement methods

Method	Wardrobe image (512 × 512)				
	PQM	RCEF	DSM	TBE_E	TBE_S
CLAHE	9.5390	4.8374	90.5060	−0.1350	−0.0082
Gamma	9.4315	3.5699	88.5773	−0.9358	−0.0363
SSR	9.4081	4.2603	99.8729	−0.0687	−0.0016
Retinex-Net	9.8339	5.5889	98.7096	−0.1383	−0.0232
SSIEN	10.2006	5.6051	107.9644	−0.1245	−0.0102
DSN	10.1140	**6.8596**	**130.0013**	**−0.0435**	**−0.0027**

4 Conclusion

In this paper, a dynamic saturated nonlinear stochastic pooling network model based on adaptive stochastic resonance is proposed, which combines the dynamic saturated nonlinear system with stochastic resonance, and it can realize adaptive iteration and enhance low-contrast images. The above experimental results show that, compared with traditional methods, the enhanced images recovered by the new method proposed in this paper not only have good visual perception, but also obtain better evaluation index values.

References

1. Ashiba, M.I., Tolba, M.S., El-Fishawy, A.S., et al.: Hybrid enhancement of infrared night vision imaging system. Multimed. Tools Appl. **79**(9), 6085–6108 (2020)
2. Zhang, W., Dong, L., Zhang, T., et al.: Enhancing underwater image via color correction and bi-interval contrast enhancement. Signal Process. Image Commun. **90**, 116030 (2021)
3. Xia, K., Zhou, Q., Jiang, Y., et al.: Deep residual neural network based image enhancement algorithm for low dose CT images. Multimed. Tools Appl., 1–24 (2021)
4. Singh, R.P., Dixit, M.: Histogram equalization: a strong technique for image enhancement. Int. J. Signal Process. Image Process. Pattern Recogn. **8**(8), 345–352 (2015)
5. Pizer, S.M., et al.: Contrast-limited adaptive histogram equalization: speed and effectiveness. In: Proceedings of the First Conference on Visualization in Biomedical Computing, p. 337 (1990)
6. Xu, G., Su, J., Pan, H.D., Zhang, Z.G., Gong, H.B.: An image enhancement method based on gamma correction. In: 2009 Second International Symposium on Computational Intelligence and Design, 1, pp. 60–63 (2009)
7. Zuiderveld, K.: Contrast limited adaptive histogram equalization. Graphics gems, pp. 474–485 (1994)
8. Huang, S.C., Cheng, F.C., Chiu, Y.S.: Efficient contrast enhancement using adaptive gamma correction with weighting distribution. IEEE Trans. Image Process. **22**(3), 1032–1041 (2012)
9. Jobson, D.J., Rahman, Z., Woodell, G.A.: Properties and performance of a center/surround retinex. IEEE Trans. Image Process. **6**(3), 451–462 (1997)

10. Jobson, D.J., Rahman, Z., Woodell, G.A.: A multiscale retinex for bridging the gap between color images and the human observation of scenes. IEEE Trans. Image Process. 6(7), 965–976 (1997)
11. Wei, C., Wang, W., Yang, W., et al.: Deep retinex decomposition for low-light enhancement. arXiv preprint arXiv (2018)
12. Zhang, Y., Di, X., Zhang, B., et al.: Self-supervised image enhancement network: training with low light images only. arXiv preprint arXiv (2020)
13. Benzi, R., Sutera, A., Vulpiani, A.: The mechanism of stochastic resonance. J. Phys. A: Math. Gen. 14, 453–457 (1981)
14. Mitaim, S., Kosko, B.: Adaptive stochastic resonance. Proc. IEEE 86(11), 2152–2183 (1998)
15. Liu, J., Hu, B., Wang, Y.: Optimum adaptive array stochastic resonance in noisy grayscale image restoration. Phys. Lett. A 383(13), 1457–1465 (2019)
16. Xiao, L., Bajric, R., Zhao, J., et al.: An adaptive vibrational resonance method based on cascaded varying stable-state nonlinear systems and its application in rotating machine fault detection. Nonlinear Dyn. 103(1), 715–739 (2021)
17. Wu, C., Wang, Z., Yang, J., et al.: Adaptive piecewise re-scaled stochastic resonance excited by the LFM signal. Europ. Phys. J. Plus 135(1), 1–14 (2020)
18. Gard, T.C.: Introduction to Stochastic Differential Equations. Monographs and Text-books in pure and applied mathematics. New York (1988)
19. Zozor, S., Amblard, P.O., Duchêne, C.: On pooling networks and fluctuation in suboptimal detection framework. Fluctuation Noise Lett. 7(01), 39–60 (2007)
20. Itzcovich, E., Riani, M., Sannita, W.G.: Stochastic resonance improves vision in the severely impaired. Sci. Rep. 7(1), 1–8 (2017)
21. Gupta, N., Jha, R.K.: Enhancement of dark images using dynamic stochastic resonance with anisotropic diffusion. J. Electron. Imaging 25(2), 023017 (2016)
22. Wang, Z., Sheikh, H.R., Bovik, A.C.: No-reference perceptual quality assessment of JPEG compressed images. In: Proceedings. International Conference on Image Processing, pp. 1: I-I. IEEE (2002)
23. Mukherjee, J., Mitra, S.K.: Enhancement of color images by scaling the DCT coefficients. IEEE Trans. Image Process. 17(10), 1783–1794 (2008)
24. Gupta, N., Jha, R.K., Mohanty, S.K.: Enhancement of dark images using dynamic stochastic resonance in combined DWT and DCT domain. In: 2014 9th International Conference on Industrial and Information Systems (ICIIS), pp. 1–6. IEEE (2014)

A Fusion Method for 2D LiDAR and RGB-D Camera Depth Image Without Calibration

Xia Hou[1], Hong Shi[1(✉)], Yuanhao Qu[2], and Minghao Yang[2]

[1] School of Computer, Beijing Information Science and Technology University, Beijing, China
{houxia,shihong}@bistu.edu.cn
[2] Institute of Automation, Chinese Academy of Science, Beijing, China
yuanhao.qu@guet.edu.cn, mhyang@nlpr.ia.ac.cn

Abstract. Two-dimensional (2D) LiDAR and RGB-D camera are two widely used sensors in various tasks of robot navigation. In spite of calibration, there are still quite a few noises such as hollows and speckle burrs in both of them caused by possibly external complex environments. This paper provides a data fusion method for 2D LiDAR and RGB-D depth images. The proposed method utilizes the phenomenon that the data provided by 2D LiDAR and RGB-D are significantly different in format but tightly relative in their depth information. With time alignment and correlation analysis, we find that the lines of 2D LiDAR are able to register to the RGB-D images in height, and conversely, the corresponding lines of RGB-D depth in height could be registered to the range of 2D LiDAR curves in width automatically, even there is no calibration of them. In experiments, we evaluate the proposed method on the Robot@Home dataset: a widely recognized in-doors open robot navigation database. The results show that the proposed method contributes to de-noise of the original data both for 2D LiDAR and RGB-D depth image simultaneously. The proposed method is also validated on the realistic navigation environments, and it could be extended to the application of more precise 2D map construction for robot navigation.

Keywords: 2D LiDAR · RGB-D · Robot Navigation

1 Introduction

In spite of various sensors in robot navigation task, 2D LiDAR and RGB-D camera are still widely used for depth information perception in robot navigation environment, such as 2D map construction [1, 2], location and environment reconstruction [3–5]. RGB-D camera depth image and 2D LiDAR provide different views of information mainly in the following two styles:

- The view and range of their depth information are different: the 2D LiDAR provides depth data in horizon level and RGB-D camera presents depth data in both horizon and vertical view of field [1].

F. Sun et al. (Eds.): ICCCS 2022, CCIS 1732, pp. 89–101, 2023.
https://doi.org/10.1007/978-981-99-2789-0_8

- Noises exist in their data but the reasons for the noises are different: noises in 2D LiDAR are shaped usually because of possible transparent glass ahead, strong light absorption for laser projection and reflections, etc. While quite a few RGB-D cameras obtain depth information depended on infrared reflection. Hollows and speckle burrs exist in the depth data of RGB-D image because of strong backlight, far or near objects out of reflection range and measurement errors caused by multiple infrared reflections [6].

There are some researchers combing these two sensors' depth information in map construction [7] and navigation location [8, 9], etc. However, there is still lack of an efficient fusion method or depth fusion strategy of them because of different principles in their signal processing. This paper provides an automatic data fusion model for 2D LiDAR and RGB-D depth image. The proposed method utilizes the phenomenon that the depth information provided by 2D LiDAR and RGB-D are tightly related. The proposed fusion method is able to de-noise the 2D LiDAR and RGB-D depth images automatically and simultaneously.

The remainders of this work are organized as follows: we first present related work in Sect. 2, and introduce the fusion model in Sect. 3. Experiments are presented in Sect. 4, and we conclude this work in Sect. 5.

2 Related Work

2D LiDAR provides horizon level depth information in front of view. It provides accessible and inaccessible navigation information in the 2D map, including real-time loop closure [10], place recognition [11, 12], obstacle object distance estimation [13, 14], etc,. Different from 2D LiDAR, RGB-D camera provides depth information for full view of observation. The RGB-D camera provides abundant three dimensional (3D) information, which supports obstacles segmentation [15], map building [2, 3], navigation [3, 4], object detection [16], etc., tasks. However, map building and navigation from RGB-D image are relatively complex since matching processing in 3D point cloud [4, 7, 9] and 3D reconstruction from RGB and depth images [3, 5, 17–19] are time consuming.

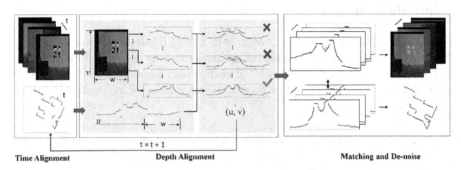

Fig. 1. The pipeline of the proposed fusion model in this work.

In spite of depth information both provided by 2D LiDAR and RGB-D camera, their image principles and formations are significantly different. The noises, such as hollows and burrs are different in the positions and ranges in 2D LiDAR and RGB-D depth data. When the 2D LiDAR and RGB-D cameras are placed in similar heights [7, 9], the depth views from similar heights help the matching processing between these two sensors. However, there are some situations that these two sensors are placed in completely different heights [1]. In these cases, it is difficult to calibrate the 2D LiDAR and RGB-D camera since the depth information in different heights make their observation in view are significant different. In these cases, an automatic method of depth information fusion for 2D LiDAR and RGB-D depth images is quite needed.

This work proposes a data fusion method for 2D LiDAR and RGB-D depth images. The proposed method utilizes the phenomenon that the data provided by 2D LiDAR and RGB-D are significantly different in format but tightly relative in depth information perception. With the proposed method, the lines of 2D LiDAR are able to register to the RGB-D depth images in height, and conversely, the corresponding lines of RGB-D depth in height could be registered to the range of 2D LiDAR curves in width automatically.

3   The Proposed Method

The work flow of the proposed fusion method is given in Fig. 1, which contains three steps:

- Align the 2D LiDAR data and RGBD depth images according to their time stamp.
- Align each line of RGB-D depth image to 2D LiDAR series along x-coordinate, and conversely, align 2D LiDAR data to RGB-D depth view along y-coordinate according to their z values, where x, y coordinates are the horizontal and vertical axis, and z present the depth information.
- Match and de-noise two kinds of depth curves after their initial alignment in horizontal and vertical.

3.1   Alignment on Time Stamp

Let D^i ($1 \leq i \leq M$) and L^j ($1 \leq j \leq N$) be the ith depth image captured by RGB-D camera and the jth 2D series obtained by 2D LiDAR sensor respectively. Considering the work frequency difference between two sensors, we align the D^i image and the L^j 2D series using $f(i) = g(j) = t$ and the functions $f(i)$ and $g(j)$ make D^t and L^t very adjacent to each other in time sequence, where $1 \leq t \leq T = f(M) = g(N)$.

3.2   Depth Alignment in the Width Range of 2D LiDAR and Height Range of RGB-d Depth Image

Supposing that W and H are the width and height of D^i, Ψ is the width of L^j, $D^t_{h,w}$ ($x^t_{h,w}, z^t_{h,w}$, $1 \leq w \leq W$) is the 2D point series along the x-coordinate at height h for D^t, and L^t_{ω} ($x^t_{\omega}, z^t_{\omega}$, $1 \leq \omega \leq \Psi$) is the 2D point series for 2D LiDAR L^t, where W and Ψ

are the length of $D_{h,w}^t$ andL^t. Let $\hat{L}_{u,\omega}^t$ ($\omega \leq u \leq \omega + W$ and$1 \leq \omega \leq \Psi$) be W points picked from L_ω^t from start point$u$, where usually $\Psi \geq W$, namely the width of 2D LiDAR data is larger than that of RGB-D depth image. Then the target of depth alignment is to find the appropriate values of (u,v), which contribute to the max value of Eq. (1), where $1 \leq u \leq \Psi - W, 1 \leq v \leq H$. Function $sim()$ in Eq. (1) is used to measure how similar of two curves, where $0 \leq sim() \leq 1.0$.

$$max[arg_{\forall(u,v)} \sum_{v=1}^{H} \sum_{u=1}^{\Psi - W} \left\| \left| sim\left(D_{v,w}^t - \hat{L}_{u,\omega}^t \right) \right| \right\|] \tag{1}$$

In history, many methods have been proposed to register one curve to another [21–23]. However, most of these methods demanded the source point set is clear in their shapes. Considering that noise and burrs both exist in $D_{v,w}^t$ and $\hat{L}_{u,\omega}^t$, we adopt Pearson correlation coefficient (PCC), normalized Cosine and Euler distance as function $sim()$ in this step, which are given as Eq. (2)–Eq. (4) respectively. For simplicity, we denote $D_{v,w}^t$ and $\hat{L}_{u,\omega}^t$ as D_v and $\hat{L}_u$ in the following discussion.

$$PCC(D_v, \hat{L}_u) = \frac{Cov\left(D_v, \hat{L}_u\right)}{\sqrt{Var(D_v)Var\left(\hat{L}_u\right)}} = \frac{E\left(D_v\hat{L}_u\right) - E(D_v)E\left(\hat{L}_u\right)}{\sqrt{E(D_v^2) - E^2\left(\hat{L}_u\right)}\sqrt{E\left(\hat{L}_u^2\right) - E^2\left(\hat{L}_u\right)}} \tag{2}$$

$$Inv_{Cos}(D_v, \hat{L}_u) = 1 - Cos(D_v, \hat{L}_u) = 1 - \frac{D_v \cdot \hat{L}_u}{||D_v||\left|\left|\hat{L}_u\right|\right|} \tag{3}$$

$$Inv_{Euler}(D_v, \hat{L}_u) = 1 - Euler(D_v, \hat{L}_u) = 1 - \frac{||D_v - \hat{L}_u||}{max(||D_v||, \left|\left|\hat{L}_u\right|\right|)} \tag{4}$$

3.3 Matching and De-noise

Matching. After obtaining the values of u and v using Eq. (1), we obtain the appropriate fusion location of x-coordinate (u) on 2D LiDAR data and the y-coordinate (v) on RGB-D depth image. In spite of hollows and burrs, the shapes of point set $D_{v,w}^t$ and $\hat{L}_{u,\omega}^t$ in (x, z) 2D plane are similar to each other since they share the same front view. Our target is to find the noise points on $D_{v,w}$ and $\hat{L}_{u,\omega}$ from their shapes. In this step, we combine Random Sample Consensus (RANSAC) [24] and homography [25] to obtain robust mapping relationship between $D_{v,w}$ and $\hat{L}_{u,\omega}$ and to decide the noise points on them.

The shape transformation for two point sets in planes which are similar to each other could be described by homography [25]. Equation (5) presents the mapping relationship using the homography method, which contributes to the converting from $m \times 3$ point set ($\hat{L}$) in 2D LiDAR plane to the $m \times 3$ points set (D) at the vertical plane of RGB-D depth image with a 3×3 homography matrix (A). In Eq. (5), (x_i^L, z_i^L) and (x_i^D, z_i^D) are the points' values in (x, z) 2D plane for 2D LiDAR and RGB-D depth images, where $1 \leq i \leq m$.

$$\hat{L}A = D \text{ where } U = D \text{ where } U = \begin{bmatrix} x_1^L & z_1^L & 1 \\ \vdots & \vdots & \vdots \\ x_m^L & z_m^L & 1 \end{bmatrix}, A = \begin{bmatrix} a_{11} & a_{12} & a_{13} \\ a_{21} & a_{22} & a_{23} \\ a_{31} & a_{32} & a_{33} \end{bmatrix}, D = \begin{bmatrix} x_1^D & z_1^D & 1 \\ \vdots & \vdots & \vdots \\ x_m^D & z_m^D & 1 \end{bmatrix} \tag{5}$$

Supposing there are α good matching pairs and β noise pairs between $D_{v,w}^t$ and $\hat{L}_{u,\omega}^t$, where $W = \alpha + \beta$ and β noise pairs are caused by burr or hollow points on either of $D_{v,w}$ and $\hat{L}_{u,\omega}$. Because of the noise pairs, an ideal strategy is to pick n pairs from α good ones and to calculate the values of homography matrix A, where $n < \text{sizeof}(\alpha) < W$. However, there is no previous label for α and β, we like to pick n pairs randomly from full W pairs, and once the picked n pairs are all good pairs, then the values of A are obtained.

Supposing $c = \alpha/(\alpha + \beta)$, then the probability of the picked n pairs are all good pairs is c^n [24]. The maximum number of attempts to find a consensus sample with 97.5% credibility ratio are about $4/c^n$ [24].

De-noise. There are mainly two kinds of noises contained in the β noise pairs: hollows and burrs. There are three cases in our de-noise step for the noises:

- A: The areas of noise pairs in 2D LiDAR and RGB-D depth data both are hollows;
- B: The areas of noise in 2D LiDAR or RGB-D depth data are hollows, and the others are not hollows;
- C: The areas of noise in 2D LiDAR or RGB-D depth data are burrs, and the others are not burrs;

It is easy to judge case A where the points in $D_{v,w}^t$ and $\hat{L}_{u,\omega}^t$ are "zero", and we can use the Eq. (6) to judge case "B" and "C". ε in Eq. (6) is a threshold to distinguish the pair' matching distances, and the points $\hat{L}_{u,w}$ or $D_{v,w}$ satisfy Eq. (6) are hollows or burrs.

$$\exists_{\beta \in W} \| \hat{L}_{u,\beta} A - D_{v,\beta} \| > \varepsilon \tag{6}$$

For the case "B" and "C", once the hollows or burrs are found in $\hat{L}$ or D (denoted as $\hat{L}_{u,\beta}$ or $D_{v,\beta}$), then we use the correspond good points in D and $\hat{L}$ to retrieve the noise data using $\hat{L}_{u,\hat{\alpha}} = D_{v,\alpha} A^{-1}$ or $D_{v,\hat{\alpha}} = \hat{L}_{u,\alpha} A$. $\hat{L}_{u,\hat{\alpha}}$ and $D_{v,\hat{\alpha}}$ are repaired data for noise points $\hat{L}_{u,\beta}$ and $D_{v,\beta}$. It is noticeable that in the case "A", both $\hat{L}_{u,\beta}$ are $D_{v,\beta}$ are zero. When this case are found, we remove the original $\hat{L}_{u,\beta}$ and $D_{v,\beta}$ points in $\hat{L}$ and D, and replace them using the good points which are adjacent to them.

4 Experiments

In experiments, we first evaluate the proposed fusion method on a well-known robot in-door navigation dataset Robot@Home [1], which contains 2D LiDAR and RGB-D depth images without calibration. After that, we continue to adopt the proposed method in a real-time navigation in-door environment with a 2D LiDAR and a RGB-D camera placed in similar and different locations on a Turtlebot-2 platform.

4.1 Robot@Home Dataset

The Robot@Home dataset is a collection of raw and processed data from five domestic settings aimed at serving as a benchmark for semantic mapping, robot localization, 2D/3D map building and object segmentation [26]. The dataset contains 87,000 + time-stamped observations gathered by a mobile robot endowed with a rig of 4 RGB-D cameras and a 2D laser scanner. In spite of a similar view in front of 2D LiDAR and RGB-D cameras, the 2D LiDAR and RGB-D depth images in Robot@Home were not calibrated.

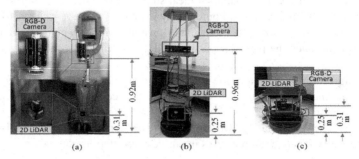

Fig. 2. The locations in height for the 2D LiDAR and RGB-D cameras in different situations; (a) the locations' configuration in Robot@Home (Fig. 1 in [1]); (b) (c) the real two experiment situations in this work.

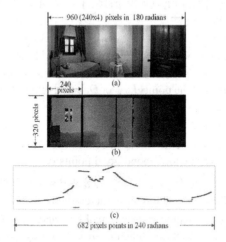

Fig. 3. A sample of RGB and depth images and 2D LiDAR curve aligned on time-stamp in Robot@Home.

4.2 Depth Alignment

Figure 3 presents the fourth RGB-D depth image and the twelfth 2D LiDAR for the "bedroom1" of the "alma-s1" apartment, where Fig. 3(a), (b) and (c) are processed RGB image, depth image and 2D LiDAR respectively. Four Asus XTion Pro Live RGB-D cameras contribute to 960 × 320 resolution images in 180° radians along horizontal direction, and 2D laser scanner URG-04LX-UG01 generates 2D curve each frame with 682 pixels in 240°. Then in depth alignment step, the 960 pixels in horizontal direction of depth image were sampled into 180 pixels, and the 682 pixels of 2D LiDAR are sampled into 240 pixels along horizontal direction. In this way, the target of depth alignment is to calculate the values of u and v simultaneously according to Eq. (1). The range of u is $\omega \leq u \leq \omega + W$, where W=240 pixels and $1 \leq \omega \leq 180$, then the value of u is about $0 \leq u \leq 60$. Theoretically, the values of v are in the range of [0, 320] pixels. However, in Robot@Home dataset, the 2D LiDAR is placed at the very low level of robot, then we set the max values of v 160 pixels in experiment.

The charts in the top row of Fig. 4 present the distributions of the values of (u,v) for $PCC(D_v, \hat{L}_u)$, $Inv_{Cos}(D_v, \hat{L}_u)$, $Inv_{Euler}(D_v, \hat{L}_u)$ respectively for $D_{v,w}^t$ and $\hat{L}_{u,\omega}^t$ given in Fig. 3. In Fig. 4(a), (c) and (e), the pixel intensity presents the matching degree of the D_v and $\hat{L}_u$. The charts in the bottom row of Fig. 4 present the corresponding 2D LiDAR and RGB-D depth curves at the max value of (u,v). We can see from Fig. 4 that the distribution of $PCC(D_v, \hat{L}_u)$ and $Inv_{Euler}(D_v, \hat{L}_u)$ are similar. Meanwhile the location of the max values of (u,v) are c lose between $PCC(D_v, \hat{L}_u)$ and $Inv_{Euler}(D_v, \hat{L}_u)$ at about (49, 21), and the two curves' fluctuations trend in Fig. 4(b) and (d) are more correlative than those in Fig. 4(f). It indicates that $PCC(D_v, \hat{L}_u)$ and $Inv_{Euler}(D_v, \hat{L}_u)$ are better than $Inv_{Cos}(D_v, \hat{L}_u)$ in the depth alignment task. In addition, we can see from Fig. 4 that the values of $PCC(D_v, \hat{L}_u)$, $Inv_{Euler}(D_v, \hat{L}_u)$ and $Inv_{Cos}(D_v, \hat{L}_u)$ are in the range of [0. 10, 0.80], [0.70, 0.93] and [0.10, 0.35] respectively. The values of $PCC(D_v, \hat{L}_u)$ are more well distributed in the range of (0,1), which contributes to better distinguish ability for u and v than those of Inv_{Euler} and Inv_{Cos}. In addition, the PCC values in the range of $(0 \leq u \leq 20)$ and $(0 \leq v \leq 160)$ are obviously smaller than those of $(40 \leq u \leq 60)$. A reason for this phenomenon is that the 2D LiDAR is located at the 0.31m height above the ground and the RGB-D camera is placed at the height of 0.92m in Robot@Home. There are some objects which are placed on the ground and are very close to the robot. In Fig. 3, a dwarf cabinet is placed at the side of the bed in the left part of the image. Most parts of cabinet do not appear in depth image, but are scanned by 2D LiDAR. Then the 2D LiDAR pixels are not able to match to the ones at the same height of RGB-D depth curves. In Fig. 4(b) and (d), the 2D LiDAR points in red in the range of $10 \leq u \leq 20$ are cabinet, while the points in blue in the same range are the front part of bed. The points in this range of 2D LiDAR and RGB-D depth images have low values of PCC. While for the points located in the right range of $60 \leq x \leq 180$ in Fig. 4(b) and (d), the points in red are closer to the points in blue than the points in the left part. It is because that the scene in the right part of Fig. 3(a) and (b) are very open, namely there is no dwarf objects located before 2D LiDAR.

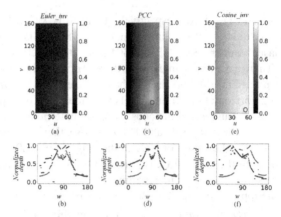

Fig. 4. The distributions of *PCC*, Inv$_{Euler}$, Inv$_{Cos}$ and the depth alignment for the 2D LiDAR (curves in red) and RGB-D depth image (curves in blue) presented in Fig. 3.

Table 1. lists the numb er of max *PCC* values which are located in the range of low (0, 0.3), middle (0.3, 0.6) and high level (0.6,1.0) respectively for the seven rooms in the biggest department sarmis-s1 of Robot@Home. We can see from Table 1 that the room "corridor1" has obviously large ratio of high level *PCC* about 66.67% than those of "bedrooms" about 18.48% and 20.10%. The reason for this phenomenon is that the scenes in the "corridor1" are obviously more open than other rooms in Robot@Home dataset.

Table 1. The number of max PCC values at the levels for the seven rooms in the biggest department sarmis-s1 of Robot@Home.

Room	Low (0.0–0.3)	Middle (0.3–0.6)	High (0.6–1.0)	High score ratio (%)
bedroom1	28	316	92	21.10
bedroom2	20	212	56	19.44
bathroom2	16	84	24	19.35
bedroom3	44	256	68	18.48
corridor1	0	72	144	66.67
kitchenroom1	4	120	20	13.89
livingroom1	28	196	76	25.33

4.3 Matching

With the depth alignment using *PCC* for D_v and $\hat{L}_u$, the values of u and v are obtained. We continue to match D_v and $\hat{L}_u$ using RANSCA and homography for the scenes with a high level of *PCC* values. According to the description and the table in §II.B of article

[22], when the value of c (w in [22]) is larger than 0.6, and each time $n=6$ sample pairs are random picked in homography calculation, then a consensus sample (6 pairs in α good pairs) will be obtained in 84 times random picking with 97.5% confidence. Then in experiment, we set the sample times as 84 and $n=6$.

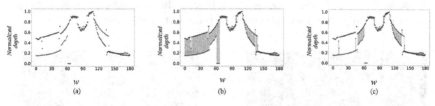

Fig. 5. The macthing procedure for the two curves after their depth alignment

Figure 5 presents the matching procedure for the scene given in Figs. 3 and 4(d), where the red curve and blue curve in Fig. 5(a) are the 2D LiDAR data and depth data for matching respectively. Figure 5(b) and (c) present all $\alpha+\beta$ pairs and the final α good pairs after 84 times samples where the noise β pairs are removed.

Figure 6 presents the recall rate for the good matching pairs and the noise pairs with different ε, where $\xi = \alpha' \cap \alpha/\alpha$ and $\eta = \beta' \cap \beta/\beta$. The α' and β' are the good and repaired pairs from noise obtained by the proposed model. In Fig. 6, the value of ε varies with different scale of E, where E is the mean error of all pairs, which is obtained by $E = \sum_{u=1}^{W} ||\hat{L}_u A - D_u||/W$. The higher values of ξ and η, the better performance of homography matrix A. In the de-noise procedure, the values of ε is set as $1.2\,E$ in experiments.

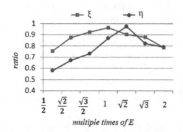

Fig. 6. The recall rate for the good matching pairs and the noise pairs with different scales.

4.4 De-noise

Figure 7 presents the de-noise results for the scene given in Fig. 3. The images in Fig. 7(a) and (b) are original and de-noise RGB-D depth images, and the ones in the bottom row are original and de-noise 2D LiDAR curves. The light green line at the bottom of the original RGB-D depth images indicates the location of depth alignment of 2D LiDAR to RGB-D depth image. The points in red block in 2D LiDAR curves are hollow, and their depth values are repaired by the corresponding information picked from RGB-D depth image, which is labeled by the arrow lines in blue. Similarly, the points in blue blocks in RGB-D depth image located on green line are hollows, and their values are repaired by the corresponding depth information picked from 2D LiDAR curve, which is labeled by the arrow lines in red. In this way, the original 2D LiDAR and RGB-D depth images are de-noise from each other. The pixels in blue in the 2D LiDAR de-noise image of Fig. 7(b) are the repaired points given by the proposed fusion model.

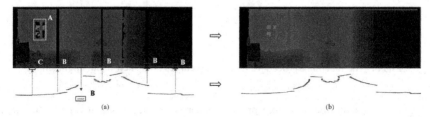

Fig. 7. De-noise results for the scene in Fig. 3.

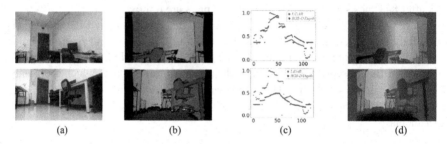

Fig. 8. The RGB images, RGB-D depth images, 2D LiDAR curves, aligned depth curves and the fusion ones for the real robot platforms given in Fig. 2(b) and (c).

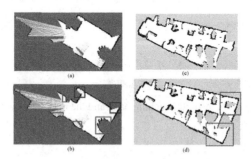

Fig. 9. The 2D map for the "bedroom1" of "alma-s1" in Robot@Home. (a) original 2D map; (b) the 2D map obtained by the proposed fusion method; (c): the 2D map for the authors' laboratory room generated by 2D LiDAR and (d) the 2D map generated with the proposed fusion method on 2D LiDAR and RGB-D depth.

Figure 9(a) and (b) present the 2D map constructed by the original and the de-noise 2D LiDAR data using the proposed fusion method for the "bedroom1" of "alma-s1" in Robot@Home. We can see from Fig. 9 that the missed boundaries in the red blocks are obviously improved and repaired by the proposed fusion model. In addition, there are some areas in blue blocks, where Robot@Home does not present the front views of them. In these areas, the depth information is both absent for 2D LiDAR and RGB-D depth images.Experiment in Real Navigation Environments

Besides Robot@Home, we evaluate the proposed model in real in-door navigation environments. We place a Hokuyo 2D LiDAR sensor and a Kinect 2 RGB-D camera on a Turtlebot-2 platform in different heights and similar heights respectively. Figure 2(b) and (c) present these two situations, where the Hokuyo and Kinect 2 are placed at the height of 0.25m and 0.96m above the ground in Fig. 2(b), and 0.31 m, 0.25 m above the ground in Fig. 2(c). The top and bottom images in Fig. 8(a)–(b) are the corresponding RGB images, original depth images, original curves of 2D LiDAR and RGB-D depth image, de-noised depth images respectively. The two curves in bottom image of Fig. 8(c) have better correlation than those images in the top Fig. 8(c). It indicates that 2D LiDAR and RGB-D camera placed at similar heights have better alignment results those cases of sensors placed in obvious different heights. In addition, Fig. 9(c) and (d) list the 2D map reconstructed for the room presented in Fig. 8(a), where the 2D map in Fig. 9(c) is generated by the 2D LiDAR without fusion. The 2D map in Fig. 9(d) is generated from the fusion results from 2D LiDAR and RGB-D depth image. We can see the 2D map in Fig. 9(d) present more complete information than Fig. 9(c) at the right-bottom and right-top corners. It is because that in these two corners, the depth information in 2D LiDAR is partially zero because of the occlusion caused by the sides of the desk. While in the depth information presented by Kinect 2 provides the full depth information including the desk and the wall. It helps to enhance the depth information of 2D LiDAR and contributes to make 2D map better from the fusion results generated by our method.

5 Conclusions

In this work, we propose a data fusion method for 2D LiDAR and RGB-D image. The proposed fusion method is evaluated on Robot@Home and in real robot in-door environment. The results show that the proposed method contributes to de-noise of the original data and provides more accuracy details information both for 2D LiDAR and RGB-D images simultaneously without of calibration of them. We believe it could be extended to more precise 2D map construction and 3D scene reconstruction for robot navigation.

Acknowledgements. This work is supported by the National Key Research & Development Program of China (No. 2018AAA0102902), the Guangxi Key Research and Development Program (AB21220038), the National Natural Science Foundation of China (NSFC) (No.61873269), Beijing Natural Science Foundation + J210012, Hebei Natural Science Foundation (F2021205014), Key Laboratory Foundation (KGJ6142210210311), the Beijing Natural Science Foundation (No: L192005), Key Laboratory Fund (KGJ6142210210311), the CAAI-Huawei MindSpore Open Fund (CAAIXSJLJJ-20202-027A), the Guangxi Key Research and Development Program (AB18221011,AB21075004, AD18281002, AD19110137), the Natural Science Foundation of Guangxi of China (No: 2020GXNSFAA297061, 2019GXNSFDA185006, 2019GXNSFDA185007), Guangxi Key Laboratory of Intelligent Processing of Computer

Images and Graphics (No GIIP201702) and Guangxi Key Laboratory of Trusted Software (NO kx201621,kx201715).

References

1. Ruiz-Sarmiento, J.R., Galindo, C., GonzáLez-JiméNez, J.: Robot@Home, a robotic dataset for semantic mapping of home environments. Int. J. Robot. Res. **36**(2), 131–141 (2017)
2. Kolhatkar, C., Wagle, K.: Review of SLAM algorithms for indoor mobile robot with LIDAR and RGB-D camera technology. In: Favorskaya, M.N., Mekhilef, S., Pandey, R.K., Singh, N. (eds.) Innovations in Electrical and Electronic Engineering. Lecture Notes in Electrical Engineering, vol. 661, pp. 397–409. Springer, Singapore (2021). https://doi.org/10.1007/978-981-15-4692-1_30
3. Tian, B., Shim, V.A., Yuan, M., Srinivasan, C., Tang, H., Li, H.: RGB-D based cognitive map building and navigation. In: 2013 IEEE/RSJ International Conference on Intelligent Robots and Systems (IROS), Tokyo, Japan, 3–7 November 2013 (2013)
4. Wellhausen, L., Ranftl, R., Hutter, M.: Safe robot navigation via multi-modal anomaly detection. IEEE Robot. Autom. Lett. **5**(2), 1326–1333 (2020)
5. Fu, Y., Yan, Q., Yang, L., Liao, J., Xiao, C.: Texture mapping for 3D reconstruction with RGB-D sensor. In: Proceedings of the IEEE Conference on Computer Vision and Pattern Recognition (CVPR) (2018)
6. Kuan, Y.W., Ee, N.O., Wei, L.S.: Comparative study of intel R200, kinect v2, and Primesense RGB-D sensors performance outdoors. IEEE Sensors J. **19**(19), 8741–8750 (2019)
7. Shin, Y.-S., Park, Y.S., Kim, A.: Direct visual SLAM using sparse depth for camera-LiDAR system. In: 2018 IEEE International Conference on Robotics and Automation (ICRA), Brisbane, Australia, 21–25 May 2018 (2018)
8. Winterhalter, W., Fleckenstein, F., Steder, B., Spinello, L., Burgard, W.: Accurate indoor localization for RGB-D smartphones and tablets given 2D floor plans. In: 2015 IEEE/RSJ International Conference on Intelligent Robots and Systems (IROS), Hamburg, Germany, 28 Sept–2 Oct 2015 (2015)
9. Song, H., Choi, W., Kim, H.: Robust vision-based relative-localization approach using an RGB-depth camera and lidar sensor fusion. IEEE Trans. Indust. Electron. **63**(6), 3725–3736 (2016)
10. Hess, W., Kohler, D., Rapp, H., Andor, D.: Real-time loop closure in 2D LIDAR SLAM. In: 2016 IEEE International Conference on Robotics and Automation (ICRA), Stockholm, Sweden, 16–21 May 2016 (2016)
11. Himstedt, M., Frost, J., Hellbach, S., Böhme, H.-J., Maehle, E.: Large scale place recognition in 2D LIDAR scans using geometrical landmark relations. In: 2014 IEEE/RSJ International Conference on Intelligent Robots and Systems (IROS 2014), Chicago, IL, USA, 14–18 September 2014 (2014)
12. Zlot, R., Bosse, M.: Place recognition using keypoint similarities in 2D lidar maps. Springer Tracts Adv. Robot. **54**(2009), 363–372 (2009)
13. Wang, D.Z., Posner, I., Newman, P.: Model-free detection and tracking of dynamic objects with 2D lidar. Int. J. Robot. Res. **34**(7), 1039–1063 (2015)
14. Arras, K.O., Mozos, O.M., Burgard, W.: Using boosted features for the detection of people in 2D range data. In: IEEE International Conference on Robotics and Automation (ICRA 2007), Roma, Italy, 10–14 April 2007 (2007)
15. Holz, D., Holzer, S., Rusu, R.B., Behnke, S.: Real-time plane segmentation using RGB-D cameras. In: Röfer, T., Mayer, N.M., Savage, J., Saranli, U. (eds.) RoboCup 2011: Robot Soccer World Cup XV. RoboCup 2011. LNCS, vol. 7416, pp. 306–317. Springer, Heidelberg (2011). https://doi.org/10.1007/978-3-642-32060-6_26

16. Wang, Y., Wei, X., Shen, H., Dinge, L., Wan, J.: Robust fusion for RGB-D tracking using CNN features. Appl. Soft Comput. **92**, 106302 (2020)
17. Zhong, J., Li, M., Liao, X., Qin, J.: A Real-time infrared stereo matching algorithm for RGB-D cameras' indoor 3D perception. Int. J. Geo-Inf. **9**(8), 472 2020
18. Zollhöfer, M., et al.: State of the art reportstate of the art on 3D reconstruction with RGB-D cameras. In: Eurographics 2018: Annual Conference of the European Association for Computer Graphics (2018)
19. Ylimaki, M., Heikkil, J., Kannala, J.: Accurate 3-D Reconstruction with RGB-D Cameras using depth map fusion and pose refinement. In: 2018 24th International Conference on Pattern Recognition (ICPR) Beijing, China (2018)
20. Chong, Z.J., Qin, B., Bandyopadhyay, T., Ang, M.H., Frazzoli, E., Rus, D.: Synthetic 2D LIDAR for precise vehicle localization in 3D urban environment. In: 2013 IEEE International Conference on Robotics and Automation (ICRA), Karlsruhe, Germany (2013)
21. Myronenko, A., Song, X.: Point set registration: coherent point drift. IEEE Trans. Pattern Anal. Mach. Intell. **32**(12), 2262–2275 (2010)
22. Lian, W.: Rotation-invariant nonrigid point set matching in cluttered scenes. IEEE Trans. Image Process. **21**(5), 2786–2797 (2012)
23. Ma, J., Zhao, J., Yuille, A.L.: Non-rigid point set registration by preserving global and local structures. IEEE Trans. Image Process. **25**(1), 53–64 (2016)
24. MA, F., RC, B.: Random Sample Consensus (RANSAC): a Paradigm for model fitting with applications to image analysis and automated cartography. In: Readings in Computer Vision, pp. 726–740 (1987)
25. Malis, S.B.: Homography-based 2D visual tracking and servoing. Int. J. Robot. Res. **26**(7), 661–676 (2007)
26. Robot@Home   (http://mapir.isa.uma.es/mapirwebsite/index.php/mapir-downloads/203-robot-at-home-dataset) (31 Oct 2017)

Shape and Pose Reconstruction of Robotic In-Hand Objects from a Single Depth Camera

Xiaoxian Jin[1], Zhen Deng[1(✉)], Zhenyu Zhang[1], Liangjun Lu[2], Ge Gao[3], and Bingwei He[1]

[1] Department of Mechanical Engineering and Automation,
Fuzhou University, Fujian 350108, China
zdeng@fzu.edu.cn
[2] Maynooth International Engineering College,
Fuzhou University, Fujian 350108, China
[3] Mech-Mind Robotics Technologies Ltd., Beijing 110000, China

Abstract. This paper tackles the task of estimating the state of an object in a robotic hand. The state of the object includes its shape and posture, which is critically important for robotic in-hand manipulation. However, in-hand objects have self-occlusion, making it challenging to perceive their complete shape and posture. To address this challenge, this work proposed a point-clouds processing framework to achieve shape completion and pose estimation of the in-hand objects. Firstly, the input point cloud are segmented based region growing algorithm to obtain the points belonging to the target object. Then, we design a neural network with the auto-encoder structure to perform shape completion and 6D pose estimation of the in-hand object. The latent feature of the network is used to regress the 6D pose, i.e., position and orientation, of the object. The effectiveness of the proposed framework is evaluated by comparison experiment and real-word experiment. Experimental results show that our approach achieves significantly high accuracy in the shape completion and pose estimation of robotic in-hand objects.

Keywords: Deep learning · Shape completion · Pose estimation

1 Introduction

Robots performing complex manipulation tasks, such as dexterous in-hand manipulation, need to perceive the shape and pose of objects. Due to insufficient knowledge of the state of the in-hand object, even simple robotic grasping and placing actions can fail. Visual information captured from vision sensors is often collected to estimate the state of the in-hand object. But, only side of the object can be perceived by vision sensors because of visual occlusion. Moreover, due to the physical contact between the robotic hand and object, the state estimation of the in-hand object remains challenging. To address the challenge, this

F. Sun et al. (Eds.): ICCCS 2022, CCIS 1732, pp. 102–114, 2023.
https://doi.org/10.1007/978-981-99-2789-0_9

work introduces a novel point-clouds processing framework for shape completion and pose estimation of the in-hand objects, which allows for better robotic in-hand manipulation.

Completing the shape of objects from partial visual information is helpful for robotic manipulation. There have been some works to complete the object shape by data fitting methods. For example, Makhal et al. [7] proposed to use the superquadratic (SQ) surface to represent the object shape. However, the SQ-based method is only suitable for symmetrical regular objects. It has poor robustness for objects with complex shapes. Dragiev et al. [2] used Gaussian process implicit surface (GPIS) to predict points located on the surface of an object for robotic grasping. Gaussian process is not limited by the number of fixed parameters. It can be used to approximate objects of arbitrary shapes. However, since GPIS needs to fit the point-to-point relationships, the local shape information of the object is easily lost in the process of shape completion. In the context of shape completion from incomplete point-clouds, more recent approaches have been developed based on a deep neural network. For example, PCN [19], TopNet [13], and PF-Net [4] both process the incomplete point cloud directly for shape completion. Varley et al. [14] proposed a 3D convolutional neural network that enabled the shape completion from a single-view depth map. However, their proposed network is unable to obtain high-resolution object shapes. Lundell et al. [6] used a pre-trained deep neural network that produced a complete three-dimensional pixel grid from the incomplete point cloud, effectively ignoring the shape uncertainty. However, the computation of the voxel grid is time-consuming. Similar to this work, Merwe et al. [8] developed a single-view reconstruction learning system, which combined shape completion with grasp planning to achieve collision-free grasping of novel objects. However, these methods do not solve the problem of how to complete the shape of the object in the robotic hand.

Apart from shape completion, estimating the 6D pose of an object, i.e., position and orientation, is also important for robotic in-hand manipulation. To track the pose of the in-hand object, Bimbo et al. [1] fused visual and haptic information to estimate the object's pose. Pfanne et al. [10] combined the position and moment measurements of the hand with visual features extracted from monocular images. Similarly, Drigalski et al. [15] proposed a particle filter-based method based on haptic information for the pose estimation of in-hand objects. Although using multiple sensors can provide rich information for pose estimation, it increases the complexity and cost of robot system. More recent works employ deep learning methods to estimate the object's pose directly from the point cloud. PoseCNN [18] is a novel neural network for object pose estimation, which estimated the 3D translation of a target object by locating its center in the image and predicting its distance from the camera. Pauwels et al. [9] used SimTrack, a method for real-time detecting and tracking the object's pose. Simtrack has high scalability in the number of observation objects and tracking methods.

The objective of this work is to achieve shape completion and pose estimation of robotic in-hand objects simultaneously from a single depth camera. We develop a point-clouds processing framework, which estimate the state of the in-hand object from partially-observable point cloud. First, object segmentation based on region growing method is performed to extract the points belonging the objects from the raw point cloud. Then, a neural network with the auto-encoder structure is employed to complete the object shape and estimate its 6D pose. Finally, extensive experiments including real-world experiment are carried out to demonstrate the performance of the proposed approach. Results show that the proposed method can effectively estimate the shape and pose of objects in the robotic hand.

2   Methodology

This section introduces the point-clouds processing framework, which consists of three main computational steps, i.e., object segmentation, shape completion and pose estimation, as illustrated in Fig. 1.

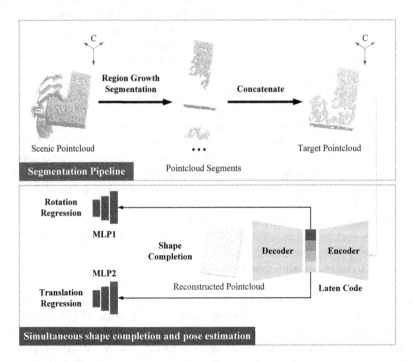

Fig. 1. Proposed point-clouds processing framework for shape completion and pose estimation of robotic in-hand objects.

2.1   Object Point-Clouds Segmentation Based on Region Growth

This subsection introduced a method that segments the point cloud of the in-hand object from the raw scenic point cloud. Considering the visual occlusion, an adapted segmentation algorithm based on region growth is proposed.

In the traditional region growth method [11], the segmentation is prone to errors due to improper selection of seed points or inaccurate feature extraction. The RANSAC algorithm is often used to obtain the seed points in 3D scattered point cloud segmentation, but it is easy to overlap the segmentation. To address the shortcomings of this method, in this work, the points with the smallest curvature are set as the seed nodes. The points with the smallest curvature in the point cloud region are selected to start growing. These points are located in the flattest region, which can reduce the total number of segments and avoid overlapping segmentation.

The point segmentation based on regional growth adopted in this paper are carried out as follows.

Step 1. The curvature of each point is calculated and ranked according to its size. The point with the smallest curvature is selected as the initial seed point for growth, the curvature K_n of a point P is computed as

$$2K_n n = \lim_{diam(A) \to 0} \frac{\nabla A}{A}, \tag{1}$$

where n is the normal vector, A is an infinitely small region around P, $diam(A)$ is the diameter of this region, and ∇ is the gradient operator about the point P. By discretizing (1), the average curvature of P_i is obtained:

$$K_n(P_i) = \frac{1}{4A_{min}} \times \sum_{j \in N_i} (\cot \alpha_{ij} + \cot \beta_{ij})(P_i - P_j) \times n, \tag{2}$$

where α_{ij} and β_{ij} are the diagonals connecting the edges of P_i and P_j, respectively.

Step 2. The search judgment of the neighboring points is performed by setting the candidate set of seed points C, clustering region Q and the spatial threshold. If the growth radius as well as the vertical distance meet the threshold condition, the point is transferred to the clustering region Q.

Step 3. Deleting the current seed point and continue to grow with the newly added seed point. Then loop through step 2 and 3 until the seed set is empty.

2.2   Shape and Pose Reconstruction

In this subsection, we detail the design of the attention-based neural network. A novel loss function is also present for network optimization.

Network Structure. The proposed network consists of three components: feature learning, pose estimation, and shape completion, as shown in Fig. 2. The first component consists of five EdgeConv [17] layers and a global feature aggregation module, which extract global features containing 6D pose information of the target object from the input point cloud. This component is the backbone of the network since the learned features are shared across all the tasks performed by the network. The second and third components will work in parallel. Both input the global features learned in the first component into a separate network to achieve the shape completion and 6D pose estimation tasks, as described next.

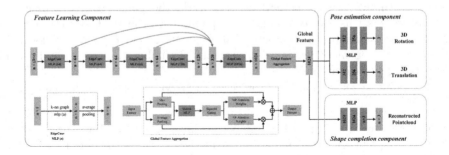

Fig. 2. Network Architecture.

In the feature learning component, the input point cloud is of dimension $n \times (3+c)$, where 3 represents the 3D coordinates of the point cloud. c represents the total number of classes of the point cloud in the dataset. This component outputs a global feature of dimension 1024. The edge features of point cloud are learned by the EdgeConv. The EdgeConv is repeated to compute the nearest neighbor graph of the feature vectors. To obtain a better global contextual aggregation, the edge feature connections of each EdgeConv are processed in the MLP layer, as shown in Fig. 2. In addition, we refine these features using an attention module. Inspired by global feature aggregation [5], this work learns the attention weights from the average-pooling and max-pooling features to scale their individual influences.

The shape completion component is responsible for generating the output point cloud from the 1024-dimensional feature vector. Our proposed component contains three fully connected layers of dimensions 1024, 1024, and $n \times 3$.

The component in the upper right corner of Fig. 2 shows the 6D pose estimation branch. Since the shape completion component is to reconstruct the point cloud segment with the same 6D pose from the global feature, the global feature contains the object 6D pose information. Therefore, the global feature is used as input to two independent networks for regression of 3D rotation and 3D translation, each network contains three MLP layers of 512, 256 and 3 dimensions, respectively.

Loss Function. A novel loss function is designed for shape completion and pose estimation. Since the point cloud is a disordered representation, a loss function for shape completion need to be independent of the relative order of the input points. Two alignment invariant metrics are used to compare disordered point cloud [3], namely Chamfer Distance (CD) and Earth Mover's Distance (EMD). CD describes the Euclidean distance between each point in the ground truth S_2

and its nearest point in the output point cloud S_1. Note that S_1 and S_2 need not be the same size to calculate CD. EMD describes the bijection between the ground truth and the output point cloud. A bijection $\emptyset : S_1 \to S_2$ is found between two-point sets to correspond the two-point sets one by one, so that the sum of the two computed Euclidean distances is minimized. Unlike CD, EMD requires S_1 and S_2 to be the same size.

$$l_{CD}(S_1, S_2) = \frac{1}{|S_1|} \sum_{x \in S_1} \min_{y \in S_2} ||x - y||_2 + \frac{1}{|S_2|} \sum_{y \in S_2} \min_{x \in S_1} ||y - x||_2. \qquad (3)$$

$$l_{EMD}(S_1, S_2) = \min_{\emptyset : S_1 \to S_2} |S_2| \sum_{x \in S_1} ||x - \emptyset(x)||_2. \qquad (4)$$

The loss function for pose estimation includes two parts: directional regression and position regression. The loss function of the rotation regression is defined using the geodesic distance in this work. We choose the axis angle $r \in \mathbb{R}^3$ as the rotation expression, which is a vector representation of the radians of rotation $\theta = ||r||_2$ around the unit vector $\frac{r}{||r||_2}$. The distance between two rotations is geometrically interpreted as the geodesic distance between two points on the unit sphere. Hence, the geodesic distance is the angle in radians between two viewpoints in exponential form [12]. The geodesic distance between the rotation matrices is used to describe the difference between the rotation matrices. The rotation loss function is defined as

$$l_r(\hat{r}, r) = \arccos\left(\frac{\text{trace}(\hat{R}R^T - 1)}{2}\right), \qquad (5)$$

where $\hat{R}$ and R are the corresponding rotation matrices for $\hat{r}$ and r respectively.

As for position regression, we use the normalized translation as the regression target. The full translation prediction $\hat{t} = t_N + t_{mean}$ is obtained by adding the mean t_{mean} of the input point cloud. The loss function for position regression is defined using L2-norm:

$$l_t(\hat{t}, t) = ||t - \hat{t}||_2, \qquad (6)$$

where t is the ground truth translation.

The total loss function for network training is an combination of the reconstruction, rotation and translation loss:

$$L_{total} = \alpha l_{CD} + \beta l_{EMD} + \gamma l_r + l_t, \qquad (7)$$

where $\alpha, \beta, \gamma \geq 0$ denotes the hyper-parameters.

3   Experiment

In this section, we first compare the proposed method with state-of-the-art methods on public datasets. Then, a real-world experiment with a four-finger robotic hand is carried out to evaluate the effectiveness of the proposed method.

3.1 Performance Evaluation

Datasets. The open-source YCB video (YCBV) dataset is used in this evaluation. YCBV is composed of 133,827 real RGB-D images intercepted from 92
RGB-D videos, and 80,000 synthetic RGB-D images. These images are composed of 21 different classes of objects, as shown in Fig. 3. Figure 3(a) shows all
21 target objects. Figure 3(b) shows the RGB image on the top and the synthe-
sized RGB image on the bottom. This work used 20% of the real RGBD images
from the YCBV dataset for network training and extracted 2949 key frames for
testing.

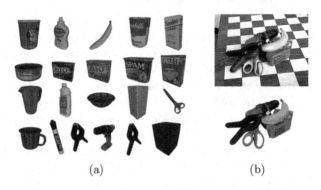

(a)   (b)

Fig. 3. YCB-Video dataset:(a) 21 different objects, (b) top is real shot, and bottom is
synthetic.

Evaluation Metric. This work used the average distance(ADD) proposed
in [20] and the average distance of a rotationally symmetric object(ADD-S)
proposed in [18] as the evaluation metrics. The ADD is defined as:

$$ADD = \frac{1}{m} \sum_{x \in \mathcal{M}} ||(Rx + t) - (\hat{R}x + \hat{t})||_2, \tag{8}$$

where R and t are the ground truth rotation and translation, and $\hat{R}$ and $\hat{t}$ are
the estimated rotation and translation. $\mathcal{M}$ denotes the set of 3D model points.
m is the number of points.

ADD-S is computed using closest point distance. It is a distance measure
that considers possible pose ambiguities caused by object rotational symmetry:

$$ADD - S = \frac{1}{m} \sum_{x_1 \in \mathcal{M}} \min_{x_2 \in \mathcal{M}} ||(Rx_1 + t) - (\hat{R}x_2 + \hat{t})||_2, \tag{9}$$

We can plot an accuracy-threshold curve for pose estimation. The area under
this curve (AUC) is calculated by setting the horizontal axis threshold to a
maximum of 0.1 m. The 6D pose estimation is considered to be accurate if the
ADD and ADD-S values are smaller than the given threshold.

During network training, we used Adam optimizer with learning rate 0.0008. The batch size is 128. The number of points of the input point cloud segments is $n = 256$. Batch normalization is applied to all layers and no dropout is used. For ICP refinement, we use the simple Point-to-Point registration provided by Open3D [12] and refine for 10 iterations. The initial search radius is 0.01 m and it is reduced by 10% after each iteration.

Table 1. The AUC of ADD-S metric on the YCB-VIDEO dataset.

	DenseFusion Iterative [16]	Ours + ICP
Object	AUC	AUC
002_master_chef_can	94.7	**95.7**
003_cracker_box	**91.0**	87.4
004_sugar_box	94.9	**98**
005_tomato_soup_can	93.7	**94.8**
006_mustard_bottle	95.6	**97.8**
007_tuna_fish_can	94.9	**97.3**
008_pudding_box	93.4	**97.3**
009_gelatin_box	97.2	**97.6**
010_potted_meat_can	89.2	**91.7**
011_banana	91.4	**97.6**
019_pitcher_base	93.7	**97.7**
021_bleach_cleanser	93.9	**95.2**
024_bowl	84.5	**97.7**
025_mug	95.3	**97.6**
035_power_drill	91.9	**97.5**
036_wood_block	**89.7**	84.7
037_scissors	90.4	**92.5**
040_large_marker	95.2	**96.3**
051_large_clamp	72	**72.3**
052_extra_large_clamp	**68.5**	48.0
061_foam_brick	92.6	**95.9**
Average	90.8	**92.0**

Results and Discussion. The proposed network is compared with the popular Densefusion network based on the YCBV dataset. The experimental results are shown in Table 1 and Fig. 4. The numbers in the table indicate the accuracy in %, and the bottom row indicates the average accuracy of all target objects under both methods. Those marked in bold are the methods with higher average accuracy under the AUC threshold. The target objects marked in red in the leftmost column indicates the symmetric objects.

From Table 1, it can be seen that compared to the DenseFusion network, the proposed network is slightly better, and the performance can be significantly

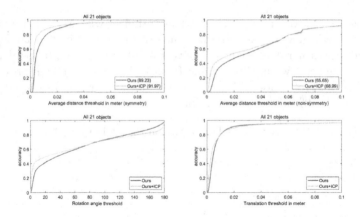

Fig. 4. Detailed results on the YCB-Video dataset.

improved by ICP optimized poses. Then, we can see that some objects are more difficult to handle and do not perform as well on the network as the Dense-Fusion, such as large clamp and extra large clamp because they look the same and the network is easily confused. For symmetrical objects, the two networks have equal processing power. In general, our proposed network has improved the generalization capability to some extent.

3.2   Performance Experiments Comparing Loss Function

The effectiveness of the loss function was also evaluated. We evaluated the loss function for shape completion. In this experiment, we used both CD and EMD for $l_1 = l_C D + l_E M D$ but only CD for $l_2 = l_C D$. Both CD and EMD can reflect the good or bad shape restoration effect of point cloud data better. We visualized the change in the value of the loss function during the training process in Fig. 5. It can be seen that the value of the loss function decreases as the number of epochs increases, and both loss functions show a better fit.

In the process of training the network, CD is usually chosen as the repair loss function because it is differentiable. Meanwhile, EMD requires more memory and longer computation time during computing, while CD is more efficient. However, from Fig. 6. we can see that the CD-trained network is prone to reconstructing point cloud with splashes and blurring at the boundaries. Therefore, a complementary EMD is introduced to correspond the reconstructed point cloud to the points in the target point cloud one by one, which makes the reconstructed point cloud more compact and effectively avoids the extreme cases. We therefore chose l_2 as the loss function for the shape completion.

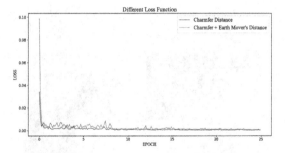

Fig. 5. The two loss curves under different loss functions during training.

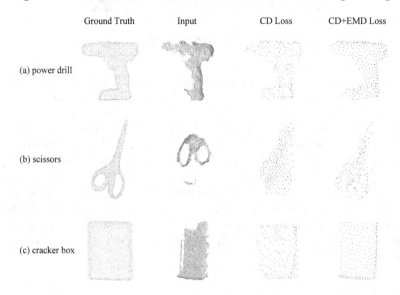

Fig. 6. The shape completion results of the network trained under different loss functions.

3.3 Real-Word Experiment

In this experiment, we evaluated the effectiveness of the proposed framework in real scenarios. The UR5 robot with the Allegro Hand as its end-effector was controlled to grasp objects, as shown in Fig. 7. We used the depth camera, i.e., the KinectV2, to capture the visual information of the scenario. After obtaining the scene point cloud data, we passed it to the segmentation pipeline to get the target object point cloud data. Finally, we input the point cloud data into the trained network to obtain the complementary shape and the corresponding 6D pose of the target object.   The target object was a real object with the category named cracker box in the YCB-V dataset. The experimental results are shown in Fig. 8. The red point cloud indicates the point cloud of the input network, which contains 1024 points. The blue part indicates the reconstructed

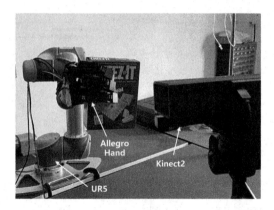

Fig. 7. Experimental setup.

point cloud, which contains 256 points. The Fig. 8(d) shows the calculated 3D rotation and 3D translation matrices of the object in this viewpoint.

We can see that the reconstruction is as expected, but there is still blurring of the edges of the reconstructed object. We use Eq. 5 to determine whether the predicted 6D pose is accurate, and when the value of ADD is less than 10% of the external sphere diameter (in meters) of the target object, the target object predicted 6D pose is accurate. The calculation results are as expected, and we use ICP refinement to make the 6D pose estimation results more accurate. By obtaining the shape and 6D pose information of this object, it is more conducive to the efficient subsequent dexterity operation. The network proposed in this chapter can estimate the shape and 6D pose of the object in hand accurately to some extent.

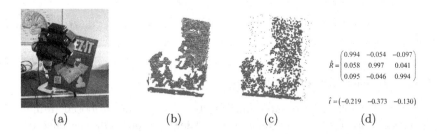

(a)   (b)   (c)   (d)

Fig. 8. Experimental Result.

4   Conclusion

This paper has proposed to complete the object shape and estimate its 6D pose from partial visual information. The information of the object shape and pose is critical important for the planning of robotic in-hand manipulation. A point-clouds processing framework with a neural network as its core component is introduced. In this framework, object segmentation based on region growing algorithm takes a point cloud captured by a depth camera as input and extract the points belonging to the objects. The neural network with the auto-encoder structure is designed to perform shape completion and 6D pose estimation of object. Experimental results demonstrated that the proposed approach achieves a high accuracy of shape completion and pose estimation. In future work. we intend to expand the proposed framework for developing robotic applications. For example, providing the object state information produced by the proposed approach for manipulation planning and control.

Acknowledgements. This work was partly supported by the National Natural Science Foundation of China (Project No. 62003089). This work was partly supported by the Natural Science Foundation of Fujian Province (Project No. 2020J01455).

References

1. Bimbo, J., Seneviratne, L.D., Althoefer, K., Liu, H.: Combining touch and vision for the estimation of an object's pose during manipulation. In: 2013 IEEE/RSJ International Conference on Intelligent Robots and Systems, pp. 4021–4026. IEEE (2013)
2. Dragiev, S., Toussaint, M., Gienger, M.: Gaussian process implicit surfaces for shape estimation and grasping. In: 2011 IEEE International Conference on Robotics and Automation, pp. 2845–2850. IEEE (2011)
3. Fan, H., Su, H., Guibas, L.J.: A point set generation network for 3D object reconstruction from a single image. In: Proceedings of the IEEE Conference on Computer Vision and Pattern Recognition, pp. 605–613 (2017)
4. Huang, Z., Yu, Y., Xu, J., Ni, F., Le, X.: PF-Net: Point fractal network for 3D point cloud completion. In: Proceedings of the IEEE/CVF Conference on Computer Vision and Pattern Recognition, pp. 7662–7670 (2020)
5. Kaul, C., Pears, N., Manandhar, S.: FatNet: a feature-attentive network for 3D point cloud processing. In: 2020 25th International Conference on Pattern Recognition (ICPR), pp. 7211–7218. IEEE (2021)
6. Lundell, J., Verdoja, F., Kyrki, V.: Beyond top-grasps through scene completion. In: 2020 IEEE International Conference on Robotics and Automation (ICRA), pp. 545–551. IEEE (2020)
7. Makhal, A., Thomas, F., Gracia, A.P.: Grasping unknown objects in clutter by superquadric representation. In: 2018 Second IEEE International Conference on Robotic Computing (IRC), pp. 292–299. IEEE (2018)
8. Van der Merwe, M., Lu, Q., Sundaralingam, B., Matak, M., Hermans, T.: Learning continuous 3d reconstructions for geometrically aware grasping. In: 2020 IEEE International Conference on Robotics and Automation (ICRA), pp. 11516–11522. IEEE (2020)
9. Pauwels, K., Kragic, D.: Integrated on-line robot-camera calibration and object pose estimation. In: 2016 IEEE International Conference on Robotics and Automation (ICRA), pp. 2332–2339. IEEE (2016)

10. Pfanne, M., Chalon, M., Stulp, F., Albu-Schäffer, A.: Fusing joint measurements and visual features for in-hand object pose estimation. IEEE Robot. Autom. Lett. **3**(4), 3497–3504 (2018)
11. Rabbani, T., Van Den Heuvel, F., Vosselmann, G.: Segmentation of point clouds using smoothness constraint. Int. Arch. Photogramm. Remote Sens. Spatial Inf. Sci. **36**(5), 248–253 (2006)
12. Salehi, S.S.M., Khan, S., Erdogmus, D., Gholipour, A.: Real-time deep registration with geodesic loss. arXiv preprint arXiv:1803.05982 (2018)
13. Tchapmi, L.P., Kosaraju, V., Rezatofighi, H., Reid, I., Savarese, S.: TopNet: structural point cloud decoder. In: Proceedings of the IEEE/CVF Conference on Computer Vision and Pattern Recognition, pp. 383–392 (2019)
14. Varley, J., DeChant, C., Richardson, A., Ruales, J., Allen, P.: Shape completion enabled robotic grasping. In: 2017 IEEE/RSJ International Conference on Intelligent Robots and Systems (IROS), pp. 2442–2447. IEEE (2017)
15. Von Drigalski, F., et al.: Contact-based in-hand pose estimation using Bayesian state estimation and particle filtering. In: 2020 IEEE International Conference on Robotics and Automation (ICRA), pp. 7294–7299. IEEE (2020)
16. Wang, C., et al.: DenseFusion: 6D object pose estimation by iterative dense fusion. In: Proceedings of the IEEE/CVF Conference on Computer Vision and Pattern Recognition, pp. 3343–3352 (2019)
17. Wang, Y., Sun, Y., Liu, Z., Sarma, S.E., Bronstein, M.M., Solomon, J.M.: Dynamic graph CNN for learning on point clouds. ACM Trans. Graph. (TOG) **38**(5), 1–12 (2019)
18. Xiang, Y., Schmidt, T., Narayanan, V., Fox, D.: PoseCNN: a convolutional neural network for 6d object pose estimation in cluttered scenes. arXiv preprint arXiv:1711.00199 (2017)
19. Yuan, W., Khot, T., Held, D., Mertz, C., Hebert, M.: PCN: point completion network. In: 2018 International Conference on 3D Vision (3DV), pp. 728–737. IEEE (2018)
20. Zhou, Q.Y., Park, J., Koltun, V.: Open3D: a modern library for 3D data processing. arXiv preprint arXiv:1801.09847 (2018)

Decision Making and Cognitive Computation

"Gongzhu" Strategy Based on Convolutional Neural Network

Licheng Wu⬥, Qifei Wu⬥, Hongming Zhong⬥, and Xiali Li⁽⊠⁾

School of Information Engineering, Minzu University of China, Beijing, China
xiaer_li@163.com

Abstract. "Gongzhu" is a very interesting and challenging poker game, which belongs to the incomplete information game. The game is divided into two stages: card-showing and card-playing. The behavior of card-showing determines the strategy of card-playing, and the whole game process is highly reversible. This paper proposes a deep learning-based game algorithm of "Gongzhu". According to the functional characteristics, the network structure of card-showing and card-playing are designed respectively. The two networks are both constructed by Convolutional Neural Network (CNN). Using the real game information of 11,000 high-level human players to generate train set and test set proportionally, the two CNN networks are trained and tested respectively. The fitting accuracy between the result predicted by the agent constructed by this algorithm and the labeled data of human players is used as the evaluation indicator. The experimental results show that the fitting accuracy of card-showing and card-playing networks can reach 88.4% and 71.4% respectively. The agent is able to generate reasonable strategies for card-showing and card-playing.

Keywords: Gongzhu · Incomplete Information · Artificial Intelligence · Deep Learning · Game · Convolutional Neural Network

1 Introduction

An incomplete information game means that the participants cannot obtain all the situation information from the game, so it is difficult to study the game algorithm. Texas Hold'em [1–3], "Doudizhu" [4–6], etc. In particular, there are a lot of related research results at home and abroad in Texas Hold'em. In 2013, Wang Xuan et al. achieved remarkable results in information representation [7], function optimization [8], game tree search [9], opponent modeling [10], Risk Model Analysis [11], etc. In the 2013 ACPC two-person limit project competition, it achieved a good fourth place [12]. In 2015, Bowling et al. proposed an improved Counterfactual Regret Minimization (CFR) algorithm, called CFR+ [13], which made significant progress in the two-player limit project, and successfully cracked the winning strategy of the project for the first time, but still could not solve the super-large-scale game problem. In 2018, Brown et al. adopted a finite depth-first approach for search [14], and the constructed agent defeated previous versions of AI programs. In 2019, Noam et al. used a self-playing method to train the agent, which is similar to the method of training AlphaGo Zero [15], AlphaZero [16], and the constructed agent Pluribus [3] defeated the human masters in six-player no-limit Texas Hold'em. In 2020, Zhang Xiaochuan et al. designed a decision-making model reward function based on the Upper Confidence Bound Apply to Tree (UCT) to update decision-making, and proposed an algorithm combining the Deep Q-Learning (DQN) and Sarsa, called DQN-S, and to improve the learning efficiency of the model. Excellent performance and won the first prize [17]. In 2021, Peng Lirong and others adopted the Aho-Corasick (AC) algorithm and introduced expert knowledge to pre-train the network parameters. The constructed agents played against other versions of Texas Hold'em agents, and the results show that the average payoff per game is more than 1 big blind [18]. In 2022, Zhang Meng et al. designed an integrated framework that includes two stages of offline training and online gaming for opponent modeling. In the face of dynamic adversary strategies, the level of the agent is improved compared with the previous method [19]. Zhou Qibin et al. reduced the availability of the strategy by considering the possible range of other players' hands, and the constructed DecisionHoldem [20] agent defeated the strongest heads-up no-limit Texas Hold'em agent Slumbot and Deepstack's advanced replication agent Openstack publicly.

There are also some literatures that study the incomplete information game problem of "Doudizhu". In 2018, Li Shuqin et al. applied the deep learning method to the "Doudizhu" playing cards; and completed the prediction of a player's single card [21]. In 2019, You Yang et al. proposed the combinatorial Q-learning (CQL) [22] algorithm, which solved the difficulty of multiple combinatorial playing methods. Jiang Qiqi et al. aimed at solving the problem that other players' playing methods and strategies could not be known, and provided the network model trained by the real game information of human players to other players for decision-making [4]. Wang Yisong et al. proposed a rule-based hand split algorithm and used the Monte-Carlo (MC) method to select the node with the largest profit as the best decision [5]. This method can better realize the self-play of "Doudizhu". In 2020, Wang Yisong and others combined the MC method with the convolutional neural network (CNN) to study the playing strategy of "Doudizhu". The agent constructed by this algorithm can obtain a relatively obvious advantage in the winning rate in the game against other agents with the existing "Doudizhu" strategy [23]. In the same year, Wang Yisong et al. used the game information collected by self-play to learn the strategy of "Doudizhu", and used a weight-based approach to overcome the problem of uneven distribution of training data, the model achieved a high winning rate in the game against real human [6]. In 2021, Zha Daochen et al. proposed a Deep Monte-Carlo (DMC) method, which used deep neural networks, action encoding, and parallel actors to improve the traditional MC method. The agent DouZero defeated all the "Doudizhu" AIs on the BotZone platform [24]. In 2022, Li Shuqin and others used the Alpha-Beta pruning algorithm to solve the endgame problem of "Doudizhu", and achieved a complete victory [25]. Yang Guan et al. adopted a perfect-training-imperfect-execution framework, in which the agent can use global information to guide policy training. Experiments show that the agent PerfectDou [26] constructed by him beats all the "Doudizhu" AIs, its performance is optimal.

"Gongzhu", also known as "Huapai", is a card game that is very popular in Chinese circles all over the world. It is a game of incomplete information. At present, there is no relevant literature on the research on "Gongzhu". Although Texas Hold'em AI can already beat professional human players, and the AI of "Doudizhu" is gradually approaching the level of human masters, they have high requirements on computing power and cannot be achieved without sufficiently powerful hardware resources. The adopted game algorithm also cannot be directly applied to "Gongzhu". This paper proposes a convolutional neural network-based game algorithm of "Gongzhu", which regards the playing and showing actions of human players in the card spectrum as correct annotations. Through supervised learning, the agent learns the strategies that human players take when showing and playing cards from real-world data.

2   "Gongzhu" Game Rules and Data Format

2.1   Game Rules

"Gongzhu" is composed of a deck of 52 cards with the big and small kings removed. There are 4 players in the game, and each player randomly gets 13 cards. In the game rules of "Gongzhu", all cards are divided into two categories: "scored cards" and "no scored cards". There are 16 cards in total, which are called scored cards, they are Spades Q ("Pig"), Diamonds J ("Sheep"), Clubs 10 ("Transformer"), and all the cards of Hearts, and the rest of the cards are no score. Table 1 shows the scored cards and their corresponding score in "Gongzhu".

Table 1. Scored cards and their corresponding score.

Cards	Hearts A	Hearts K	Hearts Q	Hearts J	Hearts (10 ~ 5)	Hearts (4 ~ 2)	Spades Q	Diamonds J
Score	−50	−40	−30	−20	−10	0	−100	+ 100

In addition, in the calculation stage of the game. For the player who gets Clubs 10, the value of other scored cards in his hand becomes twice the original value. If the player has no other scored cards in his hand, he will get +50.

There are also 4 cards that can be shown in the "Gongzhu" game, which are Clubs 10, Diamonds J, Spades Q, and Hearts A. If the Spades Q is shown, the score becomes -200; if the Diamonds J is shown, the score becomes +200; if the Hearts A is shown, the score of all Hearts × 2; Clubs 10 is shown, and the value of the other scored cards in the hand of the player who got the card × 4. If the player has no other scored cards in his hand, the score becomes +100.

Regarding the specific method of card-showing and card-playing, calculation of the scoring formula, and the final decision of the outcome. Please refer to the China Huapai Contest Rules (Trial) [27] for the gameplay.

2.2 Data Format

This paper uses a 1 × 52 one-hot matrix to represent 52 cards, and the suits are arranged in the order of Spades, Hearts, Diamonds, and Clubs. For example, the Spades K corresponds to the matrix position is 12, and Hearts 10 corresponds to the matrix position is 22. The positions of all cards in the 1 × 52 one-hot matrix are shown in Table 2.

Table 2. The position of all cards in a 1 × 52 one-hot matrix.

Cards	Spades (A ~ K)	Hearts (A ~ K)	Clubs (A ~ K)	Diamond (A ~ K)
Matrix Position	0 ~ 12	13 ~ 25	26 ~ 38	39 ~ 51

3 Card-Showing Algorithm

3.1 Card-Showing Category

There are 4 cards that can be shown in "Gongzhu". In this paper, a 1 × 4 one-hot matrix is used to encode the states and actions of the card-showing. The matrix positions represent the Clubs 10, Diamonds J, Spades Q, and Hearts A in turn. In the form of [x0 x1 x2 x3]. If a card is shown by the player, set its corresponding matrix position element to 1, otherwise, set it to 0. There are 16 kinds of card-showing categories in "Gongzhu", which can be represented by serial numbers 0 to 15. The category serial number can exactly be obtained through taking [x0 x1 x2 x3] as a Binary Code Decimal. For example, the [0 1 1 0] indicates that Diamonds 10 and Spades Q have been shown, and the sequence number of the shown card category is 6.

3.2 Card-Showing Network Design

Network Input and Output. Each player can only decide to show cards based on his or her initial hand, so the input information in the card-showing stage is a single player's 13 initial cards. In this paper, the player's initial hand card information is represented by a 1 × 52 one-hot matrix. The corresponding element of the matrix position is 1 to indicate that player has this card, and 0 to indicate no. Therefore, the one-hot matrix of each initial hand has 13 positions where the element is 1, and other position elements are 0. To facilitate the convolution operation, the 1 × 52 one-hot matrix is sequentially converted into a 4 × 13 one-hot matrix, that is, the input information of the CNN is a 4 × 13 one-hot matrix.

The output layer of the card-showing neural network consists of 16 neurons, which in turn correspond to the probabilities of outputting 16 types of card-showing, and finally, the category with the highest probability can be selected to show. When the probability of multiple categories is the same, one is randomly selected.

Network Structure. The card-showing network has 14 layers. The first layer is the input layer, and the input is a 4 × 13 matrix; the convolution kernel number of the 1-2Conv layers is 32, the size is 3 × 3, the stride is 1, and there is a ReLU layer after the Conv layer. After the first 2 × 2 Max-pooling layer, it becomes a 2 × 6 matrix; the convolution kernel number of the 3-4Conv layers is 64, the size is 2 × 2, the stride is 1, and there is a ReLU layer after the Conv layer; the convolution kernel number of the 5-6Conv layers is 128, the size is 2 × 2, the stride is 1, and there is a ReLU layer after the Conv layer; after the second 2 × 2 Max-pooling layer, it becomes a 1 × 3 matrix. After the Max-pooling layer, there is a Dropout layer, and the value of dropout is 0.2; the last layer is a Fully-connected layer, and the Softmax function is used to classify the action of card-showing, the decision to show cards is made according to the probability of each card-showing action. The CNN network model structure of the card-showing module is shown in Fig. 1.

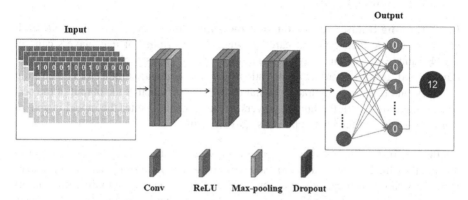

Fig. 1. The CNN model network structure of card-showing.

Loss Function. There are a total of 16 actions for card-showing, which is a multi-classification problem. The loss function used in the network structure of the card-showing in this paper is categorical cross-entropy.

For the card-showing classification problem, 16 output vector values are required. After the output vector value of each category is transformed by the Softmax function, it can be expressed as the probability that the model predicts the category, and the sum of all category probability outputs is 1.

The formula is as follows:

$$Softmax(z)_i = \frac{e^{z_i}}{\sum_{j=0}^{N-1} e^{z_j}}, i = 0, ..., N - 1 \tag{1}$$

Among them, Z_i is the original output vector value of each category, N is the number of categories of the label, and its value is 16 for the problem of card classification.

The following cross-entropy loss function is used:

$$-\log(P(y)) \tag{2}$$

Among them, y is the real label, and the card-showing categories correspond to Z_0, Z_1, ..., and Z_N, respectively.

Substituting formula (1) into formula (2), we get:

$$-\log Softmax(Z)_i = -\left(Z_i - \log\sum\nolimits_{j=0}^{N-1} e^{Z_j}\right) \tag{3}$$

If the output value of the actual corresponding category is Z_9, then the loss function is:

$$-\log Softmax(Z)_9 = -\left(Z_9 - \log\sum\nolimits_{j=0}^{N-1} e^{Z_j}\right) \tag{4}$$

3.3 Card-Showing Experiment

Card-Showing Dataset and Evaluation Indicator. Since the four players in each game can decide their types of card-showing, 11,000 games of real players can get 44,000 sample data of card-showing. In this experiment, the sample data is divided into train set and test set according to the ratio of 4:1. The evaluation indicator is the fitting accuracy, and the fitting accuracy calculation is based on whether the output of the network is consistent with the human players in the game, the better the model is able to learn the strategy of the human player's card-showing.

Training Effect. After 40 times of parameter adjustment training, the *epoch* is set to 50, and the fitting accuracy and loss values are in a state of convergence; the optimizer chooses Adam, which can adjust different learning rates according to different parameters; the output is 16 kinds of card-showing decision, that is, the loss function selects the categorical cross-entropy; the ReduceLROnPlateau function is called to optimize the learning rate; the *batch_size* is set to 128, and the input is the game information of 32 high-level human players, which satisfies the principle of not being too large or too small. Hyperparameter configuration is shown in Table 3.

The curve of the fitting accuracy of the train set and the test set of the card-showing model with the *epoch* is shown in Fig. 2.

It can be seen from Fig. 2 that when the *epoch* is 50, the model reaches a state of convergence, the fitting accuracy on the train set can reach 91.8%, and the fitting accuracy on the test set can reach 88.4%.

Table 3. Hyperparameter configuration.

Parameter	Value
Epoch	50
Batch_size	128
Loss	Categorical cross-entropy
Factor	0.5
Min_lr	0
Optimizer	Adam

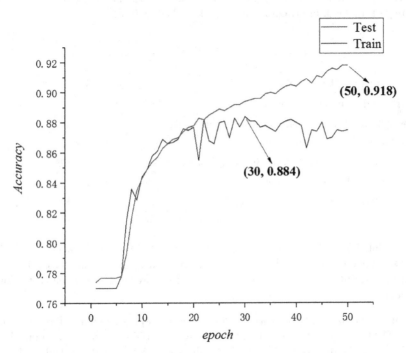

Fig. 2. Fitting accuracy of the train set and test set of card-showing.

4 Card-Playing Algorithm

4.1 Card-Playing Category

There are 52 cards in "Gongzhu" and only one card can be played at a time, so players can play cards in 52 categories. The states and actions can be encoded with a 1×52 one-hot matrix, In the form of [x0 x1 ...... x50 x51], If the corresponding element is 1, it means that this card is played, and if it is 0, it means that it cannot be played. The correspondence between cards and the position of one-hot matrix is shown in Table 2.

For example, if the Spades J is played, the one-hot matrix position corresponding to the Spades J is used as the category number, that is, 10.

4.2 Card-Playing Network Design

Network Input and Output. In this paper, the hand cards of the player in the current round, the cards in the hands of other players in the current round, the card-showing information of each player, the cards played by the other three players in the current round, and the cards played by each player in the other 12 rounds and the cards that have been collected by each player in the current round, which is respectively represented by a 1×52 one-hot matrix, in which the player's card information of the unknown round is filled with 0, that is, the input information is a 61×52 one-hot matrix. The specific meaning of the input information matrix is described in Table 4.

Table 4. Card-playing CNN input structure.

Input Information Matrix Features	Size
Player's hand for the current round	1×52
Cards in the hands of other players in the current round	1×52
Each player's hand show information	4×52
The cards of the other three players in the current round	3×52
The cards of each player in the other 12 rounds	48×52
The cards that each player has collected so far	4×52

The output layer of the card-playing neural network consists of 52 neurons, which in turn correspond to the probability of outputting 52 cards, and selecting the card with the highest probability. When the probability of multiple cards is the same, one is randomly selected.

Network Structure. The network structure of the card-playing module has a total of 46 layers. The first layer is the input layer, and the input is a 61×52 matrix; the convolution kernel number of the 1-4Conv layers is 32, the size is 5×5, and the stride is 1. After the first 2×2 Max-pooling layer, becomes a 27×26 matrix; the convolution kernel number of the 5-8Conv layers is 64, the size is 5×5, and the stride is 1. And after the second 2×2 Max-pooling layer, it becomes a 13×13 matrix; the convolution kernel number of 9-12Conv layers is 128, the size is 5×5, the stride is 1, after the third 2×2 Max-pooling layer, it becomes a 6×6 matrix; the convolution kernel number of the 13-16Conv layers is 256, the size is 2×2, and the stride is 1. After the fourth 2×2 Max-pooling layer, it becomes a 3×3 matrix, each Conv layer is followed by a ReLU layer; each Max-pooling layer is followed by 1 Batch Normalization (BN) layer and 1 Dropout layer, and the value of dropout is 0.25. The last layer is the Fully-connected layer, which uses the Softmax function to classify the card-playing actions, and makes the card-playing decision according to the probability of each card-playing action. The CNN network structure of the card-playing is shown in Fig. 3.

Loss Function. There are 52 kinds of actions in card-playing, which is a multi-classification problem. The loss function of the network structure in this paper still adopts categorical cross-entropy. For the formula, see the card-showing loss function section. Among them, since there are 52 types of cards, the N value is 52.

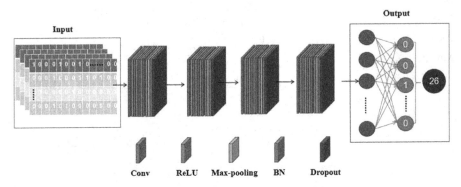

Fig. 3. The CNN model network structure of card-playing.

4.3 Card-Playing Experiment

Card-Playing Dataset and Evaluation Indicator. Every time a card is played, it can be a sample data, however, in the case of the "Gongzhu" game information used in this experiment, when all the scored cards have been played, the game will be ended directly. Therefore, only 497,020 pieces of experimental data were obtained from 11,000 games. This paper divides the experimental data into train set and test set according to the ratio of 4:1. The evaluation indicator is the fitting accuracy, and the fitting accuracy calculation is based on whether the output of the network is consistent with the human players in the game, the better the model is able to learn the strategy of the human player's card-playing.

Training Effect. After 60 times of parameter adjustment training, it is determined that the *epoch* is 50, and the fitting accuracy and loss values are in a state of convergence; the optimizer chooses Adam, which can adjust different learning rates according to different parameters; the output is 52 kinds of card-playing decisions, and the loss function chooses the categorical cross-entropy; the ReduceLROnPlateau function is called to optimize the learning rate; the *batch_size* is set to 440, that is, the single-round batch input is the card-playing information of two rounds, and the hyperparameter configuration table is shown in Table 5.

The curve of the fitting accuracy of the train set and the test set of the card-playing model with *epoch* is shown in Fig. 4.

It can be seen from Fig. 4 that when the *epoch* is 50, the model reaches a state of convergence, the fitting accuracy on the train set can reach 85.6%, and the fitting accuracy on the test set can reach 71.4%.

Table 5. Hyperparameter configuration.

Parameter	Value
Epoch	50
Batch_size	440
Loss	Categorical cross-entropy
Factor	0.5
Min_lr	0
Optimizer	Adam

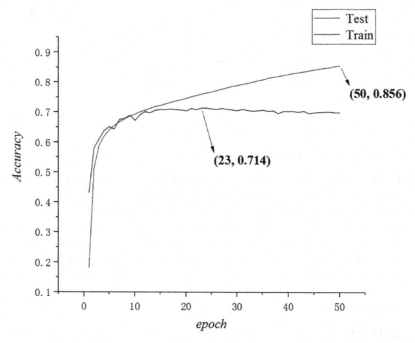

Fig. 4. Fitting accuracy of the train set and test set of card-playing.

5 Conclusion

This paper proposes a "Gongzhu" game algorithm based on CNN, using 11,000 real-person battle data for supervised learning. "Gongzhu" agent has a certain ability to show cards and play cards, but the fitting accuracy fails to reach the ideal state. The next step will be to analyze the complexity of the initial hand cards and improve the structure of CNN, or use reinforcement learning to increase the amount of training data, agent learns the strategy of card-showing and card-playing through self-play, thereby improving the fitting accuracy between high-level human players.

References

1. Blair, A., Saffidine, A.: AI surpasses humans at six-player poker. Science **365**(6456), 864–865 (2019)
2. Moravčík, M., Schmid, M., Burch, N., et al.: DeepStack: expert-level artificial intelligence in heads-up no-limit poker. Science **356**(6337), 508–513 (2017)
3. Noam, B., Sandholm, T.: Superhuman AI for heads-up no-limit poker: libratus beats top professionals. Science **359**(6374), 418–424 (2018)
4. Jiang, Q., Li, K., Du, B., et al.: DeltaDou: expert/level doudizhu AI through self-play. In: Proceedings of the 28th International Joint Conference on Artificial Intelligence (IJCAI-19), pp. 1265–1271. SweetCode Inc., Beijing (2019)
5. Peng, Q., Wang, Y., Yu, X., et al.: Monte Carlo tree search for "Doudizhu" based on hand splitting. J. Nanjing Normal Univ. (Natural Science Edition) **42**(03), 107–114 (2019). In Chinese
6. Xu, F., Wei, K., Wang, Y., et al.: "Doudizhu" strategy based on convolutional neural networks. Comput. Modernization. **11**, 28–32 (2020). In Chinese
7. Ma, X., Wang, X., Wang, X.: Information model for a class of incomplete information games. J. Comput. Res. Dev. **47**(12), 2100–2109 (2011). In Chinese
8. Wang, X., Xu, C.: The application of temporal difference in incomplete information games. In: China Machine Game Academic Symposium, pp. 16–22. Journal of Chongqing Institute of Technology, Chongqing (2007). In Chinese
9. Zhang, J., Wang, X., Yang, L., et al.: Analysis of UCT algorithm policies in imperfect information game. In: 2012 IEEE 2nd International Conference on Cloud Computing and Intelligence Systems (CCIS), pp. 132–137. IEEE, Hangzhou (2012)
10. Zhang, J.: Building Opponent Model in Imperfect Information Board Games. TELKOMNIKA Indonesian J. Electr. Eng. **12**(3), 1975–1986 (2014)
11. Zhang, J., Wang, X.: Using modified uct algorithm basing on risk estimation methods in imperfect information games. Int. J. Multimed. Ubiquitous Eng. **9**(10), 23–32 (2014)
12. Ginsberg, M.L.: Imperfect information in a computationally challenging game. J. Artif. Intell. Res. **14**, 303–358 (2001)
13. Bowling, M., Burch, N., Johanson, M., et al.: Heads-up limit hold'em poker is solved. Science **347**(6218), 145–149 (2015)
14. Brown, N., Sandholm, T., Amos, B.: Depth-limited solving for imperfect-information games. In: Proceedings of the 32nd International Conference on Neural Information Processing Systems (NIPS'18), pp. 7674–7685. Curran Associates Inc, Montréal (2018)
15. Silver, D., Schrittwieser, J., Siomonyan, K., et al.: Mastering the game of go without human knowledge. Nature **550**(7676), 354–359 (2017)
16. Silver, D., Hubert, T., Schrittwieser, J., Antonoglou, I., et al.: A general reinforcement learning algorithm that masters chess, shogi, and Go through self-play. Science **362**(6419), 1140–1144 (2018)
17. Li, Y.: Research on Intelligent Decision Model of Texas Hold'em Computer Game (2020). In Chinese
18. Li, Y., Peng, L., Du, S., et al.: A Decision Model for Texas Hold'em Game. Software Guide **20**(05), 16–19 (2021). In Chinese
19. Zhang, M., Li, K., Wu, Z., et al.: An opponent modeling and strategy integration framework for Texas Hold'em AI. Acta Automatica Sinica **48**(4), 1004–1017 (2022). In Chinese
20. Zhou, Q., Bai, D., Zhang, J., et al.: DecisionHoldem: Safe Depth-Limited Solving with Diverse Opponents for Imperfect-Information Games. arXiv preprint arXiv:2201.11580 (2022)
21. Li, S., Li, S., Ding, M., Meng, K.: Research on fight the landlords' single card guessing based on deep learning. In: Kůrková, V., Manolopoulos, Y., Hammer, B., Iliadis, L., Maglogiannis,

I. (eds.) ICANN 2018. LNCS, vol. 11141, pp. 363–372. Springer, Cham (2018). https://doi. org/10.1007/978-3-030-01424-7_36

22. You, Y., Li, L., Guo, B., et al.: Combinational Q-Learning for Dou Di Zhu. arXiv preprint arXiv:1901.08925 (2019)
23. Peng, Q.: Research on "Doudizhu" Based on Monte Carlo Tree Search (2020). In Chinese
24. Zha, D., Xie, J., Ma, W., et al.: c. In: Proceedings of the 38th International Conference on Machine Learning (ICML), pp. 12333–12344. PMLR, Virtual (2021)
25. Guo, R., Li, S., Gong, Y., et al.: Research on game strategy in two-on-one game endgame mode. Intell. Comput. Appl. **12**(04), 151–158 (2022). In Chinese
26. Yang, G., Liu, M., Hong, W., et al.: PerfectDou: Dominating DouDizhu with Perfect Information Distillation. arXiv preprint arXiv:2203.16406 (2022)
27. China Huapai Competition Rules Compilation Team.: China Huapai Competition Rules (Trial)[M]. People's Sports Publishing House (2009). In Chinese

The Survey of Self-play Method in Computer Games

Xiali Li, Zhaoqi Wang, Bo Liu, and Licheng Wu[✉]

School of Information Engineering, Minzu University of China, Beijing, China
wulicheng@tsinghua.edu.cn

Abstract. In recent years, the achievements of computer games in Go, Chess, Shogi and other chess games are amazing, which is inseparable from the development of reinforcement learning and deep reinforcement learning. DeepMind pioneered the use of deep reinforcement learning and self-play to achieve a breakthrough in the Atari arcade game. Since then, AlphaGo has also reached the level of beating the top human players. AlphaGo Zero quickly surpassed the former by just playing against itself without using human chess data. These all reflect the excellent performance of the self-play method in deep reinforcement learning. Therefore, this paper analyzes the application of self-play and deep reinforcement learning methods applied in computer games, summarizes its research status in the field of complete information games and incomplete information games and pointed out its existing problems and possible development directions.

Keywords: Computer Games · Self-play · Deep Reinforcement Learning

1 Introduction

Artificial Intelligence [1] (Artificial Intelligent, AI) was proposed by Alan Turing in the 1950s. Humans take the brain as the core and make reasonable decisions through learning, observation and thinking. How to make computers learn, think, plan and make decisions like humans is the main research content of artificial intelligence. The machine game is a widely recognized benchmark for measuring the development of artificial intelligence and even computer science [2]. Many pioneers in the computer field, such as von Neumann, Turing, Donald Knut and Alan Turing, have all explored the field of games. In 1952, Turing wrote a chess machine game program, but limited by the computing power of the computer at that time, the program failed to run successfully, but it can be called a great start of the game. Computer games are divided into two categories: incomplete information games and complete information games. This classification is based on whether the game participants can observe the completeness of the game state information, that is, whether the agent in the game process can grasp all the game information through direct acquisition or indirect calculation. Board games such as Go and chess are generally complete information games; while Landlords, Mahjong, Texas Hold'em Poker and the current hot 3D video games are incomplete information games.

F. Sun et al. (Eds.): ICCCS 2022, CCIS 1732, pp. 129–138, 2023.
https://doi.org/10.1007/978-981-99-2789-0_11

Because the incomplete information game is closer to the real-world situation, it is a hot spot in the field of artificial intelligence and computer games.

Machine games are divided into two categories according to whether the participants can observe the game state information, that is, whether the agents in the game process can grasp all the game information through direct acquisition or indirect calculation and are divided into two categories: incomplete information games and complete information games. Game. Games with complete information are mostly chess games based on chess games such as Go and Chess; while Dou Dizhu, Mahjong, Texas Hold'em and the current hot 3D video games are games with incomplete information. Since incomplete information game is closer to the real-world situation, it is a hot spot of current artificial intelligence and machine game research.

The search algorithms of traditional games include the minimax algorithm, the pruning method and their variants, etc. But they are no longer applicable when facing games with high-dimensional state spaces. Reinforcement learning methods mainly include Monte Carlo class methods and temporal difference classification methods [3]. Traditional methods have variance problems due to the use of table methods for value function evaluation and the algorithms are difficult to converge when faced with high-dimensional state-action spaces. As deep learning flourishes, computer games combine traditional reinforcement learning and deep learning techniques. The excellent feature representation ability of the deep neural network is used to fit the value function under different states and actions and it also takes into account the excellent decision-making ability of reinforcement learning.

The method of deep reinforcement learning is an important method to solve the problem of machine games at present and self-play is the most used method. The Go AI systems Alpha Go and Alpha Go Zero were developed using deep reinforcement learning and self-play. Alpha Go can defeat the top human Go players Shishi Li, Jie Ke and others through the learning of expert chess repertoire and self-play training. Alpha Go Zero was trained using only self-play without human playbooks and surpassed its predecessor, Alpha Go, in a much shorter time. It can be seen that the self-play method has played an important role in computer games.

This paper introduces the application of the self-play method in the training of computer game models. Combined with the research and development of computer games, it introduces its application and optimization in the fields of complete information games and incomplete information games. At last it also pointed out the direction of future development.

2   Self-play in Complete Information Games

Early computer games started with chess and Ludus Duodecim Scriptorum. Berliner's backgammon agent defeated world champion Villia in the 1980s and the successful construction of the supercomputer "Deep Blue" demolished the idea that robots could not beat humans in chess. However, computer games had a slowdown for a period of time. Alpha Go has given artificial intelligence the ability to "think" like humans thanks to its self-learning, self-organizing and self-computing characteristics. It has elevated deep reinforcement learning to a new level of research by taking computer games beyond simple brute-force computing. Researchers have paid close attention to its usage of self-playing deep reinforcement learning algorithms since then.

DeepMind was the first to propose self-play and deep reinforcement learning techniques and they performed several early investigations. In 2013, it was employed for the first time in a comprehensive information game. For the first time, the agent recognized the input from pure visual pictures while playing Atari through complete self-play. On Atari, the enhanced AI reached the level of a human player in 2015 [4].

DeepMind unveiled AlphaGo [5], a Go AI system that employs self-play for training and a deep reinforcement learning strategy network to enable agents to overcome top human players, in March 2016. As a result of this success, a new wave of computer games emerged. It suggested AlphaGoZero, a game based on AlphaGo, in 2017. Without human expertise, AlphaGoZero can swiftly master Go through self-play [6]. Although its training duration, computer resource consumption and learning impact have all improved significantly over its predecessors, it still relies on four TPUs to support its training. AlphaZero, a universal chess intelligence based on AlphaGoZero, was suggested in 2018 [7]. It replaces conventional hand-coding and domain expertise with DNN and tabula rasa reinforcement learning methods. It defeated the strongest professional chess AI at the time with a high winning percentage, but it did it with the help of thousands of TPUs for self-play and neural network training.

The Alpha series' high success rate is impressive, but its training requires a large number of GPUs and even TPUs, which is unmatched by normal research organizations and laboratories. As a result, academics have conducted analysis and study on the self-playing deep reinforcement learning algorithm in the hopes of optimizing it for usage in computer games.

Alexandre Laterre et al. presented the Ranked Rewards (R2) algorithm to apply self-play to the problem of single-agent games [8], which produces relative performance measurements by rating the rewards obtained by agents throughout numerous games. It simulates the strategy and value function using DNN, which overcomes the problem of self-play not being directly applicable to single-player games. Wang et al. proposed a search strategy for solitaire games that incorporated the self-playing deep reinforcement learning approach and the R2 algorithm [9].

It has been proven that the strategy works for solitary games as well. However, because of the R2 algorithm's effect, it is simple for the training to slip into the local optimum and the computational resource overhead is high.

Wang et al. investigated 12 hyperparameters in a self-play algorithm comparable to AlphaGoZero in order to better optimize the self-playing method [10]. The model training impact was shown to be very sensitive to hyperparameter selection, with too much training occasionally resulting in worse performance. They then presented a static hot-start technique employing MCTS augmentation to recreate expert data at the start of training [11], as well as a hot-start search augmentation function at the start of self-play. Based on earlier work, Hui Wang et al. suggested a technique to produce expert data with adaptive MCTS augmentation in 2021 [12]. For games with higher search levels, adaptive warm-start works effectively. This demonstrates that it may do better in larger, more complex games, but it still needs experience. Bo Ji, Hongchuan Liu and colleagues suggested a technique for measuring the quality of sample learning from self-playing games [13], which they tested on Checkers by utilizing a linear combination of sample repetition and the number of samples as a composite indicator of sample size. It reduces the computational cost of learning samples by a large amount without diminishing the learning impact, but the correctness of the experimentally determined parameters in his training still has to be validated. The researchers' examination of the parameters, expert data and samples enhanced the effectiveness of the self-play to some level, but the learning algorithm still has a lot of space for improvement.

Many academics have looked into ways to decrease the learning time of neural networks and improve resource use. To speed deep convolutional neural network models, Jiao Wang et al. suggested an efficient hardware architecture based on an FPGA platform [14]. It is also optimized using parallel processing and a pipeline architecture to achieve a good balance of processing speed and resources. The accelerator's performance can reach 3036.32 GOPS, demonstrating the benefits of parallel computing. Based on the notion of segmentation function, Li created a convolutional neural network with a maximum average output layer (MAO) and applied it to the game Go [15]. The network is sparse because the input channels are randomly selected during training, which improves generalization and learning efficiency while minimizing the number of iterations and computer resources required. Under the same settings, the Go software produced using MAO surpassed ResNet18, proving that enhanced neural networks may improve learning efficiency while reducing processing resources. Lv proposed a hybrid update deep reinforcement learning model [16]. It uses a deep neural network to combine the Sarsa(λ) algorithm with timely updates with the Q-Learning method with updates over the whole game route. The model decreases the algorithm's arithmetic requirements by lowering the number of iterations and it has been successfully applied to Go and Jiu chess (Table 1).

Table 1. Research on Self-Play in the complete information game.

Time	Game	Level	Method	Author
2013–2015	Atari	Normal human player	Input from pure visual pictures	DeepMind
2016	Go	Top human player	DRL and expert data	Deepmind
2017	Go	Over top human player	DRL but without expert data	Deepmind
2018	chess	Master human player	Replace hand-written rules with DNN and generic algorithms	Deepmind
2020	Morpion Solitaire	normal human player	Search algorithm combining R2 algorithm and DRL	Wang H, et al

3 Self-play in Incomplete Information Game

The initial solution to the incomplete information game problem was modelled using a partially observable Markovian decision-making process model, which was learned and trained using a reinforcement learning algorithm approach. Current experiments demonstrate that model-based methods perform better than model-free methods in single-agent Markov decisions. In multi-agent, model-based methods perform much worse. To this end, Liu et al. from Princeton University proposed a model-based self-play algorithm for multi-agent Markov games [17]. Its performance in two-player zero-sum games and multi-player ordinary games surpasses its predecessors and reaches an approximate Nash equilibrium. Compared to the current state-of-the-art model-free algorithms, a Markov decision can be proposed. However, these techniques have not been applied to solve incomplete game problems, and the traditional methods are only suited to small-scale games due to the high-dimensional state space characteristics of incomplete information games and the constraints of computer hardware. As a result, professionals and academics have proposed a better strategy that employs deep reinforcement learning instead of iterative reinforcement learning updates [18] and as the research proceeds, the concept of self-play deep reinforcement learning enters the researcher's vision.

Johannes Heinrich and David Silver proposed the Neural Fictitious Self-Play (NFSP) algorithm in 2016 [19], which is a self-pairing deep reinforcement learning technique that may be applied to incomplete information games. Its combination of Fictitious Self-Play (FSP) and deep neural networks allows agents without prior domain knowledge to achieve approximate Nash equilibrium through self-play. When applied to restricted Texas Hold'em Poker on a real-world scale, NFSP can achieve levels of performance that are superior to humans, but it struggles with wider search spaces or depths, such as low convergence. As a result, Li Zhang et al. developed the Monte Carlo Neural Fictitious Self-Play (MC-NFSP) and Asynchronous Neural Fictitious Self-Play (ANFSP) [20]. By using MCTS, MC-NFSP overcomes NFSP's poor performance in games with large spatial and search depth scales and the method eventually converges to an approximation of Nash equilibrium in Othello. ANFSP employs parallel actor-learners to improve slow convergence by stabilizing the training process and speeding it up. Asynchronous and parallel techniques have been shown to enhance convergence speed and reduce training time in both Texas Hold'em Poker and first-person shooter games in tests.

Card games are an important category of incomplete information games due to their high level of strategy and the complexity of the cooperative and competitive connections between players.

For the four-player card game "Big 2," Henry Charlesworth et al. devised an algorithm [21]. They employed proximal policy optimization to train deep neural networks and self-play for all training and in two days of GPU training, they outperformed amateur human players. For several agents, Jakob et al. introduced a simplified action decoder (SAD) that allows agents to learn communication protocols [22]. When playing Hanabi with two players, SAD reaches a new level, but when playing with three or more players, SAD performs significantly worse than the hat encoding approach. The game AI team at Kwai has proposed DouZero [23], an AI program for the three-player card game DouDizhu that does not require any human knowledge or abstraction of the state/action space. Deep neural networks, action encoding and parallel actors were used to improve standard Monte-Carlo approaches. It was also created using the RLcard platform, which is a reinforcement learning platform for card games. After 10 days of training with four GPUs, the augmented Monte Carlo approach achieved human-level performance. Suphx, an AI system for four-player Japanese Mahjong [24], was developed by Microsoft Research Asia. It uses a deep convolutional neural network as a model, supervised learning from data from human professional players and self-play reinforcement learning to increase performance. The training introduces global reward prediction, oracle directing and run-time policy adaptation and it can match or even outperform human players in terms of strength.

Another popular type of incomplete information game is the 3D video game. It has become a hot research issue due to its close connection to the real world, its vast action area and its heavy reliance on movement.

For the commercial game "Blade & Soul," Inseok Oh et al. created fighting AI bots [25]. They developed reward shaping and data skipping approaches to increase search space and data utilisation efficiency, as well as concurrency mechanisms to boost the model's generalisation performance. The application of these principles to the game has resulted in a high win rate. BRANDÃO et al. used proximal policy optimization more suitable for continuous environments to participate in robotic soccer VSSS (IEEE Very Small Size Soccer) [26]. He realized the training of multi-agent control and decision-making at the same time using self-play reinforcement chemistry. In the end, with an 83% win rate, he surpassed Pequi Mecânico, the best team in VSSS.By combining self-play with approximate strategy optimization algorithms and value networks based on the Atari-net architecture, Michal Warchalski et al. developed a reinforcement learning program based on the video game "Heroic - Magic Duel" [27]. It had a victory percentage of over 50% against existing AI and top-level human players, but the agents' techniques lacked long-term planning. Tencent AI Lab engineers investigated multiplayer online competitive games (MOBA) and created sophisticated AI systems that outperformed Dota AI [28]. To overcome the scalability challenge created by sophisticated training and a high number of heroes, it integrates curriculum self-play learning, policy distillation, off-policy adaptability, multi-head value estimate and Monte-Carlo tree-search.

Real-time games training necessitates a high level of computer hardware due to the huge state space, partial information and dynamic nature of the games. When solely self-play is used for training, there may be issues with the data, such as missing some categories or a lack of representative data. As a result, the self-play algorithm's convergence is slow and the training time is prolonged.

In order to speed up the process of self-play training and improve its efficiency, OpenAI used an asymmetric self-play training strategy and found that asymmetric self-play can generate a wide range of special goals [29]. The need to manually set these learning goals is greatly reduced. For large-scale multi-agent training, Trapit Bansal et al. used MuJoCo and distributed proximal policy optimization [30], which improved the variance problem to some extent. They discovered that applying additional rewards to aid in the learning of basic behaviours in the early stages of training and reducing this portion of the rewards in the later stages can ensure the model's learning effect and enhance efficiency. ReBeL is a broad paradigm suggested by Noam Brown et al., which combines self-play reinforcement learning with game search with imperfect information

[31]. They build a depth-constrained subgame rooted in the initial PBS using the broader public belief state, which is solved using an iterative equilibrium finding approach. This method requires significantly less domain knowledge than prior poker AI methods and thrives at no-limit Texas Hold'em. A self-play actor-critic (SPAC) algorithm was proposed by Yong Liu et al. [32]. The algorithm incorporates an integrated critic into the strategy gradient algorithm, allowing both sides' imperfect information to be combined. This update improves the instability problem during training by speeding up the self-play process. Antti Keurulainen et al. proposed training a meta-learner by synthesizing a collection of agents and using self-play to autonomously produce task distribution and related real-world data in the field [33]. Self-play has been shown in studies to help with task assignments. Shiyu Huang, Wenze Chen and colleagues devised a deep recurrent neural network-based offline reinforcement learning algorithm [34]. It uses a distributed learning system and its suggested offline reinforcement learning algorithm to train on a large-scale replay data set generated by a single agent's self-play. This strategy accelerates the training process and the trained model performs admirably in the GRF full game. However, there is still much potential for development in terms of robustness (Table 2).

Table 2. Research on Self-Play in the incomplete information game.

Time	Game	Level	Author
2018	Big2	Over amateur human player	Charlesworth H
2019	Othello	Amateur human player	Zhang L, et al
2019	Hanabi	Over top human player	Facebook AI Research
2020	Mahjong	Over top human player	Microsoft Research Asia
2020	Heroic-Magic Duel	Over 50% win rate against a top human player	Nordeus
2020	MOBA	Over top human player	Tencent AI Lab
2021	Dou Dizhu	Over master human player	Kwai Inc AI
2021	Blade&Magic Duel	The 50% win rate against human player	NCSoft

4 Self-play in Universal Models

In complete information games, algorithms that mix search and learning often do well. For partial information games, algorithms based on game theory reasoning and learning are better. DeepMind released the first general AI algorithm, PoG [35], in December 2021. PoG is an algorithm that produces powerful performance in both complete and imperfect information games by combining guided search, self-play and game-theoretic reasoning. Growing-Tree Counterfactual Regret Minimization (GT-CFR) and sound self-play are two aspects of PoG that teach counterfactual value and policy networks. Experiments show that as the available computation time and approximation power increase, the approach converges to perfection. It performed well in chess and Go and it defeated the top AI at No-limit Texas Hold'em Poker and Scotland Yard. When given the same resources, however, because PoG is a general-purpose algorithm, it may be weaker than specialized deep reinforcement learning algorithms like AlphaZero. At the same time, it is still an algorithm that relies on a massive number of TPUs and there's plenty of space for improvement in terms of computing resource use.

5   Summary and Prospect

In summary, while self-play deep reinforcement learning has had some success in the realm of comprehensive information games, there are still a number of issues in everyday laboratory research and applications:

The search scale is often big in the self-play training process of deep reinforcement learning and the efficiency of self-play is poor due to the search algorithm's limitations. Furthermore, some unusual moves may disappear early on, necessitating a large number of games to collect sufficient data, lengthening the training period and slowing the model convergence rate.

The training results are also influenced by the selection and optimization of DNNs. All issues to examine include how to choose an appropriate neural network for computer games, how to optimize the neural network adaptively and how to make better use of data features.

Because of their architectural design, deep reinforcement learning algorithms that use value-based functions are not suited for high-dimensional action space or continuous action space tasks and they also suffer from overestimation. High variance, sluggish convergence and a poorly established learning step size plague policy-based algorithms. In addition to online or offline learning, the employment of models, the sparsity of incentives, the clarity of reward function settings and so on, all play a role in the pace of convergence.

Using self-play deep reinforcement learning approaches to train dedicated or generic models to get outcomes that can surpass human specialists necessitates a substantial amount of processing power and computer hardware. For example, full self-gaming in an environment with a vast state or action space necessitates a huge amount of computational power, which significantly raises the cost.

Since the inception of AlphaGo, research on self-play deep reinforcement learning has progressively reached a plateau. There are numerous areas that can be investigated in the future to improve the efficiency of self-play and accelerate convergence. For example, research into reinforcement learning combination strategies, deep neural network selection, deep reinforcement learning architecture optimization, self-play game training process improvement and the use of distributed architectures to handle large state spaces will all be pursued. With the evolution of algorithms and the growth of computer hardware, optimizing the allocation of computer resources and reducing computing resource consumption remains a major challenge.

Acknowledgement. This work was supported by the project of National Science Foundation of China (Granted by 61873291 and 6177341).

References

1. Pomerol, J.C.: Artificial intelligence and human decision making. Eur. J. Oper. Res. **99**(1), 3–25 (1997)
2. Funge, J.D.: Artificial intelligence for computer games: an introduction.1st edn. AK Peters/CRC Press,63 South Avenue Natick, MA United States (2004)
3. Watkins, C., Dayan, P.: Technical note: Q-Learning. Mach. Learn. **8**(3–4), 279–292 (1992)
4. Mnih, V., et al.: Human-level control through deep reinforcement learning. Nature **518**(7540), 529–533 (2015)
5. Silver, D., et al.: Mastering the game of Go with deep neural networks and tree search. Nature **529**(7587), 484–489 (2016)
6. Silver, D., Schrittwieser, J., Simonyan, K., Antonoglou, I., Hassabis, D.: Mastering the game of Go without human knowledge. Nature **550**(7676), 354–359 (2017)
7. Silver, D., et al.: A general reinforcement learning algorithm that masters chess, shogi and Go through self-play. Science **362**(6419), 1140–1144 (2018)
8. Ranked reward: Enabling self-play reinforcement learning for combinatorial optimization, https://arxiv.org/abs/1807.01672. Accessed 23 Jun 2022
9. Wang, H., Preuss, M., Emmerich, M., Plaat, A.: Tackling morpion solitaire with alphazero-like ranked reward reinforcement learning. In: 2020 22nd International Symposium on Symbolic and Numeric Algorithms for Scientific Computing (SYNASC), pp. 149–152. IEEE, Timisoara Romania (2020)
10. Wang, H., Emmerich, M., Preuss, M., Plaat, A.: Alternative loss functions in alphazero-like self-play. In: 2019 IEEE Symposium Series on Computational Intelligence (SSCI), pp. 155–162. IEEE, Xiamen China (2019)
11. Wang, H., Preuss, M., Plaat, A.: 2020 Warm-start alphazero self-play search enhancements. In: Bäck, T., et al. (eds.) Parallel Problem Solving from Nature – PPSN XVI. PPSN 2020. LNCS, vol. 12270, pp. 528–542. Springer, Cham (2020). https://doi.org/10.1007/978-3-030-58115-2_37
12. Adaptive Warm-Start MCTS in AlphaZero-like Deep Reinforcement Learning, https://arxiv.org/abs/2105.06136, last accessed 2022/06/23
13. Ji, B., You, H., Hongxing, L., Tian, X., Liu, H.: A Quality Evaluation Method of Self-Play Chess Game Learning Samples. Small Microcomput. Syst **42**(03), 467–471 (2021). (in Chinese)
14. Li, Z.N., Zhu, C., Gao, Y.L., Wang, Z.K., Wang, J.: AlphaGo policy network: a DCNN Accel-erator on FPGA. IEEE Access **8**, 203039–203047 (2020)
15. Li, X., Lv, Z., Liu, B., Wu, L., Wang, Z.: Improved feature learning: a maximum-average-out deep neural network for the game go. Math. Probl. Eng. **2020**, 1–6 (2020)
16. Li, X., Lv, Z., Wu, L., Zhao, Y., Xu, X.: Hybrid Online and Offline Reinforcement Learning for Tibetan Jiu Chess. Complexity 2020 (2020)
17. Liu, Q., Yu, T., Bai, Y., Jin, C.: A sharp analysis of model-based reinforcement learning with self-play. In: International Conference on Machine Learning, pp. 7001–7010. PMLR (2021)
18. Wang, P.: Research on Machine Game with Incomplete Information Based on Deep Reinforcement Learning. Harbin Institute of Technology (2017)
19. Deep reinforcement learning from self-play in imperfect-information games, https://arxiv.org/abs/1603.01121. Accessed 23 Jun 2022
20. Zhang, L., et al.: A Monte Carlo Neural Fictitious Self-Play approach to approximate Nash Equilibrium in imperfect-information dynamic games. Front. Comp. Sci. **15**(5), 1–14 (2021). https://doi.org/10.1007/s11704-020-9307-6
21. Application of self-play reinforcement learning to a four-player game of imperfect information, https://arxiv.org/abs/1808.10442. Accessed 23 Jun 20223

22. Hu, H., Foerster, J.N.: Simplified action decoder for deep multi-agent reinforcement learning. In: International Conference on Learning Representations, pp. 1–14. ICRL, Addis Ababa ETHIOPIA (2020)

23. Zha, D., et al.: Douzero: mastering doudizhu with self-play deep reinforcement learning. In: International Conference on Machine Learning, pp. 12333–12344. PMLR, Vienna Austria (2021)

24. Suphx: Mastering mahjong with deep reinforcement learning, https://arxiv.org/abs/2003.13590. Accessed 23 June 2022

25. Oh, I., Rho, S., Moon, S., Son, S., Chung, J.: Creating pro-level AI for a real-time fighting game using deep reinforcement learning. IEEE Trans. Games **PP**(99), 1 (2021)

26. Brandão, B., De Lima, T.W., Soares, A., Melo, L., Maximo, M.R.: Multiagent reinforcement learning for strategic decision making and control in robotic soccer through self-play. IEEE Access **10**, 72628–72642 (2022)

27. Deep RL Agent for a Real-Time Action Strategy Game. https://arxiv.org/abs/2002.06290. Accessed 23 Jun 2022

28. Ye, D., et al.: Towards playing full moba games with deep reinforcement learning. Adv. Neural. Inf. Process. Syst. **33**, 621–632 (2020)

29. Asymmetric self-play for automatic goal discovery in robotic manipulation, https://arxiv.org/abs/2101.04882. Accessed 23 Jun 2022

30. Emergent complexity via multi-agent competition. https://arxiv.org/abs/171003748. Accessed 23 Jun 2022

31. Brown, N., Bakhtin, A., Lerer, A., Gong, Q.: Combining deep reinforcement learning and search for imperfect-information games. Adv. Neural. Inf. Process. Syst. **33**, 17057–17069 (2020)

32. Liu, S., Cao, J., Wang, Y., Chen, W., Liu, Y.: Self-play reinforcement learning with comprehensive critic in computer games. Neurocomputing **449**, 207–213 (2021)

33. Keurulainen, A., Westerlund, I., Kwiatkowski, A., Kaski, S., Ilin, A.: Behaviour-conditioned policies for cooperative reinforcement learning tasks. In: Farkaš, I., Masulli, P., Otte, S., Wermter, S. (eds.) ICANN 2021. LNCS, vol. 12894, pp. 493–504. Springer, Cham (2021). https://doi.org/10.1007/978-3-030-86380-7_40

34. TiKick: Towards Playing Multi-agent Football Full Games from Single-agent Demonstrations. https://arxiv.org/abs/2110.04507. Accessed 23 Jun 2022

35. Schmid, M., et al.: Player of games. https://arxiv.org/abs2112.03178. Accessed 23 Jun 2022

GK-Means SOM Algorithm Used to Plan the Paths for Multiple Agents Exploring Multiple Target Points

Hexing Yang[1], Qingjie Zhao[1(✉)], Lei Wang[2(✉)], Wangwang Liu[2], and Ling Chong[1]

[1] School of Computer Science and Technology, Beijing Institute of Technology, Beijing, China
zhaoqj@bit.edu.cn
[2] Beijing Institute of Control Engineering, Beijing, China
15413869@qq.com

Abstract. Multi-agent technology is widely used in many fields such as intelligent manufacturing, logistics and environment exploration. In this paper, we propose a greedy K-means self-organizing map algorithm to balance the tasks and plan the paths for multiple agents exploring multiple target points, where greedy k-means adopts greedy strategy to ensure the tasks allocated for agents tending equally, and the self-organizing map networks are used for parallel path planning to speed up the problem solved.

Keywords: Multi-agent · Path planning · Greedy K-means · Self-organizing map

1 Introduction

Multi-robot systems are widely used in intelligent manufacturing, logistics and transportation, environmental exploration and other fields, where each robot can be regarded as an agent. In most cases, reasonable task allocation and planning is required for these agents cooperating to complete a task.

Path planning for a single agent visiting multiple target points can be regarded as a Traveling Salesman Problem (TSP) [1]. The objective of TSP is to plan an optimal path to make the shortest path for traveling salesman to traverse all the cities. The optimization algorithms inspired by biological intelligence of solving such problems include Ant Colony Optimization [2], Bee Colony algorithm [3], Particle Swarm Optimization [4], Fruit Fly Optimization Algorithm [5], Self-Organizing Map (SOM) [6], etc. SOM simulates the perception principle of human cerebral cortex. For example, when the retina of the human eye receives a specific color stimulus, certain neurons in the cerebral cortex excite and neighboring neurons also respond, while neurons farther away are insensitive. SOM learns the excitatory and inhibitory mechanism of cortical neurons acquired by self-organizing. Based on this principle, SOM is applied to path planning of underwater

Pre-research Project on Civil Aerospace Technologies of China National Space Administration (Grant No. D010301).

vehicle system [7, 8]. Sun et al. proposed a SOM path planning method with locking mechanism [9].

In multi-robot cooperative exploration, it will take a long time for the robot group to finish the whole task if the loads of robots are not balanced. That is to say, the work efficiency of the robot group will be improved if the target points to be visited are evenly allocated which means the exploration tasks of all robots are balanced. Therefore, in this paper we propose a greedy K-means self-organizing map planning algorithm--GK-means SOM, where greedy K-means is firstly used to allocate balanced task for each robot, and then SOM is used to plan the path for each robot to improve the working efficiency of the robot group.

2   Multi-agent Path Planning Algorithm

2.1   Problem Description

The multi-agent multi-target path planning problem can be regarded as a Multiple Traveling Salesmen Problem (MTSP) in which multiple traveling salesmen start from different starting points, visit a series of target cities and return to the starting positions. MTSP is Descripted as below.

1. $A = \{a_1, a_2, \cdots, a_k\}$, a_i is the ith salesman. $G = \{g_1, g_2, \cdots, g_n\}$, G is the target city set. Each city is a task to visit.
2. The n tasks need to be allocated to k salesmen.
3. $G = G_1 \cup G_2 \cdots \cup G_k$, $G_i \cap G_j = \phi(i \neq j)$. k salesmen need to visit all target cities in G and there is no overlapping cities between every two salesmen.
4. For every salesman a_i in A, $cost_i$ is the cost to visit all cities in its $G_i = \{g_{i1}, g_{i2}, \cdots, g_{ig}\}$.

$$cost_i = \sum_{j=1}^{g-1} d\left(g_{ij}, g_{i(j+1)}\right) + d\left(g_{i1}, g_{ig}\right) \tag{1}$$

5. The objective is to minimize $maxspan$.

$$maxspan = max\{cost_1, cost_2, \cdots, cost_k\} \tag{2}$$

2.2   Balanced Task Allocation by Greedy K-means

Since the traditional K-means algorithm ignores the balance of clustering results in the clustering process, the uneven distribution of target points leads to unbalanced task allocation. In multi-robot cooperative explorations if allocating tasks without considering the load balancing problem, there will be heavier loads for the robots in mission intensive areas, and lighter loads for those in sparse regions, which may cause low efficiency for a multi-robot system to complete the total task.

Therefore, we introduce greedy strategy to automatically adjust the unbalanced task allocation in the iteration process to make the tasks tending to balance. Greedy strategy calculates the path cost of the current results which indicate the loads of agents, and adjusts the unbalanced loads, that is, reduce the loads with heavier values and increase those with smaller values. The steps of Greedy K-means are as follows:

1. $c = \{c_1, c_2, \cdots, c_k\}$, c_j is the center of the *jth* cluster, and there are k clusters. *Gkmeans_max_iter* is the max number of iterations. $\delta \in [0, 1]$ is the offset multiple.
2. Calculate the Euclidean distances between each target point in $G = \{g_1, g_2, \cdots, g_n\}$ and the center of each cluster. d_{ij} is the Euclidean distance of $g_i(x_i, y_i)$ and $c_j(x_j, y_j)$, $j \in \{1, 2, \cdots, k\}$.

$$d_{ij} = |g_i - c_j| = \sqrt{(x_i - x_j)^2 + (y_i - y_j)^2} \tag{3}$$

3. Each target point in $G = \{g_1, g_2, \cdots, g_n\}$ is allocated into the cluster whose center is the closest. For each cluster j, $j \in \{1, 2, \cdots, k\}$, the allocation is $C_j = \{g_1, g_2, \cdots, g_i, \cdots, g_l\}$ and the position of g_i is (x_i, y_i). l is the number of target points in C_j, $l \in [1, n]$.
4. Update each cluster center c_j to $(\bar{x}_j, \bar{y}_j)$ using the means of all target points in C_j.

$$\begin{cases} \bar{x}_j = \frac{1}{l} \sum_{i=1}^{l-1} x_i \\ \bar{y}_j = \frac{1}{l} \sum_{i=1}^{l-1} y_i \end{cases} \tag{4}$$

5. Calculate the cost of every cluster $D = \{D_1, D_2, \cdots, D_k\}$ using greedy algorithm.
6. In every iteration, the shrinking regulator *narrow*$_{iter}$ and the expanding regulator *enlarge*$_{iter}$ are used to adjust the maximum and minimum cluster space respectively.

$$narrow_{iter} = \left[\frac{(D_{max} - D_{ave})}{D_{ave}} \right] * \delta \tag{5}$$

$$enlarge_{iter} = \left[\frac{(D_{ave} - D_{min})}{D_{ave}} \right] * \delta \tag{6}$$

iter is the current number of iterations, D_{max} and D_{min} are the max and min cost of clusters in D. D_{ave} is the average cost.

7. Use *narrow*$_{iter}$, *enlarge*$_{iter}$ and the target points in the max cost cluster and min cost cluster to produce virtual task locations. For the max cost cluster, the virtual position is:

$$\begin{cases} x_i = x_i - narrow_{iter} \times (\bar{x}_{max} - x_i) \\ y_i = y_i - narrow_{iter} \times (\bar{y}_{max} - y_i) \end{cases} \tag{7}$$

And for the min cost cluster the virtual position is

$$\begin{cases} x_i = x_i + enlarge_{iter} \times (\bar{x}_{min} - x_i) \\ y_i = y_i + enlarge_{iter} \times (\bar{y}_{min} - y_i) \end{cases} \tag{8}$$

$(\bar{x}_{max}, \bar{y}_{max})$ is the center of the max cost cluster while $(\bar{x}_{min}, \bar{y}_{min})$ is the center of the min cost cluster.

8. Finally, if *iter* does not reach the maximum number of iterations, return to the second step for the next iteration according to the generated virtual task target position, otherwise output the final task assignment.

2.3 Self-organizing Map for Path Planning

The architecture of SOM is shown in Fig. 1. SOM relies on competitive learning among neurons to optimize network parameters step by step in which neurons at the input layer simulate sensory organs receiving input data, and neurons at the output layer simulate cerebral cortex responding to the input data. And each input can find a neuron most similar to it in the output layer through competition, which is called the winning neuron, whose neighborhood node also updates its weight appropriately according to its activation distance.

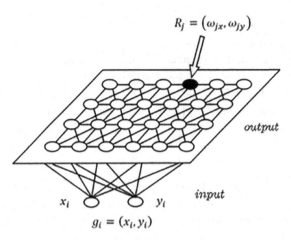

$$R_j = \left(\omega_{jx}, \omega_{jy}\right)$$

output

input

$$g_i = (x_i, y_i)$$

Fig. 1. The architecture of SOM.

The learning steps of SOM are as follows:

1. Initialize the number M of output layer neurons and the weight matrix.
2. Competition. The closest neuron of output layer is the champion when a target point g_i is input into SOM. d_{ij} is the Euclidean distance of g_i whose position is (x_i, y_i) and the *jth* neuron of output layer R_j while $\left(w_{jx}, w_{jy}\right)$ is the weight. Ω is the value set of output neurons. $I(t)$ is the champion in the tth iteration.

$$d_{ij} = |g_i - R_j| = \sqrt{\left(x_i - w_{jx}\right)^2 + \left(y_i - w_{jy}\right)^2} \tag{9}$$

$$I(t) = min\{d_{ij}\}, I \in \Omega \tag{10}$$

3. Cooperation. Once the champion is found, the neurons in its neighborhood need to confirm and their weights will be updated using the field function which determines the impact to the neighborhood according to the rule that the farther the neuron is the less important it is. We use Gaussian field function:

$$G_{j,I(t)}(t+1) = e^{-d_j^2/2\sigma^2(t)} \tag{11}$$

$$\sigma(t+1) = M * (1 - \alpha)^t \tag{12}$$

$G_{j,I(t)}$ is the Euclidean distance between the jth output neuron and $I(t)$. The range of the champion's neighborhood is $\sigma(t)$ and the decay rate of neighborhood is $\alpha \in [0, 1]$.

4. Adaption. The following strategy is used to update the weights to push the neurons in the neighborhood towards to the champion.

$$w_j(t+1) = w_j(t) + \eta(t) \times G_{j,I(x)} \times \left(g_i - R_j\right) \tag{13}$$

$$\eta(t+1) = \eta_0(1 - \beta)^t \tag{14}$$

5. The algorithm goes to the second step till $\eta(t) \leq \eta_{min}$ which is the minimize of learning rate.

The output layer is set a ring structure for the traveling salesman problem. After iterations, it will appear as an elastic ring, constantly approaching the cities and finally arriving them. We also use the decay of learning rate to avoid large offset to ensure the convergence. After the training, the city coordinates are associated with the winning neurons.

The above SOM path planning method is suitable for single agent situation. Combined with the greedy k-means algorithm mentioned above, it can be used for multi-agent multi-target point path planning problems.

2.4 Multi-agent Multi-target Path Planning Algorithm

The multi-agent multi-target path planning algorithm includes two stages: firstly, the tasks are clustered by GK-means and the path length of each cluster in the current iteration is estimated by GK-means, which is used to adjust the clustering results and balance task assignments. Then, parallel multiple SOMs are used to plan paths for robots in the system (As shown in Algorithm 1).

Algorithm 1. Multi-agent multi-target path planning algorithm

---

① Initialize the number of agents k, the maximal number of iterations
   $Gkmeans\_iter$, the offset multiple δ and the center of clusters
② **while** $iter < Gkmeans\_iter$ **do**

   Calculate all Euclidean distances between target points and cluster centers

   Allocate the target points into their closest clusters

   Update the centers of clusters
   Calculate $\{D_1, D_2, \cdots, D_k\}$ using greedy algorithm
   Adjust the tasks using $narrow_{iter}$ and $enlarge_{iter}$ according to Equation (5)-(8)

   end while
③ **for** each agent **do**
   Initialize the weights of output neurons
   while $\eta(t) > \eta_{min}$ **do**
   Determine the winner neuron according to the distances
   Update the neighborhood range of the winner neuron
   Update the learning rate and calculate the neighborhood weights
   end while
④ Output the paths of agents

---

3 Experiments

The computing device is Apple M1 with macOS 11 and 16 GB memory. The test dataset
TSPLIB [10] is an international standard database for traveling salesman problems,
which contains multiple data sets. In the following we give the experimental results on
kroA100, tsp225 and u574.

3.1 Parameter Selection

Experiments show that with the increase of δ $maxspan$ tends to be large at both ends and
small in the middle. The reason is that, when δ is small, the regulating is not obvious;
when δ is large, K-means has poor clustering effect; and the best case is $\delta = 0.1$. Also
based on the experiments and analysis, we select the neighborhood decay rate $\alpha =
0.9997$, the initial learning rate $\eta_0 = 0.8$, the learning rate decay rate $\beta = 0.99997$, and
the maximum iteration $kmeans_{iter} = 100$.

   The number of output neurons has an influence on the execution time and the quality
of final path planning result. Extensive tests have been carried out on the datasets with
different sizes where the output neuron multiple θ is respectively set as 1, 2, 3, 5, 7 times
of the number of tasks. After GK-means SOM is independently run for 20 times and the
average result is showed as in Fig. 2.

   It's obvious that when the number of output neurons is large the iteration times will
increase which means the convergence will be slow. From Fig. 2, $\theta = 3$ is the best
selected, so the output neurons multiple θ is set to three in our following experiments.

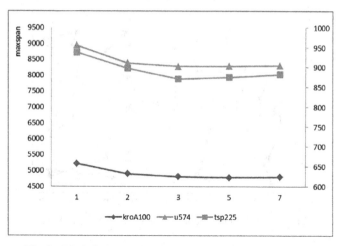

Fig. 2. The influence of output neuron multiple θ on *maxspan*.

3.2 Comparison Experiments

In order to verify the advantages of GK-means, we compare the planning results of GK-means SOM and K-means SOM on the datasets with same parameters. Each group of experiments is run for 20 times, and the average is taken as the results.

The results are shown in Table 1, where display the optimal value, average value, worst value and standard deviation of agent path cost on different scale datasets with different scales. We can find *maxspan* is smaller when using GK-means. The standard deviation is used to evaluate the load balance. As the results shown, GK-means SOM can avoid the load imbalance among multiple agents and has stronger robustness.

Table 1. Results of K-means SOM and GK-means SOM

dataset	method	Optimal value	Average value	Worst value	standard deviation
kroA100	K-means SOM	5631.43	5938.61	6320.47	984.62
	GK-means SOM	**4794.16**	**5056.83**	**5330.42**	**383.67**
tsp225	K-means SOM	911.03	931.65	950.18	266.82
	GK-means SOM	**840.28**	**868.06**	**889.52**	**51.33**
u574	K-means SOM	9238.03	9592.88	10279.28	1111.23
	GK-means SOM	**7472.99**	**8210.37**	**8524.98**	**443.10**

Taking the result on krA100 as an example, five agents are required to complete 100 tasks. The path planning result of K-means SOM and GK-means SOM is separately shown in Fig. 3 and Fig. 4. The costs of agents are shown in Fig. 5.

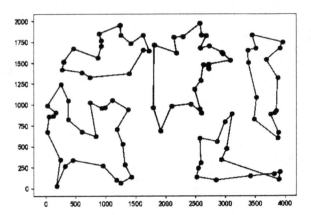

Fig. 3. Path planning result of K-means SOM on krA100.

The maximal cost of GK-means SOM is 5141, while that of K-means SOM is 6178, which shows GK-means SOM makes the loads of agents tend to balance. As shown in Fig. 3, K-means SOM avoids conflicts among agents, but the task allocation is unbalanced which can be observed in Fig. 3 that the third agent has heavier load, which will reduce the working efficiency of the whole system. GK-mean SOM well avoid the task imbalance among the agents (see Fig. 4 and Fig. 5).

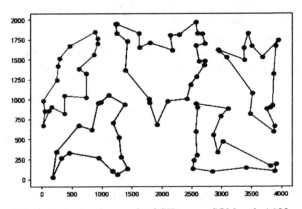

Fig. 4. Path planning result of GK-means SOM on krA100.

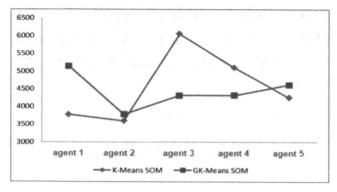

Fig. 5. Costs of agents of two algorithms.

References

1. Niendorf, M., Kabamba, P.T., Girard, A.R.: Stability of solutions to classes of traveling salesman problems. IEEE Trans. Cybern. **46**(4), 973–985 (2016)
2. Dorigo, M., Birattari, M., Stutzle, T.: Ant colony optimization. IEEE Comput. Intell. Mag. **1**(4), 28–39 (2006)
3. Karaboga, D., Akay, B.: A comparative study of artificial bee colony algorithm. Appl. Math. Comput. **214**(1), 108–132 (2009)
4. Han, H., Bai, X., Han, H., Hou, Y., Qiao, J.: Self-adjusting multi-task particle swarm optimization. IEEE Trans. Evol. Comput. https://doi.org/10.1109/TEVC.2021.3098523
5. Pan, W.T.: A new fruit fly optimization algorithm: taking the financial distress model as an example. Knowl.-Based Syst. **26**(2), 69–74 (2012)
6. Kohonen, T.: The self-organizing map. Proc. IEEE **78**(9), 1464–1480 (1990)
7. Zhu, D., Huang, H., Yang, S.X.: Dynamic task assignment and path planning of multi-AUV system based on an improved self-organizing map and velocity synthesis method in three-dimensional underwater workspace. IEEE Trans. Cybern. **43**(2), 504–514 (2013)
8. Zhu, D., Cao, X., Sun, B., et al.: Biologically inspired self-organizing map applied to task assignment and path planning of an AUV system. IEEE Trans. Cognit. Dev. Syst. **10**(2), 304–313 (2017)
9. Sun, W., Zhang, F., Xue, M., et al.: An SOM-based algorithm with locking mechanism for task assignment. In: 2017 IEEE International Conference on Cybernetics and Intelligent Systems (CIS) and IEEE Conference on Robotics, Automation and Mechatronics (RAM), pp. 36–41. IEEE (2017)
10. Reinelt, G.: TSPLIB—A traveling salesman problem library. ORSA J. Comput. **3**(4), 376–384 (1991)

Association Analysis of Gene Expression and Brain Image Identifies Gene Signatures in Major Depression Disorder

Wei Liu$^{(\boxtimes)}$, Jian-po Su, and Ling-Li Zeng

College of Intelligence Science and Technology, National University of Defense Technology,
Changsha 410073, Hunan, People's Republic of China
Liuwei314@nudt.edu.cn

Abstract. Major depression disorder (MDD) usually comes with structural and functional alterations of the brain, which are determined by altered gene expression patterns. We extracted threefunctional metrics, including Amplitude of Low Frequency Fluctuation, fractional Amplitude of Low Frequency Fluctuation, Regional Homogeneity from fMRI data, and one structural metrics, grey matter density from MRI data, to explore inter-group differences between MDD patients and healthy controls. Based on the association analysis of gene expression and brain imaging, we found 796 gene signatures whose expression was significantly correlated with functional or structural alteration of brain images in MDD. To understand how these genes affect the development of human brains, we used a measure, DPM, to quantitatively estimate the temporal and spatial specificity of MDD-related genes expressed in a region during a period, based on the gene expression dataset of developing human brains from the Allen Brain Atlas. We found MDD-related genes displayed more spatial specificity and less temporal specificity than control genes, suggesting MDD may occur at all periods, but only cause functional and structural changes in specific brain regions. Furthermore, MDD-related genes were found significantly overexpressed in ventrolateral prefrontal cortex, posterior superior temporal cortex, inferior parietal cortex, orbital frontal cortex, primary somatosensory cortex, primary motor cortex, primary auditory cortex, mediodorsal nucleus of the thalamus and cerebellar cortex mainly during adolescence and adulthood, associated with a decrease of grey matter density and a compensatory increase of functional activities affected by MDD. These findings can provide new insights into temporal-spatial expression pattern of MDD-related genes in human brains.

Keywords: major depression disorder · gene signature · temporal-spatial expression characteristics

1 Introduction

Major Depressive Disorder (MDD) is a common psychiatric illness, affecting over 350 million people worldwide and takes an immense personal toll on patients and their

F. Sun et al. (Eds.): ICCCS 2022, CCIS 1732, pp. 148–159, 2023.
https://doi.org/10.1007/978-981-99-2789-0_13

families, placing a vast economic burden on society [1]. This disorder is of complex origin, being determined by the interplay of a multitude of environmental factors and genetic variations [2–4]. Neuroimaging of large cohorts can identify characteristic of MDD, and brain structural and functional alterations in MDD have been reported using different modalities of magnetic resonance imaging (MRI) [5–7].

Recent advances in comprehensive brainwide gene expression atlases such as the Allen Human Brain Atlas (AHBA) [8], have provided the possibility of linking spatial variations in gene expression to macroscopic neuroimaging phenotypes. Considering the highly conserved transcriptional architecture of the cerebral cortex across individuals [9], several transcription neuroimaging association studies have explored the molecular mechanisms underlying MRI and resting state functional magnetic resonance imaging (fMRI) in healthy subjects by correlating brain imaging derived from living human brains to gene expression data derived from postmortem human brains [10–13]. The analytical approach to combine imaging and genomic data has been methodologically established [14, 15] and applied in the context of neuropsychiatric disorders [16–19]. For example, integrative omics analysis was reported to identify transcriptome signatures associated with regional grey matter volume variations in MDD [5]. Functional connectivity of the prefrontal cortex and amygdala, the right inferior frontal gyrus and orbitofrontal cortex was found related to depression status in major depressive disorder [20, 21].

Temporal-spatial variation of gene expression can happen extensively among tissues, developmental stages, physiological conditions and individuals [22, 23]. The variation is believed to link with gene function and pathology. The temporal-spatial preference of these pattern genes carries crucial information of what the genes do and how they work together to execute certain physiological functions. However, there is still a lack of understanding about the temporal and spatial expression characteristics of MDD-related genes in human brain. Analyzing temporal-spatial dynamic pattern of MDD-related gene expression in human brain is essential for a comprehensive understanding of the occurrence mechanism.

The work described here had two major objectives. The first aim is to develop an approach based on combined analysis of gene expression and brain imaging for identifying gene signatures related with MDD. The second aim is to explore the temporal-spatial expression specificity of MDD-related genes. To this end, we performed a comprehensive correlation analysis of gene expression and brain imaging to identify gene signatures and analyzed their expression specificity in different brain regions at different periods. Firstly, we detected inter-group difference between MDD patients and healthy control based on fMRI and MRI data. Then, we extracted sample-wise gene expression data across six postmortem adult human brains from AHBA. Subsequently, we used correlation analysis to identify genes whose transcriptional profiles were significantly linked to the case-control differences in functional and structural metrics. Finally, temporal-spatial specificity of MDD-related genes in expression were analyzed based on the gene expression dataset of developing human brains from the Allen Brain Atlas, to explain how these genes play crucial roles in regulating MDD.

2 Result

2.1 Identification of Functional and Structural Alterations in MDD Based on Brain Imaging Analysis

The whole-brain resting-state functional connectivity metrics and voxel-based morphometry analysis was used to determine the pattern of functional and structural changes in MDD relative to healthy controls. For fMRI data, we extracted three functional metrics, including Amplitude of Low Frequency Fluctuation (ALFF), fractional Amplitude of Low Frequency Fluctuation (fALFF), Regional Homogeneity (ReHo) [24], to represent the functional characteristics of the brain. For structural MRI data, the normalized grey matter was extracted by CAT12 and inter-group difference of ALFF, fALFF, ReHo and grey matter density (GMD) between MDD patients and healthy control were determined by two-sample t-test.

We mapped the t statistics for the case-control differences in functional metrics and GMD at each cortical area (Fig. 1). A positive t statistic means that the corresponding metrics increased in patients, whereas a negative t statistic means that corresponding metrics decreased in patients. MDD patients had decreased ALFF and fALFF in the left fusiform gyrus, left angular gyrus, left posterior cingulate gyrus. The MDD group showed significantly lower GMD than the healthy control group in the right middle frontal gyrus, right superior frontal gyrus, left inferior frontal gyrus and left superior frontal gyrus. However, the healthy control group showed significantly lower GMD than the MDD group in the right precuneus, left anterior central gyrus and right anterior cingutate.

We found significant positive correlations between the three functional metrics (all R > 0.9 and P < 0.001), and significant negative correlations between the functional

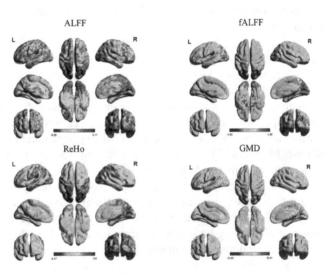

Fig. 1. Visualization of discriminative brain regions between patients with MDD and healthy controls in two-sample t-test. T-values are indicated by colors. Positive values indicate higher metrics in patients and negative values indicate lower metrics in patients.

metrics and GMD (all R < -0.5 and P < 0.001). The discriminative brain regions between patients with MDD and healthy control indicated by the three functional metrics are basically consistent. This may be related to the decrease of GMD and the compensatory increase of functional activities in brain regions associated with MDD.

2.2 Identification of MDD-Related Genes Based on the Association Analysis of Gene Expression and Brain Imaging

Integrating the expression data of AHBA and functional and structural metrics of neu-roimaging, we identified 796 MDD-related genes, correlated with case-control differences in functional metrics and GMD. We found 781 gene signatures correlated with functional alterations in MDD, including 289 genes positively correlated with functional metrics and 492 genes negatively correlated with functional metrics, with all Spearman correlation R > 0.2 or R < − 0.2 and P < 0.001 for ALFF, fALFF and ReHo. Mean-while, we found 218 gene signatures significantly correlated with GMD alterations in MDD, including 112 genes positively correlated with GDM and 106 genes negatively correlated with GMD, with R > 0.15 or R < − 0.15 and P < 0.001. The expression of 203 genes was simultaneously correlated with functional and structural alterations in MDD, including 101 genes positively correlated with functional metrics and negatively correlated with GMD and 102 genes negatively correlated with functional metrics and positively correlated with GMD. All these genes correlated with functional or structural alterations in MDD are considered as MDD-related genes, as shown in Fig. 2a. As an

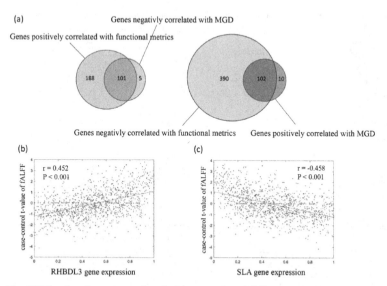

Fig. 2. The MDD-related genes correlated with case-control differences in functional or structural metrics. (a) The overlap of genes correlated with functional and structural alterations in MDD. (b) The expression of RHBDL3 is significantly positively correlated with fALFF alterations in MDD. (c) The expression of SLA is significantly negatively correlated with fALFF alterations in MDD. For convenience, gene expression levels of RHBDL3 and SLA were normalized to between 0 and 1.

illustration, the expression of two typical genes, including Rhomboid-related protein 3 (RHBDL3) and Homo sapiens Src-like-adaptor (SLA), were given in Fig. 2bc, which were significantly correlated with fALFF alterations in MDD.

2.3 Functional Annotation Analysis of MDD-Related Genes

We analyzed the function annotations and biological pathways of MDD-related genes using the gene annotation and analysis resource Metascape [25]. Gene ontology (GO) was used to enrich these genes for specific molecular functions, biological processes and cellular components, and Kyoto encyclopedia of genes and genomes (KEGG) was used to identify related biological pathways. Accumulative hypergeometric P-values and enrichment factors were calculated and used for filtering. The most enriched functional annotation terms of MDD-related genes are shown in Fig. 3.

Enrichment analysis showed that these gene signatures were mainly associated with ion transport and voltage gated channels. The genes negatively correlated with functional alterations and positively correlated with GMD alterations were enriched for the GO biological processes of positive regulation of dendrite extension ($P = 2.04 \times 10^{-5}$) and the KEGG pathway of neuroactive ligand-receptor interaction ($P = 5.37 \times 10^{-9}$). The genes positively correlated with functional alterations and negatively correlated with GMD alterations were enriched for the GO biological processes of inorganic cation transmembrane transport ($P = 4.47 \times 10^{-6}$) and voltage gated potassium channels ($P = 7.76 \times 10^{-7}$).

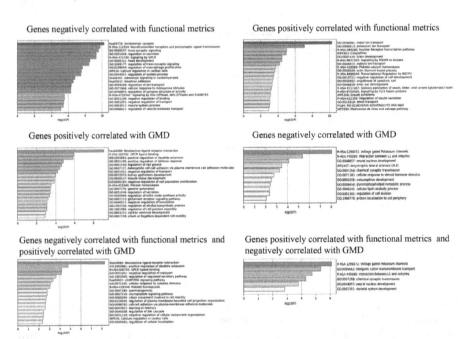

Fig. 3. Enriched ontology clusters of MDD-related genes.

2.4 Temporal-Spatial Expression Specificity Analysis of MDD-Related Genes

To investigate the temporal-spatial expression characteristics of MDD-related genes, we downloaded the gene expression dataset of developing human brains from the Allen Brain Atlas [26] and obtained 547 transcriptome samples in 16 brain regions during 8 periods (see Materials and Methods). Then, we used the measure DPM to evaluate the expression specificity of MDD-related genes in different periods or regions (see the section of Materials and Methods) [27]. As shown in Fig. 4a, we counted the mean temporal and spatial DPM values of the MDD-related genes, respectively. Taking the genes not correlated with the functional or structural alterations in MDD as the control, MDD-related genes were found to have a higher spatial DPM value but a lower mean temporal DPM value (Two sample t-test $P < 0.001$). That means MDD-related genes display more spatial specificity and less temporal specificity than control genes, implying that MDD may occur at multiple periods, but only cause functional and structural alterations in specific brain regions.

We investigated the periods and brain regions in which MDD-related genes were significantly highly expressed compared with control genes based on two sample t-test, as shown in Fig. 4b. The genes positively correlated with functional alterations and negatively correlated with GMD alterations in MDD were significantly overexpressed in VFC, STC, IPC, OFC, S1C, M1C, A1C, MD and CBC than control genes ($P < 0.001$). There may be a decrease of GMD and a compensatory increase of functional activities in these brain regions affected by MDD. The genes positively correlated with functional alterations and negatively correlated with GMD alterations in MDD, were significantly highly expressed in CBC and V1C from childhood to adulthood, but highly expressed in other related brain regions mainly during adolescence and adulthood. This result implied that early phenotypes of MDD may appear in CBC and V1C. The genes negatively correlated with functional alterations and positively correlated with GMD alterations in MDD, were significantly overexpressed in MFC, ITC, HIP and AMY, which has been widely reported by previous investigations [5, 28]. It suggested that the level of functional activity in these brain regions may decline affected by MDD.

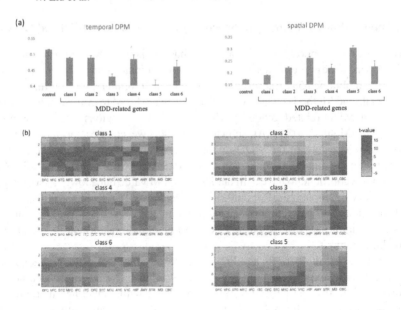

Fig. 4. The temporal and spatial specificity of MDD-related genes. (a) The mean temporal and spatial DPM values of MDD-related genes. (b) The periods and regions in which MDD-related genes significantly overexpressed. The t statistic of the expression of MDD-related genes compared with that of control genes was mapped to the periods and brain regions. Class 1: genes negatively correlated with functional metrics; class 2: genes positively correlated with functional metrics; class 3: genes negatively correlated with GMD; class 4: genes positively correlated with GMD; class 5: genes positively correlated with functional metrics and negatively correlated with GMD; class 6: genes negatively correlated with functional metrics and positively correlated with GMD. M1C: primary motor cortex; S1C: primary somatosensory cortex; A1C: primary auditory cortex; V1C: primary visual cortex; DFC: dorsolateral prefrontal cortex; VFC: ventrolateral prefrontal cortex; OFC: orbital frontal cortex; IPC: inferior parietal cortex; STC: posterior superior temporal cortex; ITC: inferolateral temporal cortex; MFC: medial prefrontal cortex; HIP: hippocampus; AMY: amygdala; STR: striatum; MD: mediodorsal nucleus of the thalamus; CBC: cerebellar cortex.

3 Discussion

In this paper, we revealed several important aspects of gene signatures associated with brain structural and functional alterations in MDD, from their molecular function and biological process involved, and temporal-spatial expression specificity. Firstly, we simultaneously detected inter-group difference between MDD patients and healthy control based on fMRI and MRI data and identified gene signatures correlated with functional or structural alterations in MDD by the association analysis of gene expression and brain imaging. Functional analysis showed that these gene signatures were mainly associated with ion transport and voltage gated channels. Secondly, we evaluated the temporal and spatial specificity of MDD-related genes in 8 periods and 16 brain regions by introducing the specific measure of gene expression. Compared with other genes, MDD-related genes display higher spatial specificity and lower temporal specificity in expression.

This result reminds us that MDD-related genes have unique temporal-spatial expression specificity, and the abnormal expression of genes affected by MDD may cause serious symptoms in specific brain regions.

The development of MDD is accompanied by a series of brain structural and functional alterations. Previous analyses usually focused only on one aspect of the symptoms caused MDD, such as structural alterations or functional alterations. In this paper, we simultaneously examined the structural and functional alterations between the MDD patients and the control based on fMRI and MRI. We found that MDD can cause more extensive functional alterations than structural alterations and there is a close correlation between structural and functional alterations in MDD. We found significant positive correlations among the three functional metrics, and significant negative correlations between the functional metrics and GMD. The decrease of GMD in specific brain regions is usually accompanied by the enhancement of functional activity, while the increase of GMD is accompanied by the weakening of functional activity.

Different to the PLS analyses used to identify a linear combination of genes that had a similar cortical pattern of expression to the map of volumetric changes in MDD in previous research [5], we correlated the structural and functional alterations in MDD with the gene expression profiles in normal human brain samples to in order to investigate the association between each gene and MDD. We identified MDD-related genes and separated them by directional effects, such as positively or negatively correlated expression. As a result, we found 203 genes simultaneously correlated with functional and structural alterations in MDD. There is no intersection between the genes negatively correlated with functional metrics and those positively correlated with structural GMD or between the genes positively correlated with functional metrics and those negatively correlated with GMD. These findings can provide insights into understanding the genetic determinants of structural and functional variations in MDD.

Understanding temporal-spatial characteristics of gene expression in human brain can help explain the neurodevelopment, as well as our increased susceptibility to certain brain disorders [22]. Based on the developmental data of human brains, we analyzed the periods and regions in which MDD-related genes significantly overexpressed. The MDD-related genes were found significantly overexpressed in VFC, STC, IPC, OFC, S1C, M1C, A1C, MD and CBC than control genes during adolescence and adulthood. Especially, most MDD-related genes have higher expression levels in adulthood than in adolescence, which may explain the different symptoms of MDD occurred in adolescence and adulthood [29, 30]. These findings may provide important implications for the development of targeted diagnosis and treatment methods for MDD patients.

There are several limitations to this work. First, the use of gene expression profiles from the healthy human brain in AIBS to explain structural and functional alterations is limited to the extent that transcription in patients could be different from those in healthy brains. Second, limited by the sampling accuracy of the BrainSpan atlas, we only analyzed the temporal-spatial expression characteristics of MDD-related genes in 8 periods and 16 brain regions. Therefore, our findings need to be validated through other transcriptome atlases when the expression datasets with higher resolution of sample collection are available. Finally, we should note that how these genetic signatures work together in the process of MDD remains to be further studied, considering MDD is a complex polygenic disorder.

4 Materials and Methods

4.1 Brain Imaging Data of MDD

The brain imaging data utilized in the research come from the REST-meta-MDD project (http://rfmri.org/REST-meta-MDD) [31]. The project are initialized to explore the functional and structural alteration in the brain caused by MDD based on big data of brain images. The consortium contributed 2428 previously collected datasets (1300 MDDs and 1128 NCs) from 25 research groups in 17 hospitals. By far, it is the largest datasets for neuroimaging research of MDD. Detailed imaging and demographic information can be found in the website.

4.2 Allen Human Brain Atlas

Publicly available gene expression data from six human postmortem donors, aged 24–57 years of age, were obtained from the AHBA, downloaded after the updated microarray normalization pipeline (http://human.brain-map.org) [8]. Specifically, AHBA offers comprehensive coverage of nearly the entire brain, consisting of normalized expression data of 20,737 genes with unique Entrez IDs detected by 58,692 probes taken from 3,702 spatially distinct tissue samples. Spatial correspondence between gene expression and neuroimaging phenotypes could be established in MNI space.

We combined the gene expression dataset across all six brains using the typical workflow for processing brain-wide transcriptomic data described in previous paper [14]. By mapping the samples to cerebral cortex of human brains, we obtain an integrated dataset including 15,745 genes expressed in 1280 samples. The following probe filtering criteria were applied during the process: 1) the probe-to-gene annotations were updated using Re-Annotator package; 2) probes where expression measures do not exceed the background in more than 50% samples were removed; 3) a representative probe for a gene was selected based on the highest intensity; 4) applying limma batch effect removal on cross-subject aggregated data, followed by the scaled robust sigmoid normalization. After normalization, samples have no inter-individual differences in gene expression.

4.3 BrainSpan Atlas

We downloaded the gene expression dataset of developing human brains from the Allen Brain Atlas (http://www.brainspan.org) [26]. Only brains from clinically unremarkable donors with no signs of large-scale genomic abnormalities were included in the study (N = 41; age, 8 post-conceptual weeks (pcws) to 40 years; sex, 22 males and 19 females). We created a 8-period system spanning from embryonic development to adulthood (Table 1) and used 547 transcriptome samples in 16 brain regions for investigation, including the cerebellar cortex, mediodorsal nucleus of the thalamus, striatum, amygdala, hippocampus, and 11 neocortex (NCX) areas. Information about data preprocessing and normalization is available on the Brainspan Atlas website (http://help.brain-map.org//display/devhumanbrain/Documentation).

Table 1. Periods of human development and adulthood as defined in this study.

Period	Description	Age
1	Early fetal	Age < 14pcw
2	Middle fetal	14pcw ≤ Age < 24pcw
3	Late fetal	24pcw ≤ Age < Birth
4	Infancy	Birth ≤ Age < 1yrs
5	Early childhood	1yrs ≤ Age < 6yrs
6	Middle and late childhood	6yrs ≤ Age < 12yrs
7	Adolescence	12yrs ≤ Age < 20yrs
8	Adulthood	20yrs ≤ Age

4.4 The Measure for Evaluating Temporal and Spatial Specificity of Genes

The specificity measure (SPM) was used to quantitatively estimate the relative expression specificity of a gene in a sample [27]. Each gene expression profile was first transformed into a vector X:

$$X = (x_1, x_2, \ldots, x_i, \ldots, x_{n-1}, x_n) \tag{1}$$

where n is the number of samples in a profile. At the same time, a vector Xi was created to represent the gene expression in sample i:

$$X_i = (0, 0, \ldots, x_i, \ldots, 0, 0) \tag{2}$$

The SPM of a gene in a sample was then determined by calculating the cosine value of intersection angle θ between vector X_i and X in high-dimension feature space:

$$SPM_i = \cos\theta = X_i \cdot X / |X_i| \cdot |X| \tag{3}$$

where $|X_i|$ and $|X|$ are the length of vector X_i and X, respectively. The higher the SPM value is, the more specific the gene expression is in a sample.

Based on SPM, we used the parameter of dispersion measure (DPM) to evaluate the specificity degree of a gene expression profile in a period or region. The gene expression profile (X) was converted to its corresponding SPM profile (X_{SPM}):

$$X_{SPM} = (SPM_1, SPM_2, \ldots, SPM_i, \ldots, SPM_{n-1}, SPM_n) \tag{4}$$

The DPM was then determined by

$$DPM = \sqrt{\frac{\sum_{i=1}^{n} \left(SPM_i - \overline{SPM}\right)^2}{n-1}} \cdot \sqrt{n} \tag{5}$$

where n is the sample number, and $\overline{SPM}$ is the mean of SPMs in a gene expression profile. unlike conventional SD analysis, DPM is independent of gene expression level and sample number by scaling into a region of 0–1.0 as above. The temporal and spatial specificity of genes can be quantitatively estimated by taking the gene expression of samples in a period and a region as a profile, respectively. The higher the DPM value is, the more specific the gene expression is in a period or region. Therefore, DPM can serve as a good indicator in quantitative description and identification of specific genes overexpressed in a period or a region.

References

1. Mill, J., Petronis, A.: Molecular studies of major depressive disorder: the epigenetic perspective. Mol. Psychiatry **12**(9), 799–814 (2007)
2. Alshaya, D.S.: Genetic and epigenetic factors associated with depression: An updated overview. Saudi J. Biol. Sci. **29**(8), 103311 (2022)
3. Wray, N.R., Ripke, S., Mattheisen, M., et al.: Genome-wide association analyses identify 44 risk variants and refine the genetic architecture of major depression. Nat. Genet. **50**, 668–681 (2018)
4. Jansen, R., Penninx, B.W., Madar, V., et al.: Gene expression in major depressive disorder. Mol. Psychiatry **21**(3), 444 (2016)
5. Sha, Z., Banihashemi, L.: Integrative omics analysis identifies differential biological pathways that are associated with regional grey matter volume changes in major depressive disorder. Psychol. Med. **52**(5), 924–935 (2022)
6. Shen, X., Reus, L.M., Cox, S.R., et al.: Subcortical volume and white matter integrity abnormalities in major depressive disorder: findings from UK Biobank imaging data. Sci. Rep. **7**(1), 5547 (2017)
7. van Tol, M.J., van der Wee, N.J., van den Heuvel, O.A., et al.: Regional brain volume in depression and anxiety disorders. Arch. Gen. Psychiatry **67**(10), 1002–1011 (2010)
8. Jawrylycz, H.M., Lein, E.S., Guillozet-Bongaarts, A.L., et al.: An anatomically comprehensive atlas of the adult human brain transcriptome. Nature **489**(7416), 391–399 (2012)
9. Hawrylycz, M., Miller, J., Menon, A.V., et al.: Canonical genetic signatures of the adult human brain. Nat. Neurosci. **18**(12), 1832–1846 (2015)
10. Richiardi, J., Altmann, A., Milazzo, A.C., et al.: Correlated gene expression supports synchronous activity in brain networks. Science **348**(6240), 1241 (2015)
11. Wang, D., Liu, S., Warrell, J., et al.: Comprehensive functional genomic resource and integrative model for the human brain. Science **362**(6420), eaat8464 (2018)
12. Zhu, D., Yuan, T., Gao, J., et al.: Correlation between cortical gene expression and resting-state functional network centrality in healthy young adults. Hum. Brain Mapp. **42**(7), 2236–2249 (2021)
13. Han, K.M., De Berardis, D., Fornar, M., Kim, Y.K.: Differentiating between bipolar and unipolar depression in functional and structural MRI studies. Prog. Neuropsychopharmacol Biol. Psychiatry **91**, 20–27 (2019)
14. Arnatkeviciute, A., Fulcher, B.D., Fornito, A.: A practical guide to linking brain-wide gene expression and neuroimaging data. NeuroImage **189**, 353–367 (2019)
15. Markello, R.D., Arnatkevičiūtė, A., Poline, J.B., et al.: Standardizing workflows in imaging transcriptomics with the abagen toolbox. Elife **10**, e72129 (2021)
16. Liu, H., Xu, L., Fu, J., et al.: Prefrontal granule cell-related genes and schizophrenia. Cereb Cortex **31**(4), 2268–2277 (2021)
17. Morgan, S.E., Seidlitz, J., Whitaker, K.J., et al.: Cortical patterning of abnormal morphometric similarity in psychosis is associated with brain expression of schizophrenia-related genes. Proc. Natl. Acad. Sci. U. S. A. **116**(19), 9604–9609 (2019)

18. Li, M., Santpere, G., Imamura Kawasawa, Y., et al.: Integrative functional genomic analysis of human brain development and neuropsychiatric risks. Science **362**(6420), eaat7615 (2018)

19. Grothe, M.J., Sepulcre, J., Gonzalez-Escamilla, G., et al.: Molecular properties underlying regional vulnerability to Alzheimer's disease pathology. Brain **141**(9), 2755–2771 (2018)

20. Zhang, A., Yang, C., Li, G., Wang, Y., et al.: Functional connectivity of the prefrontal cortex and amygdala is related to depression status in major depressive disorder. J. Affect Disord. **274**, 897–902 (2020)

21. Rolls, E.T., Cheng, W., Du, J., et al.: Functional connectivity of the right inferior frontal gyrus and orbitofrontal cortex in depression. Soc. Cogn. Affect. Neurosci. **15**(1), 75–86 (2020)

22. Kang, H.J., Kawasawa, Y.I., Cheng, F., et al.: Spatio-temporal transcriptome of the human brain. Nature **478**(7370), 483–489 (2011)

23. Zhu, Y., Sousa, A.M.M., Gao, T., et al.: Spatiotemporal transcriptomic divergence across human and macaque brain development. Science **362**(6420), eaat8077 (2018)

24. Zang, Y., Jiang, T., Lu, Y., He, Y., Tian, L.: Regional homogeneity approach to fMRI data analysis. Neuroimage **22**(1), 394–400 (2004)

25. Zhou, Y., Zhou, B., Pache, L., et al.: Metascape provides a biologist-oriented resource for the analysis of systems-level datasets. Nat. Commun. **10**(1), 1523 (2019)

26. Carlyle, B., et al.: A multiregional proteomic survey of the postnatal human brain. Nat. Neurosci. **20**(12), 1787–1795 (2017). https://doi.org/10.1038/s41593-017-0011-2

27. Pan, J.B., Hu, S.C., Wang, H., Zou, Q., Ji, Z.L.: PaGeFinder: quantitative identification of temporal-spatial pattern genes. Bioinformatics **28**(11), 1544–1545 (2012)

28. Amidfar, M., Quevedo, J., Réus, Z.G., Kim, Y.K.: Grey matter volume abnormalities in the first depressive episode of medication-naïve adult individuals: a systematic review of voxel based morphometric studies. Int. J. Psychiatry Clin. Pract. **25**(4), 407–420 (2021)

29. Zhang, F.: Resting-state functional connectivity abnormalities in adolescent depression. EBioMedicine **17**, 20–21 (2017)

30. Cullen, K.R., Westlund, M., Klimes-Dougan, B., et al.: Abnormal amygdala resting-state functional connectivity in adolescent depression. JAMA Psychiatry **71**(10), 1138–1147 (2014)

31. Yan, C.G., et al.: Reduced default mode network functional connectivity in patients with recurrent major depressive disorder. Proc. Natl. Acad. Sci. U. S. A. **116**(18), 9078–9083 (2019)

Transfer Learning Based Seizure Detection: A Review

Xiaonan Cui[1], Jiuwen Cao[1,2(✉)], Tiejia Jiang[3], and Feng Gao[3]

[1] Machine Learning and I-health International Cooperation Base of Zhejiang Province, Hangzhou Dianzi University, Hangzhou, China
{xncui,jwcao}@hdu.edu.cn
[2] School of Automation, Hangzhou Dianzi University, Hangzhou, China
[3] Department of Neurology, The Children's Hospital, Zhejiang University School of Medicine, National Clinical Research Center for Child Health, Hangzhou 310003, China
{jiangyouze,epilepsy}@zju.edu.cn

Abstract. Seizure detection automatically recognizes Electroencephalogram (EEG) signals in epileptic seizure states through machine learning, time-frequency analysis, statistical test, etc., which provides an objective reference for epilepsy diagnosis, treatment and evaluation. Traditional seizure detection is based on the assumption that the training and testing data are equally distributed. However, the characteristics of EEG signals are highly influenced by various individual differences such as age and gender, which seriously affect the generalization performance and re-usability of traditional models. To overcome individual differences and reduce the requirements for training data, transfer learning has been studied in seizure detection, which utilizes knowledge of related domains, such as data characteristics, model parameters, etc., to assist learning in new domain. This paper details three main methods to practical application of transfer learning in seizure detection. The future research directions in the field of seizure detection based on transfer learning are also pointed out.

Keywords: Seizure detection · Transfer learning · Electroencephalogram

1 Introduction

Epilepsy, a common neurological syndrome caused by abnormal firing of brain cells, is thought to be a hypersynchronous state caused by an imbalance between

This work was supported by the National Key Research and Development Program of China (2021YFE0100100, 2021YFE0205400), the National Natural Science Foundation of China (U1909209), the Open Research Projects of Zhejiang Lab (2021MC0AB04), the Key Research and Development Program of Zhejiang Province (2020C03038), and Zhejiang Provincial Natural Science Foundation (LBY21H090002).

F. Sun et al. (Eds.): ICCCS 2022, CCIS 1732, pp. 160–175, 2023.
https://doi.org/10.1007/978-981-99-2789-0_14

the excitatory and inhibitory inputs of neuronal networks in the brain [54]. The clinical manifestations of epilepsy are diverse and complicated due to the differences in the initial location and propagation mode of abnormal brain discharge activities [2]. Repeated seizures can cause severe and irreversible damage to the brain, even life-threatening. Therefore, the research on the diagnosis and treatment of epilepsy has very important clinical significance. Electroencephalogram (EEG) is an electrophysiological monitoring method for non-invasive recording of brain electrical activity [49]. It records the electrical signals generated by synchronized neuronal activity in the brain over a period of time, which is essential in the diagnosis of epilepsy. Generally, artificial epilepsy detection requires prolonged visual inspection of EEG signals by neurologists. EEG signals have noise and latency, and epilepsy diagnosis relies heavily on the subjective judgment of neurologists, which results in low efficiency of manual inspection. To overcome the limitations of traditional diagnostic methods and improve medical outcomes, intelligent recognition of epileptic EEGs is a promising option, and many machine learning techniques have been proposed [5, 19, 28, 29].

At present, seizure detection mostly focuses on patient-dependent scenarios that rely on the patient history records. An important reason for the high accuracy of patient-dependent seizure detection models is that the training and testing data have the same feature distribution. Actually, EEG signals are nonstationary and nonlinear, and are greatly affected by individual differences such as age and mentality. Therefore, it is challenging to build a general machine learning model in seizure detection that is effective for different subjects, devices and tasks. Generally, some labeled data of new subjects is required to calibrate the training model, which is time-consuming and user-unfriendly.

To address the above issues, transfer learning utilizes the labeled data or knowledge structure of the relevant domain to complete or improve the model learned on the target domain, which has been used for this purpose [55, 70]. For the general automatic epilepsy detection model, the data used for training model and the actual test data may come from different subjects, which leads to a large difference in data distribution. Therefore, the generalization performance and accuracy of the model can be improved through transfer learning. In addition, the problem of insufficient data makes it difficult to train deep networks with complex structures in seizure detection. The transfer learning model can still maintain the learning ability based on the prior knowledge learned in related tasks without requiring a large amount of data [50].

The advantages of transfer learning in seizure detection are as follows: *1) Match individual differences:* Inconsistencies in subjects, sampling time, and task objectives may lead to huge differences between training and testing data, which increases the difficulty of modeling [42]. Transfer learning can make the model flexible to match different individuals and tasks by adjusting. It has become a new trend to study cross-subject seizure detection algorithms adapted to different subjects. *2) Reduce data requirements:* Based on the prior knowledge learned in similar domains, transfer learning aims to tune the classifier with a small amount of target data to reduce the requirement for available data.

Transfer learning is increasingly used in seizure detection, and specific algorithms have also been proposed. This paper attempts to briefly summarize the relevant literature on the application of transfer learning algorithms in seizure detection for the reference of researchers in related fields. The rest of this paper is organized as follows. Section 2 introduces the basic concepts of transfer learning. Section 3 provides a detailed division and description of transfer learning methods used in seizure detection. Finally, Sect. 4 discusses the future directions of transfer learning in seizure detection.

2 Preliminaries

2.1 Definition of Transfer Learning

Before introducing transfer learning algorithms for seizure detection, some basic definitions and categories of transfer learning are introduced in this section.

Domain: A domain $\mathcal{D}$ consists of two parts, the feature space $\mathcal{X}$ and the probability distribution $P(X)$ that generates these data, in other words, $\mathcal{D} = \{\mathcal{X}, P(X)\}$. $X = \{\mathbf{x}_1, \mathbf{x}_2, \ldots, \mathbf{x}_n\} \in \mathcal{X}$ represents a set of instances of the domain. In transfer learning, the domain containing known knowledge is called the source domain, which is usually represented by $\mathcal{D}_S$. The domain containing unknown knowledge to be learned is called the target domain and is usually represented by $\mathcal{D}_T$. According to the domain definition $\mathcal{D}_S = \{\mathcal{X}_S, P(X_S)\}$ and $\mathcal{D}_T = \{\mathcal{X}_T, P(X_T)\}$, $\mathcal{D}_S \neq \mathcal{D}_T$ means $\mathcal{X}_S \neq \mathcal{X}_T$ and/or $P(X_S) \neq P(X_T)$. That is, if the source and target domains are different, then their feature spaces are different or obey different marginal probability distributions.

Task: A task $\mathcal{T}$ consists of the label space $\mathcal{Y}$ and the decision function $f(\cdot)$, i.e. $\mathcal{T} = \{\mathcal{Y}, f\}$. According to the definition of the task, the label spaces of the source and target domains are denoted as $\mathcal{Y}_S$ and $\mathcal{Y}_T$, respectively.

Transfer learning: When there are differences between data and tasks in different domains, but similar to a certain extent, the knowledge learned in the source domain can be transferred to the target domain through transfer learning. Given a source domain $\mathcal{D}_S$ and the corresponding task $\mathcal{T}_S$, a target domain $\mathcal{D}_T$ and the corresponding task $\mathcal{T}_T$, transfer learning is the process of using the relevant information of $\mathcal{D}_S$ and $\mathcal{D}_T$ to improve the target prediction function $f_T(\cdot)$, where $\mathcal{D}_S \neq \mathcal{D}_T$ or $\mathcal{T}_S \neq \mathcal{T}_T$. In this definition, transfer learning approach attempts to reduce the differences between domains or tasks to ensure that the source and target domains have similar feature or label spaces. Examples of $\mathcal{D}_S \neq \mathcal{D}_T$ and $\mathcal{T}_S \neq \mathcal{T}_T$ in seizure detection are given in Fig. 1.

2.2 Categories of Transfer Learning

Space-Setting-Based Categorization. So far, there is no unified standard for the categories of transfer learning. A common categorization is based on the consistency between the feature space of source and target domains, where the case of $\mathcal{X}_S = \mathcal{X}_T$ is defined as homogeneous transfer learning, and the case of

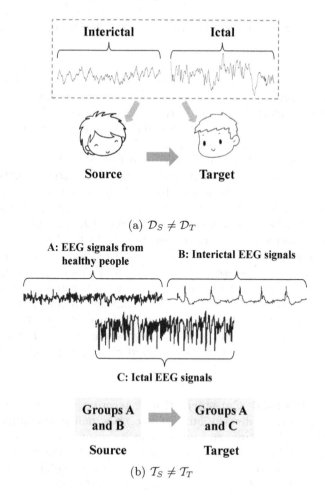

(a) $\mathcal{D}_S \neq \mathcal{D}_T$

(b) $\mathcal{T}_S \neq \mathcal{T}_T$

Fig. 1. Examples of $\mathcal{D}_S \neq \mathcal{D}_T$ and $\mathcal{T}_S \neq \mathcal{T}_T$ in seizure detection.

$\mathcal{X}_S \neq \mathcal{X}_T$ is defined as heterogeneous transfer learning. In homogeneous transfer learning, the semantics and dimensions of the feature space in the source and target domains are the same. In contrast, in heterogeneous transfer learning, the semantics and dimensions of feature space in the source and target domains are not exactly the same. Returning to the example of seizure detection, the classification of EEG signals from different subjects is a typical homogeneous transfer learning task. This is because feature extraction is usually performed in the same way for different subjects, and individual differences lead to inconsistent probability distributions for training and testing data. Signals collected by different portable devices may trigger heterogeneous transfer learning tasks. Saeed et al. pointed out that the sequence and number of channels in different EEG acquisition protocols are not the same, resulting in heterogeneous EEG datasets [45]. It is difficult to aggregate different heterogeneous datasets to augment the training data.

Label-Setting-Based Categorization. From the aspect of label setting, transfer learning problems can be classified into three types, that is, transductive transfer learning, inductive transfer learning, and unsupervised transfer learning. In simple terms, transductive transfer learning refers to the fact that label information only comes from the source domain. For example, Xia et al. proposed a cross-domain epileptic EEG classification model based on knowledge utilization maximization, exploiting the global data structure of source and unlabeled target domains [58]. If the target domain has a small amount of label information available, the scenario can be considered as inductive transfer learning. In [26], a specific epilepsy EEG recognition model was built with partially labeled target-domain samples and learned source parameters. If the label information of both the target and source domains is unknown, this situation is called unsupervised transfer learning. Unsupervised transfer learning focuses on solving unsupervised learning problems in the target domain, such as clustering, dimensionality reduction, density estimation, etc. Transfer component analysis (TCA) [35] is a classical unsupervised transfer learning method that maps the source and target domains to a domain-invariant space through a unified transformation. Jiang et al. applied TCA in seizure detection to reduce individual differences among patients [22].

Learning-Style-Based Categorization. According to the learning style, transfer learning can be divided into two categories: online and offline. In offline transfer learning, the source and target domains are fixed, and the purpose is to perform a knowledge transfer to complete the establishment or adjustment of the model. Most seizure detection literature is based on offline transfer. After training the transfer model with offline data, the model cannot perform real-time updates on newly added target data, which may result in poor performance of the model on the new datasets. Online transfer learning updates the model in real time with dynamically added data, and the model obtained in this learning mode has stronger applicability. However, the transmission and processing of dynamic data results in that the distribution difference between the source and target domains is difficult to measure, thus it is more challenging than offline transfer learning.

3   Transfer Learning Methods in Seizure Detection

This section introduces the commonly used transfer learning methods for seizure detection in this survey, starting from two broad categories of shallow and deep transfer. Based on what to transfer, shallow transfer learning is also divided into feature-representation transfer and parameter transfer. For better understanding, Table 1 represents the mentioned categorizations of transfer learning and the corresponding literature.

Table 1. Transfer learning methods in seizure detection.

Transfer learning methods	Algorithm	Literature
Feature-Representation-Based transfer	Data-centered	$[10,22,25,38,60]$
		$[24,59,65–67]$
	Subspace-centered	$[13,17,37]$
Parameter-Based transfer	Continuously updated	$[7,52]$
	Regularization-based	$[8,9,12,26]$
Deep transfer networks	Fine-tuning	$[27,34,36,39,40,48,57,62]$
		$[3,4,11,14,20,21,31,41,43,46]$
		$[1,6,32,33,47,51,53,61]$
	Deep network adaptation	$[18]$
	Generative adversarial network transfer	$[63,68]$
	Others	$[56,69]$

3.1 Feature-Representation-Based Transfer Learning

Feature-representation-based transfer, also known as domain adaptation, requires only one weak assumption: the distributions of the source and target domains are similar. In this case, a feasible solution is to minimize the distribution mismatch between the source and target domains by transferring or mapping the features to another space. In general, there are two kinds of feature-based transduction transfer learning methods: data-centered methods and subspace-centered methods. The data-centered methods seek a uniform transformation that projects the source and target domains into a domain-invariant space to minimize the distribution mismatch between domains. The goals of constructing new feature representations include minimizing marginal and conditional distribution differences, preserving the properties and underlying structure of the data. How to effectively measure the distribution difference between the source and target domains is an important issue. The maximum mean discrepancy (MMD) [16] is a widely used measure in transfer learning, which can be expressed as:

$$
\text{MMD}\left(X^S, X^T\right) = \left\| \frac{1}{n^S} \sum_{i=1}^{n^S} \phi\left(\mathbf{x}_i^S\right) - \frac{1}{n^T} \sum_{j=1}^{n^T} \phi\left(\mathbf{x}_j^T\right) \right\|_{\mathcal{H}}^2 , \tag{1}
$$

where n^S and n^T represent the number of samples in source and target domains, and $\phi\left(\cdot\right)$ represents the mapping function. Intuitively, MMD quantifies distribution differences by computing the distance of the instance mean in the reproducing kernel Hilbert space (RKHS) space. In seizure detection, the initial consideration is the alignment of the marginal distributions of the source and target domains. Yang et al. applied large-margin projection transduction support vector machine (LMPROJ) to epileptic EEG recognition [60]. LMPROJ is a transductive transfer learning method that introduces MMD on the basis of SVM to reduce the differences between interictal and ictal EEG signals. Takagi-Sugeno-Kang (TSK) fuzzy system is widely used in medical diagnosis due to its strong interpretability [23,64]. In [25], the TSK fuzzy system and marginal distribution minimization are implemented in one framework, and semi-supervised learning is introduced to exploit the data structure of the unlabeled target domain. It exhibits good robustness and interpretability in dealing with epilepsy EEG signal differences. Jiang et al. proposed a cross-subject seizure detection framework based on TCA to reduce individual differences between patients [22]. Similarly, Prabhakar and Lee applied the sparse representation model in combination with TCA to the Bonn dataset [38].

In recent years, researchers have argued that in addition to marginal distribution differences, there are also conditional distribution differences in epileptic EEG signals. Deng et al. attempted to combine knowledge of marginal distribution and conditional distribution to reduce the difference between epileptic EEG signals, and also adopted TSK fuzzy system as a base classifier [10]. It should be pointed out that the label information of target domain is often limited or even unknown. The measurement of conditional distribution differences requires the label of the target domain, which is difficult to estimate. To address this problem, most methods resort to pseudo-labeling strategies, which assign pseudo-labels to unlabeled target-domain instances. Therefore, an iterative joint knowledge strategy is also introduced in [67] to enhance the transfer learning ability. Xie et al. introduced the generalized hidden-mapping to unify the representation of some classical models, such as fuzzy systems, neural networks, etc., while considering the distribution drift between domains, which has good adaptability in epileptic EEG recognition [59]. Zhang et al. adopted clustering technology to select source instances online, which reduced the computational complexity of joint distribution adaptive classifier [65]. Ref. [66] proposed a regularization for knowledge transfer from the perspective of error consensus, which was unified with joint distribution adaptation to achieve epileptic EEG classification. Jiang et al. considered multi-source transfer learning in epileptic EEG recognition, and proposed a domain entropy weighting index to evaluate the significance of each source domain [24].

Subspace-centered methods attempt to reduce domain shift by manipulating the subspaces of the source and target domains. Gu et al. combined dictionary and transfer learning to construct an epileptic EEG recognition system, which learns a shared dictionary to connect source and target domains by projecting all data into a common subspace [17]. Dong et al. attempted to the problem of epileptic seizure classification with the framework of transfer learning and non-negative matrix factorization. The model reconstructs EEG signals from different domains to find shared hidden features between auxiliary and target domains [13]. Besides non-negative matrix factorization, affinity propagation congregation based mutual information (APCMI) was also used to extract sparse representation features of EEG signals to absorb the shared knowledge of source and target domains [37].

3.2 Parameter-Based Transfer Learning

Parameter-based transfer learning focuses on utilizing the model parameters of the source domain to improve the construction of the prediction function of the target domain. Assuming that some parameters and prior distributions are shared between the decision functions of the source and target tasks, parameter-based transfer learning aims to transfer these shared parameters or prior knowledge into the target prediction function, thereby reducing the classification error. Thomas et al. originally developed a patient-independent neonatal seizure detection system that learns a global classifier with source-domain data and then iteratively updates the weights using a small amount of labeled data from target patients [52]. Not only considering offline data, Ref. [7] established an adaptive seizure detection model by updating the classifier parameters with the patient-specific features acquired in real time.

Most parameter-based transfer learning methods have a general regularization:

$$\left\| \theta - \beta \tilde{\theta} \right\|_F^2 , \tag{2}$$

where $\tilde{\theta}$ and θ represent the model parameters of the source and target domains, respectively, and β are the weighted parameters that control the degree of transfer. Thomas et al. trained an adaptive SVM classifier with a bit of labeled target data and reference classifier parameters learned from source domain to achieve personalized heart rate-based seizure detection [8]. Not only that, they treated seizure detection as anomaly detection problem and proposed an adaptive one-class SVM classifier in the presence of only interictal samples [9]. Similarly, Dhulekar et al. built an epileptic EEG anomaly detection model based on autoregression (AR) and used the prediction error as decision parameter [12]. The AR model was constructed by extracting features from the EEG signals as well as the time-varying EEG synchronization graph, and the concept of transfer learning was introduced to use the model parameters learned in the source domain to improve the prediction in target domain. To solve the problem of insufficient training data and lack of interpretability of the model, Jiang et al. combined gamma-least squares regression with transfer learning, and proposed a new knowledge and label space induction transfer learning algorithm for multi-class EEG signals recognition [26]. The model achieves good classification performance on the target domain without using kernel tricks by transferring the knowledge and generalized label space from source domain to target domain.

3.3 Deep Transfer Networks

Deep learning is especially popular in the field of machine learning. Data dependence is one of the most important problems in deep learning, because deep networks require large amounts of instances to understand the underlying patterns of the data. In seizure detection, the construction of high-quality annotated datasets is complex and expensive, thus insufficient training data is an unavoidable problem. In addition, individual differences between subjects can lead to inconsistent data distribution. Even if high-quality annotated data is obtained, it cannot be effectively applied to new tasks. Transfer learning relaxes the assumption that the training and testing data are independent and identically distributed. The parameters learned in the source domain initialize the target-domain model, which significantly reduces the demand for training data and time. There are three main approaches to the application of deep transfer learning in seizure detection, namely fine-tuning, deep network adaptation, and generative adversarial network transfer, as illustrated in Fig. 2.

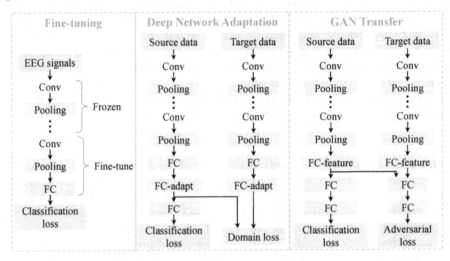

Fig. 2. Structures of Deep Transfer Networks.

Fine-tuning. The most commonly used deep transfer method in seizure detection is to fine-tune pre-trained deep neural networks. Deep networks can be fine-tuned in two aspects, one is fine-tuning the whole network and the other is fine-tuning certain layers. Song et al. constructed a single-channel seizure detection system based on the brain rhythm recurrence biomarker (BRRM) reflecting the nonlinear dynamics of raw EEG signals, and the transfer model ONASNet was used to extract the features of different brain rhythms in BRRM [48]. In Ref. [36], a spatiotemporal CNN trained on the human action recognition (HAR) dataset was transferred to video clip-based epilepsy detection. Similarly in [27], a network pre-trained on HAR dataset was used to extract the spatiotemporal features of epileptic seizures, and a long short-term memory (LSTM) fully connected classifier was constructed for prediction. Seizure prediction algorithms implemented in wearable headband also employ transfer models [34,40]. In Ref. [39], the enhanced convolutional neural network (ECNN) model was done via the following steps, first the optimized CNN model obtained on the UCI machine learning repository was selected as the initial point and then the model is fine-tuned for obtaining least error. The conventional approach to apply fine-tuning in seizure detection is to convert the EEG signals into spectrograms [11,31,57,62], continuous wavelet transform scales [44], power spectral density energy maps

[14], etc. [4,20,21,46] as the input of the deep network, and then fine-tune the deep transfer model to construct an epileptic seizure detection system. Several successful network structures have been widely used, such as Resnet50, Inceptionv3, Vgg19 and Densenet201 [3,33,41,43,47,51].

In addition to transferring networks trained on large datasets such as ImageNet, some studies have also attempted to initialize target models by learning network parameters on similar tasks. Yu et al. fused handcrafted features and hidden deep features extracted based on prior knowledge, and constructed an epileptic seizure prediction system by transferring the weights of the base model trained on the EEG data of all patients to the target patient [61]. Abdelhameed et al. used the deep convolutional autoencoder to learn discriminative spatial features from multi-channel unlabeled raw EEG signals and employed bidirectional LSTM for seizure prediction. After training and testing the system on selected patients, transfer learning was performed using the weights of the pretrained encoder to initialize another similar supervised network for prediction of other patients [1]. Ref. [32] initialized the seizure detection model for preterm infants with network weights trained on the full-term infant dataset. Thuwajit et al. trained an end-to-end multi-scale convolutional neural network model on source subjects to initialize cross-subject model parameters [53]. Daoud et al. connected the bidirectional LSTM recurrent neural network with DCNN for seizure detection, and trained an autoencoder-based semi-supervised model to initialize the parameters of DCNN to improve the convergence speed [6].

Deep Network Adaptation. Fine-tuning can save training time of deep networks and improve learning accuracy. However, it is less effective when the source and target domains have different distributions. Recently, researchers try to develop adaptive layers to make the data distribution of the source and target domains closer. The cost function of the original network is adjusted by increasing the distance loss to measure the distribution of source and target domains. Hang et al. proposed an end-to-end convolutional neural network for cross-subject EEG signal recognition, which minimizes the distribution difference of deep features between the source and target domains using maximum mean discrepancy [18].

At present, some state-of-the-art deep transfer learning techniques have also been applied to seizure setection. Wu et al. proposed a new two-stage training scheme based on end-to-end knowledge distillation, which extracts informative features from multiple subjects through a pre-trained general model, and then a patient-specific model was obtained with the extracted knowledge and additional personalized data [56]. Specifically, a pooling network (P-network) was pre-trained in the first stage, which modeled the mechanisms of all possible seizures with source subjects, regardless of individual differences. In the second stage, a custom network (C-network) with initial random conditions was introduced, and the P-network and C-network were iteratively optimized under the consistency constraint. Zhu et al. established a multi-class epilepsy detection system based on meta-transfer learning to reduce the influence of changes in seizure patterns among patients [69].

Generative Adversarial Network Transfer. Generative adversarial networks (GAN), which is composed of generative network and discriminant network for unsupervised learning, was first proposed in 2014 [15]. The fake samples are generated by generators according to the given data and estimated by discriminators to distinguish the source of them. In recent years, researchers have applied the principles of GAN to transfer learning. In generative adversarial network transfer, the features of the source and target domains learned by the generator are sent to the discriminator, and the discriminator judges the source of the features and feeds the results back to the generator until it is impossible to indistinguishable. Zhang et al. proposed an adversarial representation learning framework to build a robust patient-independent seizure detection algorithm that removes the effects of inter-patient noise [63]. The framework utilized deep generative model and convolutional discriminative model to decompose EEG signals into seizure-related and patient-related components. Meanwhile, attention mechanism was introduced to automatically explore the importance of each EEG channel, thus bringing fine-grained analysis to epilepsy diagnosis. Zhu et al. proposed an unsupervised adversarial domain adaptation approach, in which features from different subjects were mapped to a subject-invariant space through autoencoder and adversarial network, and the cross-subject seizure detection model was learned in the subject-invariant feature space [68].

4   Conclusion and Future Direction

The application of transfer learning in the field of seizure detection enables it to deal with more complex real-world scenarios. This paper reviews the application of commonly used transfer learning methods for seizure detection: feature-representation-based transfer, parameter-based transfer, and deep transfer networks. According to the above survey, combined with the transfer learning methods and the actual application scenarios of seizure detection, this paper looks forward to the future research directions in this field from the following aspects.

Feature Transferability. In seizure detection based on transfer learning, it can also be classified into two categories: manual features combined with shallow transfer learning methods and deep transfer learning. In the application of shallow transfer learning, the extraction of handcrafted features relies on the experience of previous tasks. These features may perform well for specific tasks, but whether they are suitable for transfer is unknown. The quality of the features seriously affects the final classification performance. Therefore, it is a potential research direction in this field to introduce the idea of transfer learning into the classical feature selection method and derive new algorithms based on the distinguishing ability and transferability of features as common evaluation objectives.

Negative Transfer. In cross-subject seizure detection studies, the source domain usually consists of data from multiple individuals. Assuming that there is little similarity between the source subject and the target subject, negative transfer is easy to occur. Negative transfer means that the knowledge learned in the source domain has a negative impact on target task due to data differences or unreasonable transfer methods. To avoid negative transfer and reduce the requirement of computing power, it is necessary to correctly analyze and evaluate the similarity of the source and the target subjects to ensure the correct selection of data sources before establishing an effective model. In most cases, the lack of available labels in the target domain results in a relatively brute-force selection of the source domain. Therefore, it is necessary to develop a strong adaptive source selection method.

Deep Transfer Networks. Compared with traditional models, deep networks have significant advantages in processing multi-channel data. The application of deep transfer learning in seizure detection focuses on fine-tuning. Although the data requirements are reduced, the problem of distribution differences has not been effectively solved. Therefore, deep network adaptation and generative adversarial network transfer are expected to further improve the robustness of deep seizure detection models. Adaptive weighting of multi-source domains is also worth considering in deep transfer learning.

In a more demanding case where target domain is completely inaccessible, how to use only the feature information of the source domain for model learning. Domain generalization [30] combines data from multi-source domains to learn a shared model, or integrates independent models trained in each domain, which provides a new solution for building models with good generalization performance when the target-domain data cannot be accessed.

ETHICAL STANDARDS. This study has been approved by the Second Affiliated Hospital of Zhejiang University and registered in Chinese Clinical Trail Registry (ChiCTR1900020726). All patients gave their informed consent prior to their inclusion in the study.

References

1. Abdelhameed, A.M., Bayoumi, M.: Semi-supervised deep learning system for epileptic seizures onset prediction. In: 2018 17th IEEE international conference on machine learning and applications (ICMLA). pp. 1186–1191. IEEE (2018)
2. Beghi, E.: The epidemiology of epilepsy. Neuroepidemiology **54**(2), 185–191 (2020)
3. Caliskan, A., Rencuzogullari, S.: Transfer learning to detect neonatal seizure from electroencephalography signals. Neural Computing and Applications **33**(18), 12087–12101 (2021). https://doi.org/10.1007/s00521-021-05878-y
4. Cao, J., Hu, D., Wang, Y., Wang, J., Lei, B.: Epileptic classification with deep transfer learning based feature fusion algorithm. IEEE Trans. Cogn. Dev. Syst. **14**, 684–695 (2021)
5. Chandaka, S., Chatterjee, A., Munshi, S.: Cross-correlation aided support vector machine classifier for classification of EEG signals. Expert Syst. Appl. **36**(2), 1329–1336 (2009)
6. Daoud, H., Bayoumi, M.A.: Efficient epileptic seizure prediction based on deep learning. IEEE Trans. Biomed. Circuits Syst. **13**(5), 804–813 (2019)

7. De Cooman, T., Kjær, T.W., Van Huffel, S., Sorensen, H.B.: Adaptive heart rate-based epileptic seizure detection using real-time user feedback. Physiol. Meas. **39**(1), 014005 (2018)
8. De Cooman, T., et al.: Personalizing heart rate-based seizure detection using supervised SVM transfer learning. Front. Neurol. **11**, 145 (2020)
9. De Cooman, T., Varon, C., Van de Vel, A., Ceulemans, B., Lagae, L., Van Huffel, S.: Semi-supervised one-class transfer learning for heart rate based epileptic seizure detection. In: 2017 Computing in Cardiology (CinC). pp. 1–4. IEEE (2017)
10. Deng, Z., Xu, P., Xie, L., Choi, K.S., Wang, S.: Transductive joint-knowledge-transfer tsk fs for recognition of epileptic EEG signals. IEEE Trans. Neural Syst. Rehabil. Eng. **26**(8), 1481–1494 (2018)
11. Desai, S.A., Tcheng, T., Morrell, M.: Transfer-learning for differentiating epileptic patients who respond to treatment based on chronic ambulatory ecog data. In: 2019 9th International IEEE/EMBS Conference on Neural Engineering (NER). pp. 1–4. IEEE (2019)
12. Dhulekar, N., Nambirajan, S., Oztan, B., Yener, B.: Seizure Prediction by Graph Mining, Transfer Learning, and Transformation Learning. In: Perner, P. (ed.) MLDM 2015. LNCS (LNAI), vol. 9166, pp. 32–52. Springer, Cham (2015). https://doi.org/10.1007/978-3-319-21024-7_3
13. Dong, A., Li, Z., Zheng, Q.: Transferred subspace learning based on non-negative matrix factorization for EEG signal classification. Front. Neuro. **15**, 647393 (2021)
14. Gao, Y., Gao, B., Chen, Q., Liu, J., Zhang, Y.: Deep convolutional neural network-based epileptic electroencephalogram (EEG) signal classification. Front. Neurol. **11**, 375 (2020)
15. Goodfellow, I., Pouget-Abadie, J., Mirza, M., Xu, B., Warde-Farley, D., Ozair, S., Courville, A., Bengio, Y.: Generative adversarial nets. Advances in neural information processing systems 27 (2014)
16. Gretton, A., Borgwardt, K., Rasch, M., Schölkopf, B., Smola, A.: A kernel method for the two-sample-problem. Advances in neural information processing systems 19 (2006)
17. Gu, Xiaoqing, Shen, Zongxuan, Qu, Jia, Ni, Tongguang: Cross-domain EEG signal classification via geometric preserving transfer discriminative dictionary learning. Multimed. Tools and Appl. **81**, 41733–41750 (2021). https://doi.org/10.1007/s11042-021-11244-w
18. Hang, W., Feng, W., Du, R., Liang, S., Chen, Y., Wang, Q., Liu, X.: Cross-subject EEG signal recognition using deep domain adaptation network. IEEE Access **7**, 128273–128282 (2019)
19. Hu, D., Cao, J., Lai, X., Liu, J., Wang, S., Ding, Y.: Epileptic signal classification based on synthetic minority oversampling and blending algorithm. IEEE Trans. Cogn. Dev. Syst. **13**(2), 368–382 (2020)
20. Hu, D., Cao, J., Lai, X., Wang, Y., Wang, S., Ding, Y.: Epileptic state classification by fusing hand-crafted and deep learning EEG features. IEEE Trans. Circuits Syst. II: Express Briefs **68**(4), 1542–1546 (2020)
21. Ilakiyaselvan, N., Khan, A.N., Shahina, A.: Deep learning approach to detect seizure using reconstructed phase space images. J. Biomed. Res. **34**(3), 240 (2020)
22. Jiang, X., Xu, K., Chen, W.: Transfer component analysis to reduce individual difference of eeg characteristics for automated seizure detection. In: 2019 IEEE Biomedical Circuits and Systems Conference (BioCAS). pp. 1–4. IEEE (2019)
23. Jiang, Y., Deng, Z., Chung, F.L., Wang, G., Qian, P., Choi, K.S., Wang, S.: Recognition of epileptic EEG signals using a novel multiview tsk fuzzy system. IEEE Trans. Fuzzy Syst. **25**(1), 3–20 (2016)

24. Jiang, Y., et al.: Smart diagnosis: a multiple-source transfer tsk fuzzy system for eeg seizure identification. ACM Trans. Multimed. Comput. Commun. and Appl. (TOMM) **16**(2s), 1–21 (2020)

25. Jiang, Y., et al.: Seizure classification from EEG signals using transfer learning, semi-supervised learning and tsk fuzzy system. IEEE Trans. Neural Syst. Rehabil. Eng. **25**(12), 2270–2284 (2017)

26. Jiang, Z., Chung, F.L., Wang, S.: Recognition of multiclass epileptic EEG signals based on knowledge and label space inductive transfer. IEEE Trans. Neural Syst. Rehabil. Eng. **27**(4), 630–642 (2019)

27. Karácsony, T., Loesch-Biffar, A.M., Vollmar, C., Noachtar, S., Cunha, J.P.S.: A deep learning architecture for epileptic seizure classification based on object and action recognition. In: ICASSP 2020–2020 IEEE International Conference on Acoustics, Speech and Signal Processing (ICASSP). pp. 4117–4121. IEEE (2020)

28. Kharbouch, A., Shoeb, A., Guttag, J., Cash, S.S.: An algorithm for seizure onset detection using intracranial EEG. Epilepsy & Behavior **22**, S29–S35 (2011)

29. Li, Y., Liu, Y., Cui, W.G., Guo, Y.Z., Huang, H., Hu, Z.Y.: Epileptic seizure detection in EEG signals using a unified temporal-spectral squeeze-and-excitation network. IEEE Trans. Neural Syst. Rehabil. Eng. **28**(4), 782–794 (2020)

30. Muandet, K., Balduzzi, D., Schölkopf, B.: Domain generalization via invariant feature representation. In: International Conference on Machine Learning. pp. 10–18. PMLR (2013)

31. Nogay, H.S., Adeli, H.: Detection of epileptic seizure using pretrained deep convolutional neural network and transfer learning. European Neurol. **83**(6), 602–614 (2020)

32. O'Shea, A., et al.: Deep learning for EEG seizure detection in preterm infants. Int. J. Neural Syst. **31**(08), 2150008 (2021)

33. Ouichka, O., Echtioui, A., Hamam, H.: Deep learning models for predicting epileptic seizures using ieeg signals. Electronics **11**(4), 605 (2022)

34. Page, A., Shea, C., Mohsenin, T.: Wearable seizure detection using convolutional neural networks with transfer learning. In: 2016 IEEE International Symposium on Circuits and Systems (ISCAS). pp. 1086–1089. IEEE (2016)

35. Pan, S.J., Tsang, I.W., Kwok, J.T., Yang, Q.: Domain adaptation via transfer component analysis. IEEE Trans. Neural Netw. **22**(2), 199–210 (2010)

36. Pérez-García, F., Scott, C., Sparks, R., Diehl, B., Ourselin, S.: Correction to: Transfer Learning of Deep Spatiotemporal Networks to Model Arbitrarily Long Videos of Seizures. In: de Bruijne, M., et al. (eds.) MICCAI 2021. LNCS, vol. 12905, pp. C1–C1. Springer, Cham (2021). https://doi.org/10.1007/978-3-030-87240-3_80

37. Prabhakar, S.K., Lee, S.W.: ENIC: Ensemble and nature inclined classification with sparse depiction based deep and transfer learning for biosignal classification. Appl. Soft Comput. **117**, 108416 (2022)

38. Prabhakar, S.K., Lee, S.W.: Improved sparse representation based robust hybrid feature extraction models with transfer and deep learning for eeg classification. Expert Syst. Appl. **198**, 116783 (2022)

39. Prathaban, B.P., Balasubramanian, R., Kalpana, R.: A wearable foreseiz headband for forecasting real-time epileptic seizures. IEEE Sensors J. **21**(23), 26892–26901 (2021)

40. Prathaban, B.P., Balasubramanian, R., Kalpana, R.: Foreseiz: An iomt based headband for real-time epileptic seizure forecasting. Expert Syst. Appl. **188**, 116083 (2022)

41. Raghu, S., Sriraam, N., Temel, Y., Rao, S.V., Kubben, P.L.: EEG based multi-class seizure type classification using convolutional neural network and transfer learning. Neural Netw. **124**, 202–212 (2020)
42. Ramadan, R.A., Vasilakos, A.V.: Brain computer interface: control signals review. Neurocomputing **223**, 26–44 (2017)
43. Rasheed, K., Qadir, J., O'Brien, T.J., Kuhlmann, L., Razi, A.: A generative model to synthesize EEG data for epileptic seizure prediction. IEEE Trans. Neural Syst. Rehabil. Eng. **29**, 2322–2332 (2021)
44. Roy, A.D., Islam, M.M.: Detection of epileptic seizures from wavelet scalogram of eeg signal using transfer learning with alexnet convolutional neural network. In: 2020 23rd International Conference on Computer and Information Technology (ICCIT). pp. 1–5. IEEE (2020)
45. Saeed, A., Grangier, D., Pietquin, O., Zeghidour, N.: Learning from heterogeneous eeg signals with differentiable channel reordering. In: ICASSP 2021–2021 IEEE International Conference on Acoustics, Speech and Signal Processing (ICASSP). pp. 1255–1259. IEEE (2021)
46. Sarvi Zargar, B., Karami Mollaei, M.R., Ebrahimi, F., Rasekhi, J.: Generalizable epileptic seizures prediction based on deep transfer learning. Cogn Neurodyn. **17**, 1–13 (2022)
47. Singh, R., Ahmed, T., Singh, A.K., Chanak, P., Singh, S.K.: Seizsclas: an efficient and secure internet-of-things-based EEG classifier. IEEE Internet of Things J. **8**(8), 6214–6221 (2020)
48. Song, Z., Deng, B., Wang, J., Yi, G., Yue, W.: Epileptic seizure detection using brain-rhythmic recurrence biomarkers and onasnet-based transfer learning. IEEE Trans. Neural Syst. Rehabil. Eng. **30**, 979–989 (2022)
49. Subha, D.P., Joseph, P.K., Acharya U, R., Lim, C.M., et al.: EEG signal analysis: a survey. J. Med. Syst. **34**(2), 195–212 (2010)
50. Tan, C., Sun, F., Kong, T., Zhang, W., Yang, C., Liu, C.: A Survey on Deep Transfer Learning. In: Kůrková, V., Manolopoulos, Y., Hammer, B., Iliadis, L., Maglogiannis, I. (eds.) ICANN 2018. LNCS, vol. 11141, pp. 270–279. Springer, Cham (2018). https://doi.org/10.1007/978-3-030-01424-7_27
51. Tang, Y., Li, W., Tao, L., Li, J., Long, T., Li, Y., Chen, D., Hu, S.: Machine learning-derived multimodal neuroimaging of presurgical target area to predict individual's seizure outcomes after epilepsy surgery. Front. Cell and Dev. Biol. **9**, 3859 (2021)
52. Thomas, E., Greene, B., Lightbody, G., Marnane, W., Boylan, G.: Seizure detection in neonates: Improved classification through supervised adaptation. In: 2008 30th Annual International Conference of the IEEE Engineering in Medicine and Biology Society. pp. 903–906. IEEE (2008)
53. Thuwajit, P., et al.: EEGWaveNet: Multiscale CNN-based spatiotemporal feature extraction for EEG seizure detection. IEEE Trans. Ind. Inf. **18**(8), 5547–5557 (2021)
54. Trinka, E., Cock, H., Hesdorffer, D., Rossetti, A.O., Scheffer, I.E., Shinnar, S., Shorvon, S., Lowenstein, D.H.: A definition and classification of status epilepticus-report of the ilae task force on classification of status epilepticus. Epilepsia **56**(10), 1515–1523 (2015)
55. Wan, Z., Yang, R., Huang, M., Zeng, N., Liu, X.: A review on transfer learning in EEG signal analysis. Neurocomputing **421**, 1–14 (2021)
56. Wu, D., Yang, J., Sawan, M.: Bridging the gap between patient-specific and patient-independent seizure prediction via knowledge distillation. J. Neural Eng. **19**, 036035 (2022)

57. Wu, Z., Zhou, S.: Epileptic seizure detection by transfer learning considering ictal-like non-ictal signals in electroencephalogram. In: 2021 International Conference on e-Health and Bioengineering (EHB). pp. 1–4. IEEE (2021)

58. Xia, K., Ni, T., Yin, H., Chen, B.: Cross-domain classification model with knowledge utilization maximization for recognition of epileptic EEG signals. IEEE/ACM Trans. Comput. Biol. Bioinf. **18**(1), 53–61 (2020)

59. Xie, L., Deng, Z., Xu, P., Choi, K.S., Wang, S.: Generalized hidden-mapping transductive transfer learning for recognition of epileptic electroencephalogram signals. IEEE Trans. Cyber. **49**(6), 2200–2214 (2018)

60. Yang, C., Deng, Z., Choi, K.S., Jiang, Y., Wang, S.: Transductive domain adaptive learning for epileptic electroencephalogram recognition. Artif. Intell. Med. **62**(3), 165–177 (2014)

61. Yu, Z., Albera, L., Le Bouquin Jeannes, R., Kachenoura, A., Karfoul, A., Yang, C., Shu, H.: Epileptic seizure prediction using deep neural networks via transfer learning and multi-feature fusion. Int. J. Neural Syst. **32**, 2250032–2250032 (2022)

62. Zhang, B., Wang, W., Xiao, Y., Xiao, S., Chen, S., Chen, S., Xu, G., Che, W.: Cross-subject seizure detection in EEGS using deep transfer learning. Comput. Math Methods. Med. **2020** (2020)

63. Zhang, X., Yao, L., Dong, M., Liu, Z., Zhang, Y., Li, Y.: Adversarial representation learning for robust patient-independent epileptic seizure detection. IEEE J. Biomed. Health Inf. **24**(10), 2852–2859 (2020)

64. Zhang, Y., Dong, J., Zhu, J., Wu, C.: Common and special knowledge-driven tsk fuzzy system and its modeling and application for epileptic eeg signals recognition. IEEE Access **7**, 127600–127614 (2019)

65. Zhang, Y., Zhou, Z., Bai, H., Liu, W., Wang, L.: Seizure classification from EEG signals using an online selective transfer tsk fuzzy classifier with joint distribution adaption and manifold regularization. Front. Neuro. **14**, 496 (2020)

66. Zhang, Y., et al.: Epilepsy signal recognition using online transfer tsk fuzzy classifier underlying classification error and joint distribution consensus regularization. IEEE/ACM Trans. Comput. Bio. Bioinf. **18**(5), 1667–1678 (2020)

67. Zheng, Z., Dong, X., Yao, J., Zhou, L., Ding, Y., Chen, A.: Identification of epileptic eeg signals through tsk transfer learning fuzzy system. Front. Neurosci. **15**, 738268 (2021)

68. Zhu, B., Shoaran, M.: Unsupervised domain adaptation for cross-subject few-shot neurological symptom detection. In: 2021 10th International IEEE/EMBS Conference on Neural Engineering (NER). pp. 181–184. IEEE (2021)

69. Zhu, Y., Saqib, M., Ham, E., Belhareth, S., Hoffman, R., Wang, M.D.: Mitigating patient-to-patient variation in eeg seizure detection using meta transfer learning. In: 2020 IEEE 20th International Conference on Bioinformatics and Bioengineering (BIBE). pp. 548–555. IEEE (2020)

70. Zhuang, F., Qi, Z., Duan, K., Xi, D., Zhu, Y., Zhu, H., Xiong, H., He, Q.: A comprehensive survey on transfer learning. Proceedings of the IEEE **109**(1), 43–76 (2020)

Cognitive Computing and System Analysis of Seven Times Pass Method Applications and Its Significance

Xueqiu Wu[1] and Xiaohui Zou[2(✉)] (iD)

[1] Shuangyashan Radio and Television University, Shuangyashan 155100, Heilongjiang, China
[2] Searle Research Center, No. 100 Renshan Road, Hengqin 519000, Guangdong, China
zouxiaohui@puk.org.cn

Abstract. The aim is to further explore the cognitive computing and system design of the Seven Times Pass method application by combining the learning method with the eight-person group management method through years of accumulated teaching practice and management practice. The method is: first, systematically review and summarize multiple series of Seven Times Pass Method application, and then extract the cognitive content of interpersonal communication and further development, finally, focus on the characteristics of the era of human-computer interaction and even collaboration, which are characterized by both personalization and standardization. The results not only summarize and review the various cognitive methods of the seven-pass learning method and the eight-person group training management method, but also further explore the cognitive computing and system design optimization methods of the seven-pass application, and further enrich the teaching and learning practice. The way in which the management practice is active. The significance is that this not only has a positive effect on the teaching, review and examination quality of national education system at all levels, but also has a significant effect on the quality education, vocational and lifelong education have positive and ongoing effects.

Keywords: Learning Method · Interpersonal Communication · Human-Computer Interaction, Cognitive Computing · System Design · National Education System · Training Management Method

1 Introduction

The aim is to further explore the cognitive computing and system design of the Seven Times Pass Method application by combining the learning method with the Eight-Person Group management method through years of accumulated teaching practice and management practice.

The specific discussion covers the following aspects: the method part includes, overview of Seven Times Pass method, classification and clustering principles of knowledge modules, unique collaborative structure; the results part is outstanding, the synergistic structure possessed by Seven Times Pass cognitive computing intelligent system; the conclusion part further summary and discussion, "Seven Times Pass" and "Eight-Person Group" teaching mode structure [1–25].

2   Method

The method is: first, systematically review and summarize multiple series of Seven Times Pass method application, and then extract the cognitive content of interpersonal communication and further development, finally, focus on the characteristics of the era of human-computer interaction and even collaboration, which are characterized by both personalization and standardization.

2.1   Overview of Seven Times Pass Method

Seven Times Passes is a multi-faceted, complex and modular teaching and learning model in which everyone, even man-machine are synergistically constructed to guide smart education with socialized systems engineering. It could not be understood like that it is just superimposed in quantity. It implements or carries out dual-subject teaching by taking students as the main body with a desire to learn independently, teachers as the main body that guides students, and AI-assisted teaching with the Internet and campus network. On the basis of certain teaching method guidance, learning method combination, and algorithm analysis, it strives to build a teaching mode that optimizes students' learning literacy through smart education.

Table 1. Seven Times Pass methods expressions both in Chinese and English.

Typical	Seven Times Pass method expression both in Chinese and English in Rongzhixue
1st-level	听、说、读、写、译、述、评
	Listening, Speaking, Reading, Writing, Translating, Narrating, Commenting
2nd-level	懂、会、熟、巧、用、分、合
	Understand&Knowing, Able, Familiar, Skillful, Using, Divide, Combine
3rd-level	图、纲、线、块、基、点、题
	Textbook & Knowledge Graph, Brief & Detailed Outlines, Main & Auxiliary Lines, Large & Small Blocks, Basic Concepts & Principles & Methods, Key Points & Difficulties& Blind Spots, Examples & Exercises & Test Questions
4th-level	晨、上、中、下、晚、习、夜
	Morning, First Half Day, Noon, Second Half Day, Evening, Self-Study, Night
5th-level	床、凳、桌、窗、台、廊、院
	Bed, Stool, Desk&Table, Window, Balcony, Corridor, Courtyard

From Table 1, it can be seen that the Chinese characters are very concise, but the corresponding English expressions are very direct. The two languages and their cognitive or thinking expressions have their own advantages and disadvantages. Therefore, they can complement each other.

It focuses on the student-centered for Chinese and foreign languages study that is first pass the text seven times: Listening, Speaking, Reading, Writing, Translating, Narrating,

Commenting; And then do it in terms of skills seven times: Understand & Knowing, Able, Familiar, Skillful, Using, Divide, Combine; For the processing of subject knowledge modules, emphasis is placed on seven times posses like going through: textbook and knowledge graph, brief and detailed outlines, main and auxiliary lines, large and small blocks, basic concepts and principles and methods, key points and difficulties and blind spots, examples and exercises and test questions; A day during school is distribution seven times through: morning, first half day, noon, second half day, evening, self-study, night; At home (indoor, outdoor or community courtyard and nearby parks, etc.) or school (classroom, study room, library, dormitory, campus) space distribution seven times through: bed, stool, desk&table, window, balcony, corridor, courtyard; The teacher teaching process is completed seven times: guiding direction, path guidance, learning, thinking, practicing, testing, evaluating and other various methods. Combining the unique theoretical internal logic structure characteristics of each discipline, give full play to the quality of students' own learning initiative, take thinking training as the center, stimulate learning motivation as the premise, based on the knowledge module structure, combine organic modern educational information technology, cultivate students self-learning and group learning and hands-on ability.

Consolidate the foundation for large-scale production of knowledge-based economy and realize the World Declaration on Education for All: Education is the basic right of all people and all peoples in the world; education helps to ensure a safer, healthier, more prosperous, and a better protected world contributes to social, economic and cultural progress, tolerance and international cooperation. Education is the key to personal and social progress; basic education must be universal and its quality improved. The World Declaration on Higher Education: In an economy characterized by the emergence or transformation of new modes of production based on knowledge and its applications or information processing, … Rapid breakthroughs in new information and communication technologies, Fundamentally changing the way knowledge is developed, acquired and transmitted…. the link between education and the field of work or society should be built on a completely new basis.

As well as what Chinese leader Jinping Xi expected in his congratulatory letter to the International Conference on Educational Informatization: Building a learning society in which everyone can learn, can learn everywhere, and can learn from time to time, and cultivate a large number of innovative talents is a common challenge or major issue for all mankind.

2.2 Classification and Clustering Principles of Knowledge Modules

When we use the word knowledge module to express, it means *the existence of classification and clustering*. There is a Chinese saying, *Things are gathered by like, and people are divided by groups*. In smart education information technology, *Classification*, for a classifier, you usually need to tell it *"this thing is classified into such and such"* as some examples, ideally, a classifier will *"learn"* from the training set it gets, so as to have the ability to perform analysis on unknown data. The ability to classify, *the process of providing training data is often called supervised learning*. *Clustering*, simply put, is to group similar things into a group. When clustering, we don't care what a certain class is. The goal we need to achieve is to group similar things together. Therefore, *a clustering algorithm* usually only needs to know *how to calculate the similarity to start cooperating*, so clustering usually does not need to use training data for learning, which is called *unsupervised learning in Machine Learning*.

For classification, when classifying a data set, we know how many types this data set has, which is the same as the concept of species in ordinary logic. For clustering, when operating on a data set, we do not know how many classes the data set contains. What we need to do is to summarize similar data in the data set, which is similar to the concept of genus in ordinary logic that is similar. Through their distance similarity measure, density similarity measure, connectivity similarity measure, and concept similarity measure, four similarity comparison methods, such as what people often say "smells congruent", are clustered into n types or sets, which is clustering.

The seven-pass classification and clustering principles of knowledge modules are more complex and changeable. While the classification is realized through the direct or indirect formalized sequence-position logical architecture system of eight forms: character&word, formula, diagram&graph, table, sound, image, three-dimensional, and living body. And also through the extraction of multiple complex structures to form a single set, a hierarchical set, and a classification set from the heterogeneous set &miscellaneous set to achieve personalized and collaborative clustering to adapt to data collection and intelligent analysis under a variety of algorithms. There are two main principles of cognitive computing and system analysis formed by it:

(1) Classification must clearly know the information of each category in advance, and can judge that all items to be classified have a corresponding category.
(2) Clustering must not only satisfy the most typical K-means algorithm, but also satisfy various algorithms such as partitioning, hierarchical method, density-based, grid-based, and a certain model-based algorithm. Through the establishment of linked correspondence, various algorithms can automatically or semi-automatically match and process feedback when collecting and analyzing massive data information.

2.3   Unique Collaborative Structure

In 2019, Stanford University established the "Human-centered AI" Institute, which aims to promote the rapid and reasonable development of artificial intelligence under the correct guidance of human beings, and realize human-machine conjugation and symbiosis. It can be seen that human-machine collaboration is the focus of the future development of artificial intelligence. Based on the logic of artificial intelligence "machines" and the synergy of "human" consciousness, it promotes the continuous development of educational informatization towards a high-end form, that is, human-machine collaborative intelligent education, which promotes innovation and change in the whole process of education. It can be seen that human-computer collaboration in the future artificial intelligence era, as a new form of human-computer interaction, will be the object of scientific philosophy research. Human-machine collaboration will definitely reshape the social form, and make human society evolve from the current violent, dominant and exchange-based social form to a collaborative social form. Existing system structures such as technology and social values will inevitably undergo deconstruction and reconstruction. If we observe and analyze from the perspective of the three dimensions of ontology, epistemology and axiology of human-machine collaboration:

First, the collaboration between human and computer is realized based on the input and output of information, its essence is the calculation process of jointly operating and processing information. The production of rules is effectively coordinated to form a complex intelligent system. This is the ontological basis of human-machine collaboration.

Second, in philosophical epistemology, human is always the subject, and others are objects. The free and conscious activity of understanding the world is a unique feature of human beings, human has always been the subject. The artificial intelligence of the computer is an artificial thing of human beings, and it is the infusion of intelligence by human beings. The generation and acquisition of knowledge are completed by the joint reasoning and calculation of the human-machine dual intelligence. The subject, the "machine" that was the object will also be transformed into the subject, and the subjectivization of the object will occur, that is, intersubjectivity. According to the dissipative structure theory, material and energy are constantly exchanged between humans and machines, and the state of the human-machine collaborative education system moves away from or approaches the threshold as the parameters change to form a new state, which will remain in order or place.

Third, the value of human-machine collaboration (HMC) is that the HMC and mutual promotion form a positive feedback relationship: on the one hand, the HMC system can replace part of human work, saving human time and energy; Invest more energy in innovating "machine" intelligence, externalizing human intelligence into "machine", thereby enhancing intelligence, and freeing human beings to devote energy to other more creative work. At the same time, the individual value of the HMC system will accelerate the generation of social and cultural values, which is one of the ways to externalize knowledge. As a tool, "machine" can not only reduce the cognitive load of people and enhance people's learning ability, but also can connect individuals to carry out cross-disciplinary and cross-field collaboration, thereby promoting the coordinated development of human society.

3   Result

The results not only summarize and review the various cognitive methods of the seven-pass learning method and the eight-person group training management method, but also further explore the cognitive computing and system design optimization methods of the seven-pass application, and further enrich the teaching and learning practice. The way in which the management practice is active.

The Human-Machine Collaboration (HMC) The current research on HMC in the field of smart education mainly focuses on the following three aspects:

One is how to establish an intelligent architecture for HMC. Existing scholars have discussed from multiple perspectives, such as HMC intelligence hierarchical structure model and intelligence classification, such as hard intelligence, soft intelligence, smart intelligence and cognitive intelligence, emotional intelligence, hobby intelligence and innovative intelligence.

The second is how to realize the transformation of the role of teachers under HMC. Existing scholars have thought and analyzed from multiple perspectives, such as the collaborative relationship between artificial intelligence and teachers, the formation of "new subject teachers" by human teachers and intelligent tutors, and the new "double-teacher classroom" supported by artificial intelligence educational robots.

The third is how to apply it in all-category education and whole-process teaching. Existing scholars put forward the precise teaching mode of HMC, pointing out that the link of "diagnosis-feedback-intervention-reflection" of HMC runs through the entire teaching process.

The Seven Times Pass (STP)   The Seven Times Pass (STP) cognitive computing intelligent system has a unique collaborative structure, that is, through the behavioral coordination of group members in the form of "group of eight" between individuals, the three intelligences and double integration between human and machines, ssequential coordination, multi-machine coordination between machines. It constitutes individuals entity space collaborative autonomous learning model, a human-machine hybrid space teaching and learning model, and a machine-machine virtual space deep learning model. These three modes interrelate and closely couple the elements of teaching and learning through the embodied interaction between teachers and students and the teaching situation. At the same time, they obtain the physiological characteristics of teachers and students through machines, and realize human-computer dynamic interaction based on cognitive awareness teaching. For example, applying high-fidelity and low-latency interactive technologies, through visual, auditory, tactile and other multi-sensory interactions, digitize knowledge that is difficult to obtain in physical space and abstract knowledge, and transform it into close-range and intuitive in virtual space form, presented to the learner. Or teachers' guidance, introduction, learning, thinking, practice, examination, evaluation and artificial intelligence to analyze and clarify the learning status of each individual learner, use the subject knowledge to pass seven times; it accurately pushes personalized learning resources, realizes human-machine collaboration, provides an interactive learning experience for learners' perception and machine structured reflection, enhances the learner's sense of presence promotes the deep learning and transfer application of knowledge, and realizes the functions of seven times of skills: understanding, knowing, familiar, skillful, using, dividing, integrating. In addition, the human-machine collaborative virtual teaching model breaks through the constraints of traditional time and space. It can pass seven times through time distribution and space distribution, to realize that everyone can learn, can learn everywhere, and can learn from time to time, which plays an important role in promoting the development of lifelong education and a learning society.

The synergistic structure possessed by Seven Times Pass cognitive computing intelligent system will reshape the teacher-student relationship in the traditional teaching model. Under the influence of the double-teacher directly both by the human teacher and indirectly by the AI intelligent tutor in the virtual space, learners can obtain direct and indirect knowledge experiences. This human-machine collaborative teaching mode makes the human brain, body, environment and technology of teachers and students inseparable, and effectively integrates learner intelligence, machine intelligence, and teacher intelligence, forming a new form of education and teaching in which humans and machines are the main body.

Verified by Cognitive Computing (Known & Unknown)

$$7^x + 8^y \leq 15^z \tag{1}$$

As can be seen from Eq. (1), there is a big difference between $7 + 8 \leq 15$ and $7^x + 8^y \leq 15^z$. If it is calculated only as a number, then it is a very simple arithmetic problem. Either a natural person or a computer can be instantly verified by cognitive computing; however, if it is the Seven-Times Pass described in this paper, with the help of the structured expression in Table 1, even if the social person is to be divided into several groups of Eight Person Group, how many knowledge modules or how many drawable knowledge graphs are there in Table 1? Humans are no more than computers. Why? Where exactly is the gap between these two cognitive computing and systems? It is worth our further study. Therefore, this ternary linear equation is not easy to solve because of three unknowns! The unknowns (x, y, z) in the formula are added, multiplied, squared, exponential or logarithmic, how to calculate?

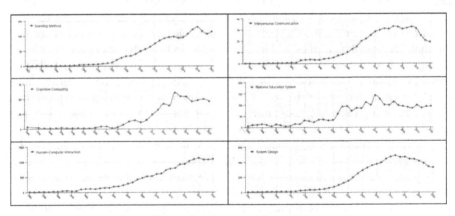

Fig. 1. Comparison of six trends.

As can be seen from Fig. 1, although there are differences in the six aspects of research or paper publication, learning methods, interpersonal communication, cognitive computing, national education system, human-computer interaction, and system design, the general trend is amazing and ground similar. From this we get enlightenment, the key lies in the three aspects of learning methods, cognitive computing, and human-computer interaction. In short, this is an area where we (especially the Seven-Times-Pass method) can play a big role.

4   Conclusion

The significance is that this not only has a positive effect on the teaching, review and examination quality of national education system at all levels, but also has a significant effect on the quality education, vocational and lifelong education. Have positive and ongoing effects.

The Intelligent Evaluation Direction of Promoting Education by Evaluation
"Seven times pass" and "Eight people group" are an interactive teaching mode based on the network environment of students and students, teachers and students, students and teachers, all of them with the help of machines. The assessment consists of three teaching modules, that are interconnected, motivated and complement each other. Among them, teacher guidance is an important part of the teaching process. Through guided learning, students can master a certain knowledge module from point to line, from line to surface, from surface to body or system, which is a gradual and progressive learning process. It can be seen that the driving force of this teaching model is the combined force of students' autonomous learning ability and teachers' guiding ability.

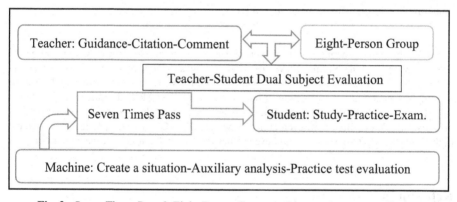

Fig. 2. Seven-Times-Pass & Eight-Person Group teaching mode structure diagram

It can be seen from Fig. 2 that Teacher's Guidance-Citation-Comment and Machine's Create a situation-Auxiliary analysis-Practice test evaluation both serve Student's Study-Practice-Exam. It is worth emphasizing: Seven-Times-Pass and Eight-Person Group (student's own combination of 2, 4, 6, 8 and their series) and the Teacher-Student Dual Subject-Evaluation.

In this teaching mode, it is guided by constructive theory, introduces modern educational technology means, realizes human-machine collaborative interaction. It emphasizes the main role of students and the leading role of teachers. The process of perception of problems $\rightarrow$ abstraction $\rightarrow$ fixed situation is the process of students' active cognition. In this process, teachers are guides and facilitators to help students realize knowledge internalization.

The "machine" in this teaching mode is the abbreviation of modern educational information technology. As a teaching medium, it is not only a tool to assist teaching, but also a cognitive tool for students to learn independently. It participates in the teaching process and becomes an integral part of it with a different functional focus. "Machine" is used by teachers to create situations in the teaching process and as a cognitive tool for students' active learning and collaborative exploration.

Through human-machine interaction, it provides favorable conditions for students' perception and cognition. Through self-regulation and control feedback, students' creative thinking is stimulated, and students are encouraged to think more actively, explore actively, and discover actively, and promote students' active development.

The "interaction" in this teaching mode refers to the role and communication of "students-students, teachers-students, students-teachers" in the whole process of teaching. It includes the communication between teachers and students, students and students, teachers and students, and all of them with the "machine". Teaching information is transmitted in multiple directions and acts on each other.

The main body or system of this teaching model is people, which determines the combination of various modules in the teaching model. Compared with people, educational technology plays an auxiliary role for people, and its role in the model depends on its role in teaching auxiliary level. The main body of teaching activities is "students", which are the bearers, initiators, and defenders of learning activities. Both "teacher" and "machine" play the role of assisting students.

Acknowledgments. Thanks to Beijing Normal University Future Education Advanced Innovation Center and Beijing Normal University Modern Educational Technology Research Institute "Basic Education Leap-forward Development Innovation Experimental Research" 2022 Annual Meeting for awarding the application of our Seven Times Pass Method to the first senior high school in Shuangyashan City, Heilongjiang Province the prize of the "Human-Computer Interaction Teaching Design Based on Seven Times Passes and Eight Groups", the practical achievement of the Eight-Person Group theory and Seven Times Pass method guiding the teaching reform!

References

1. Agresti, A., Kateri, M.: Classification and clustering. In: Foundations of Statistics for Data Scientists (2021)
2. Mavridis, C.N., Baras, J.S.: Online deterministic annealing for classification and clustering. IEEE Trans. Neural Netw. Learn. Syst. (2022)
3. Mohammad, W., et al.: Diagnosis of breast cancer pathology on the Wisconsin dataset with the help of data mining classification and clustering techniques. Appl. Bionics Biomech. **2022** (2022)
4. Gupta, S., et al.: Store-n-learn: classification and clustering with hyper- dimensional computing across flash hierarchy. ACM Trans. Embed. Comput. Syst. (TECS) **21**, 1–25 (2022)
5. Srivastava, S., et al.: ExpertNet: a symbiosis of classification and clustering. arXiv abs/2201.06344 (2022)
6. BalajiK., M., Subbulakshmi, T.: Malware analysis using classification and clustering algorithms. Int. J. Collab. **18**, 1–26 (2022)
7. Silver, D., et al.: Mastering the game of Go without human knowledge. Nature **550**, 354–359 (2017)

8. Tshitoyan, V., et al.: Unsupervised word embeddings capture latent knowledge from materials science literature. Nature **571**, 95–98 (2019)
9. Perkel, J.M.: Workflow systems turn raw data into scientific knowledge. Nature **573**, 149–150 (2019)
10. Schrittwieser, J., et al.: Mastering Atari, Go, Chess and Shogi by planning with a learned model. Nature **588**(7839), 604–609 (2020)
11. Willett, F.R., et al.: High-performance brain-to-text communication via handwriting. Nature **593**(7858), 249–254 (2021)
12. Jumper, J.M., et al.: Highly accurate protein structure prediction with AlphaFold. Nature **596**, 583–589 (2021)
13. Gotian, E.N.: Better together: collaborative spaces can inspire scientists of all ages. Nature (2021)
14. Senior, A.W., et al.: Improved protein structure prediction using potentials from deep learning. Nature **577**, 706–710 (2020)
15. Perkel, J.M.: Synchronized editing: the future of collaborative writing. Nature **580**, 154–155 (2020)
16. Kinzler, K.D., Shutts, K.: Ways to promote and foster collaborative research in your lab. Nature **560**, 673 (2018)
17. Cruz-Benito, J., et al.: A deep-learning-based proposal to aid users in quantum computing programming. In: HCI (2018)
18. DeFalco, J.A., Sinatra, A.M.: Adaptive instructional systems: the evolution of hybrid cognitive tools and tutoring systems. In: HCI (2019)
19. Fulbright, R.: Calculating cognitive augmentation - a case study. In: HCI (2019)
20. Rutkowski, L., et al.: Artificial Intelligence and Soft Computing. LNCS. Springer, Cham (2019). https://doi.org/10.1007/978-3-030-20912-4
21. Chengalur-Smith, S.: Human Centered Computing. LNCS. Springer, Cham (2018). https://doi.org/10.1007/978-3-030-15127-0
22. Zou, X.: Innovation management in educational technology. In: Proceedings of the 7th International Conference on Social Science and Higher Education (ICSSHE 2021) (2021)
23. Zou, X., et al.: Smart system studied: new approaches to natural language understanding. In: Proceedings of the 2019 International Conference on Artificial Intelligence and Computer Science (2019)
24. Zou, S., et al.: How to do knowledge module finishing. In: IFIP TC12 ICIS (2018)
25. Xu, Z., et al.: How to understand the basic graphic analysis method. In: ICCSIP (2018)

Cognitive Computing and Systems Analysis of Alumni Economic Theory and Practice

Shijie Lv[1] and Xiaohui Zou[2,3](✉) [iD]

[1] China Alumni Economic Research Institute, No. 160, Fuxingmennei Street, Beijing, China
[2] Searle Research Center, No. 100, Renshan Road, Hengqin 519000, Guangdong, China
[3] Peking University Alumni Association, Beijing 100871, China
zouxiaohui@puk.org.cn

Abstract. It aims to inspire domestic and foreign alumni to participate in the construction of the Second Centenary National Demonstration Zone of the Communist Party of China by reviewing the cognitive computing and system analysis of the latest progress in the establishment of alumni economic theory and its practice. The method is as follows: First, review the historical process that alumni economic theory and practice have gone through, and then, through cognitive computing and systematic analysis of alumni economic theory and practice, provide more specific examples for further development and expansion of its application scenarios, and finally, through the establishment of the China Alumni Economic Theory and Practice Research Institute, especially in combination with the characteristics of the times, it highlights the basic topics of cognitive computing and systematic analysis of alumni economic theory and practice. The results are beneficial for the economic practice of domestic and foreign alumni to further rise to the height of theoretical research, and to achieve sufficient interpersonal communication and high interpersonal collaboration. Its significance lies in: Chinese and foreign alumni economic talents, with the help of cognitive computing and systematic analysis of alumni economic theory and practice, not only better participate in the construction of the national demonstration zone, but also an international and domestic demonstration base for the practice of alumni economic theory in the Hengqin Guangdong-Macao Deep Cooperation Zone, and extended to at home and abroad. Taking advantage of the location advantage, the International Research Institute of Alumni Economics was established.

Keywords: Cognitive Computing · System Analysis · Alumni Economic Theory

1 Introduction

It aims to inspire domestic and foreign alumni to participate in the construction of the Second Centenary National Demonstration Zone of the Communist Party of China by reviewing the cognitive computing and system analysis of the latest progress in the establishment of alumni economic theory and its practice.

As can be seen from Fig. 1, the development trend of alumni economic theory research shows that there have been significant developments and changes in the past

F. Sun et al. (Eds.): ICCCS 2022, CCIS 1732, pp. 186–195, 2023.
https://doi.org/10.1007/978-981-99-2789-0_16

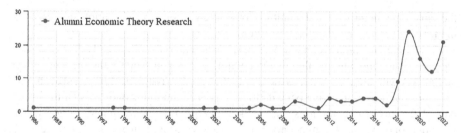

Fig. 1. The development trend of alumni economic theory research.

two or three years. Under this situation, we have not only established the China Alumni Economic Research Institute, but also are actively preparing for the International Alumni Economic Research Institute and the Outstanding Achievement Award Fund for famous alumni.

2 Method

The method is as follows: First, review the historical process that alumni economic theory and practice have gone through, and then, through cognitive computing and systematic analysis of alumni economic theory and practice, provide more specific examples for further development and expansion of its application scenarios, and finally, through the establishment of the China Alumni Economic Theory and Practice Research Institute, especially in combination with the characteristics of the times, it highlights the basic topics of cognitive computing and systematic analysis of alumni economic theory and practice.

2.1 Review Alumni Economic Theory and Practice

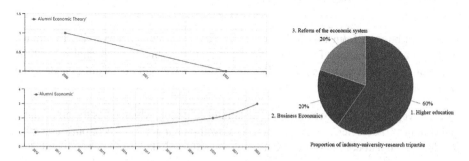

Fig. 2. The further subdivision and research

As can be seen from Fig. 2, further subdivision and research can find that the trend and ratio between the alumni economy and its theoretical research are not proportional. At the same time, in terms of the relationship between production, education and research, higher education is in a dominant position.

The alumni economy refers to the exchange of material, culture and talents between the alma mater and alumni, alumni and alumni, alumni and society, and the alma mater and society with the alma mater as the core in the social activities of alumni and economic activities that bring objective benefits to society. The alumni economy is an important part of the modern economic system. Like the digital economy, green economy, and sharing economy, the alumni economy is also a new form of business, applicable to all industries. The alumni economy is to maintain close contact with alumni entrepreneurs, establish good cooperative relations, smooth the information exchange between alumni local enterprises, government departments, and promote all parties in industrial development, infrastructure, scientific and technological talents, ecological environment, poverty alleviation. The value of substantive cooperation in various fields such as tackling tough problems, education and medical care is reflected. The alumni economy generated by alumni groups is a common strategic resource and valuable wealth for localities and universities. More and more universities and cities have begun to pay attention to this "development boosting force" that cannot be ignored. The alumni economy has quietly detonated, becoming a new outlet. The exchange of material, culture and talents between the alma mater, alumni and society has become a new format. This theoretical concept was put forward in 1997 by the scholar Professor Shijie Lu [1–4].

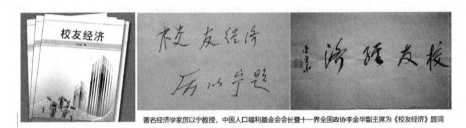

著名经济学家厉以宁教授，中国人口福利基金会会长暨十一界全国政协李金华副主席为《校友经济》题词

Fig. 3. Professor Yining Li and Jinhua Li wrote an inscription for "Alumni Economy"

As can be seen from Fig. 3, Professor Li Yining, a famous economist, Li Jinhua, the chairman of the China Population Welfare Foundation and the vice chairman of the National Committee of the Chinese People's Political Consultative Conference, wrote an inscription for "Alumni Economy".

The Concept of Alumni Economy: The alumni economy, which takes the alma mater and alumni as the carrier, aims to promote the development of related economic activities in a long-term and continuous cycle through the natural link between alumni and their alma mater. Alumni resources are important resources available in school and social development, and are a major part of forming and measuring the alumni economy. Alumni economic development work is mainly responsible for alumni management agencies such as alumni associations, government agencies and the leaders of the alma mater. All social forces need to be mobilized to participate.

2.2   Cognitive Computing and Systematic Analysis

Attributes of Alumni Economy (Extraction of basic attributes)

Table 1. Cognitive Computing and System Analysis of Attributes of Alumni Economy.

ID	Chinese	English	Attributes of Alumni Economy
1	多元性	Diversity	Attributes of alumni resources development
2	广泛性	Extensiveness	Attributes of development of the school with alumni
3	综合性	Comprehensive	Attributes of school public image construction
4	动态性	Dynamics	Attributes of a virtuous cycle of continuous
5	自发性	Spontaneity	Attributes of alumni activities
6	潜在性	Potentiality	Attributes of the biggest features
7	优质性	Quality	Attributes of a high-quality alumni community

From Table 1, it can be seen that the author of "Alumni Economy" and the authors of this article have summarized seven characteristics or attributes, which can help readers instantly understand a basic portrait or overview of what we call alumni economy, which is beneficial to university leaders, enterprise managers and party/government leaders or cadres, especially all top leaders, decision makers, always keep in mind: China, as a major country in higher education, is the most important human resource and its basic attributes or characteristics.

According to Aristotle's formal logic, concept = connotation (mainly attributes) + extension (mainly distinguishing formalizable classes).

$$A^x + E^y \leq C^z \tag{1}$$

It can be seen from the formula that the attribute (connotation) and the extension (embodied in the formalized class), as two aspects or parts of the concept, according to Aristotle's system view which is good for our cognitive calculation and system analysis that the whole is greater than the sum of the parts, and with introducing the theory and method of the circular logarithmic function, we can use this developing formula to enrich and complete our research on alumni economic theory.

Characteristics of Alumni Economy (from 1 to 7 here)

(1) *Diversity (Alumni resources as some attributes that can be calculated and analyzed).*
    The alumni resources often have diversity in industry, region, level, age gradient, and social influence after years of development.

(2) *Extensiveness (development of the school with alumni).* The school graduates thousands or even tens of thousands every year. With the development of the school, the number of graduates will become larger and larger.

(3) *Comprehensive (school public image construction).* Alumni resources can provide the school with comprehensive contributions in terms of people, finances, materials, and information. If the alumni resources can be fully utilized, they will receive comprehensive support in financial support, participation in the teaching and research process, employment placement services, school public image construction, and campus culture construction.

(4) *Dynamics (a virtuous cycle of continuous).* The resources of alumni are not static, but a virtuous cycle of continuous development in terms of diversity, breadth and comprehensiveness over time.

(5) *Spontaneity (particularly alumni activities).* In general, alumni activities are more common in spontaneity, with a single effective purpose strong sustainability, and no dependence.

(6) *Potentiality (the biggest features).* Potential alumni resources are diverse, extensive, comprehensive, and dynamic, one of the biggest features is potentiality. Many alumni resources are difficult to fully utilize/develop due to various restrictions on location and age as that the waste of precious resources in the development and construction of colleges and universities. Therefore, the alumni economy developed on the basis of alumni platform resources also presents diversity, extensiveness, comprehensiveness, dynamics and potentiality.

(7) *Quality (the attributes of a high-quality alumni community).* Using the school platform as the carrier, to gather energy and build a high-quality alumni community is the main feature of the alumni economy. The most important feature of the "alumni economy" is the linkage effect of the trinity of schools, alumni and society. The school cultivates talents, alumni use the school as a link to realize their own development, and the society uses the power of alumni to release economic vitality.

2.3   Research Institutes Both at Home and Abroad

Since the above-mentioned characteristics of the alumni economy and its rich contents still need to be verified by practice and theoretical research in all aspects, we have not only organized a large number of alumni economic practice activities (the alumni associations of various colleges and universities have jointly organized many Alumni Economic Activities) and also established the China Alumni Economic Research Institute (Beijing on August 12, 2022). A demonstration base for intelligent/smart system studied (learning and research) that combines the Department of Civil Law and the Department of Law of the Sea, and the second leap in human cognition to theory and practice, and organizes the International Alumni Economic Research Institute actively, as well as supporting alumni who have made outstanding contributions both at home and abroad a comprehensive reward fund for continuous encouragement. Its characteristic is to use wisdom to control money or economy. Its theoretical basis is: the essence of finance is ZhiRong (integration of human wisdom); the nature of raising funds is RongZhi (intelligence of all aspects of integration). The corresponding practical basis is that the Hengqin Guangdong-Macao Deep Cooperation Zone is not only a capital depression, but also a financial highland.

In 1997, Shijie Lv, then secretary-general of the Beijing Alumni Association of Central University of Finance and Economics, visited many schools and began to study the economy of alumni. He realized that the school, as a "special factory", created talents and alumni. Once alumni enter the society, they begin to use the foundation of learning in their alma mater to serve the society. Many successful people are proud of graduating from a certain school, and their alma mater is also proud of having outstanding alumni. This natural connection enables alumni to take practical actions to feed back to their alma mater and offer suggestions for accelerating the construction of their alma mater; the connection between alumni deepens the natural relationship between alumni, and mutual help and mutual learning have also become an important connection between alumni factor. In this way, the alumni economy will naturally form. Alumni, like comrades-in-arms and co-workers, are known as an important link in the big social family, and play an important role in social progress and unity. In the same year, the alumni economy was proposed.

3 Result

The results are beneficial for the economic practice of domestic and foreign alumni to further rise to the height of theoretical research, and to achieve sufficient interpersonal communication and high interpersonal collaboration. Reflecting the blueprint of the General Secretary, with the chief executive of Macao as the main body, and the leaders of participating departments and provinces, municipalities and autonomous regions as representatives, taking advantage of the location advantage of the Hengqin Guangdong-Macao Deep Cooperation Zone, establishing the International Research Institute of Alumni Economics to further develop the capital's economy circle, East China and the Greater Bay Area, especially all kinds of talents, especially the researchers and practitioners of alumni economic theory, play a common role.

Table 2. Classification of the main roles of the alumni economy.

ID	对谁而言?	For whom?	Classification of the main roles
1	对校友而言	for alumni	Mutual support and help between alumni
2	对学校而言	for schools	As a huge wealth: the resources of alumni
3	对城市而言	for the city	The dynamic and competitive resource
4	对企业而言	for Enterprises	For development such as wisdom and talents
5	对事业而言	for Institutions	For various development such as achievements

It can be seen from Table 2 that the main functions of the alumni economy are classified. Among them, 1–3 are summarized by the author of "Alumni Economy", and

4–5 are added in this article. This highlights the main role of the alumni economy in five aspects.

The Main Role of the Alumni Economy (1–3 in "Alumni Economy")

For alumni (1). For alumni, the reason why "alumni" becomes "alumni" is that they have basic feelings for cultivating their alma mater and nurturing their city, and the nurturing and development of such feelings depend on the cultural nourishment during their school days. Mutual support and help is the closest relationship between alumni. After graduation and becoming an "alumni", it is possible to have "home care and feedback" to the alma mater.

For alumni (2). For schools, alumni are playing an increasingly important role. As a huge wealth of colleges and universities, the school should take the initiative to stretch out its hands, revitalize the resources of alumni, and provide more support for the development of the school.

For a city (3). For a city, a city with all kinds of colleges and universities can fully and reasonably develop the dynamic and competitive resource of the alumni economy, which is bound to inject inexhaustible water into the development of the city. After all, the "alumni economy" not only means capital, but also contains important elements for various development such as wisdom, achievements, and talents.

With the progress of the society and the development of the economy, the power of alumni has become more and more influential in our country's economic environment. More and more universities and cities have begun to pay attention to this development boosting force that cannot be ignored. The alumni economy is an important force for future development. Do a good job of alumni work, create a good alumni culture, and give full play to the role of alumni. The development of colleges and cities will inevitably usher in the spring.

The alumni economy as far as the current situation is concerned, the work of alumni in most domestic colleges and universities is still in its infancy. The alumni organization lacks norms, and the communication between the school and alumni and between alumni and alumni is mostly accidental, and it is in a low-level stage of disorder, inefficiency and irregularity.

The urban alumni economy, which relies on the work of various college alumni, is also in the initial stage of development because the alumni culture is not strong enough [5–9].

The International Alumni Economic Innovation and Development

The International Alumni Economic Innovation and Development Funds, the International Alumni Economic Research Institutes and the cooperation of relevant universities can set up prestigious professorships for a small number of scholars with significant achievements or contributions. This is a new kind of formal sustainable development that is more sustainable than simply conferring academic achievement awards. It is a good way to gather wisdom from the combination of the old and the new, the inheritance and the development, and the collective and the individual.

As can be seen from Fig. 4, the International Alumni Economic Research Institute echoes the China Alumni Economic Research Institute through top awards (co-organized

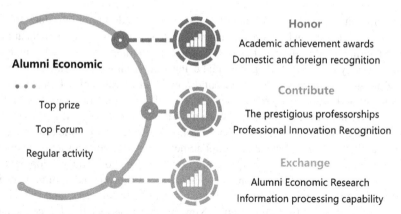

Fig. 4. International and Chinese (co-organized) Alumni Economic Research Institute

with funds), top-level forums (co-organized with universities) and regular events (co-organized with various alumni associations). It is characterized by the highest honor and academic achievement award recognized at home and abroad, continuous contribution and honorary professorship with professional innovation recognition, continuous exchange and alumni economic research with information processing capabilities, three aspects are integrated to bring substantial contributions to all aspects.

4 Conclusion

Its significance lies in: Chinese and foreign alumni economic talents, with the help of cognitive computing and systematic analysis of alumni economic theory and practice, not only better participate in the construction of the national demonstration zone, but also an international and domestic demonstration base for the practice of alumni economic theory in the Hengqin Guangdong-Macao Deep Cooperation Zone, and extended to at home and abroad.

Ping Hao's alumni economy must maintain the ability to continue learning. Ping Hao, president of Peking University, believes that the alumni economy in the information age has come to fruition [10].

Yixin Chen blew the "New Era Alumni Economy" development rally. He was then Deputy Secretary of the Hubei Provincial Party Committee and Wuhan Municipal Party Committee, and is currently the Secretary-General of the Central Political and Legal Committee. Yixin attended the meeting and pointed out that: We will work together to fight the successful implementation of investment promotion projects, sound the rallying call for the development of the "New Era Alumni Economy". It is necessary to make "alumni investment promotion" the top priority of investment promotion, and accelerate the development of "alumni economy in the new era". Focusing on continuous innovation of "alumni investment promotion", we will deeply explore alumni resources in all aspects, regions and fields, and promote project signing and project implementation, domestic "double recruitment and double introduction" to international "double recruitment and double introduction", and "alumni and alumni introduction". "Economy" to "Academician Economy", material ties to spiritual ties, "investment attraction" to "cultivating &strengthening business", project investment to city marketing and investment, policy investment to optimizing business environment, centralized assault to normal long-term effect, and strive to build a national investment hub city with high-end elements and a model city for the development of the "New Era Alumni Economy" [11].

Xian Shi Alumni economy is both a platform and a booster. Xian, vice president of Nanjing University of Finance and Economics, believes that the alumni economy is both a platform and a booster. Alumni come from major colleges and universities with high literacy and can provide high-quality services for economic development. Nanjing University of Finance and Economics has vigorously promoted alumni work in recent years, set up alumni back-to-school days, created alumni cases, and regularly shared the experience of outstanding alumni. Now the alumni association has covered overseas [12].

Zhimin, Luo a professor at Yunnan University, believes that foreign universities take donations as an important part, while Chinese universities only serve as necessary supplements. Regardless of whether they are active or passive, it is an inevitable trend for universities to absorb social donations and diversify the sources of school-running funds. Nowadays, financial investment can only be To ensure that the school "will not starve to death", it cannot guarantee "a good life". The school depends on the government, but also on the society. China is supporting a group of universities in their efforts to build world-class universities. The overall plan for the construction of universities and first-class disciplines points out that "colleges and universities should continuously expand funding channels, actively attract social donations, expand social cooperation, improve long-term social support mechanisms, pool resources through multiple channels enhance self-development capabilities" [13].

In addition to the above-mentioned calls from university leaders, several features of our online virtual exploration of the International Alumni Economics Institute are worthy of further exploration or trial use: first, the application of Qibantong and the Eight-Person Group for teaching, management and learning and using as of socialized systems engineering method, finishing knowledge modules can help train the backbone of alumni, so as to better give back to the alma mater in the construction of advantageous disciplines, the continuous improvement of teaching quality, and the combination of production, education and research; second, direct application of new mathematics such as AI factor space and circle mathematics tools to dig deep into the basic attributes and main functions of the alumni economy and several sub-fields it covers, that is: not only discovering but also using cognitive computing and system analysis to analyze various characteristics, attributes, and the role or landmark achievement realizes human-machine collaboration or automated processing [14–20].

Then there is the comparison and analysis of various typical examples of economic research and practice of alumni at home and abroad, which must play the positive role of the above two aspects (on the one hand, highlighting the intellectual advantages of human beings, on the other hand, highlighting the advantages of AI).

References

1. China News. The rise of the "alumni economy" is the right time for overseas returnees to start businesses-Chinanews.com. 2021–01–19
2. Xinhuanet. The alumni economy is becoming a new outlet. How to make good use of alumni resources? 2021–01–22
3. Daily, X.: "Alumni Economy", a new outlet in the industry. Nanjing Agricultural University, 2021–01–22
4. People's Daily. Are these huge signings fulfilled? Don't let alumni be fooled! - Finance, People's Daily. 2021–01–23
5. People's Daily. To dig deep into the alumni economy, first focus on the alumni culture - local, People's Daily. 2021–02–17
6. China Economic. To dig deep into the alumni economy, first focus on the alumni culture. China Economic Net reference date 2021–02–17
7. Tianjin CPPCC. Reply to the proposal on "Proposals to vigorously develop the alumni economy and promote the high-quality economic development of our city" - Proposal work - Tianjin CPPCC. 2021–02–18
8. Xinhuanet. Chinese universities' "alumni economy" has set a record, but it is still hard to beat world famous schools. Xinhuanet. 2021–02–16
9. China Economic Net. To dig deep into the alumni economy, first focus on the alumni culture. China Economic Net. 2021–01–30
10. Xinhuanet. 500 global alumni of Peking University gathered in Wuhan, Wuhan "alumni economy" to a new height. Xinhuanet. 2020–07–15
11. Chen, Y.: Sounded the "New Era Alumni Economy" development. Wuhan Civilization Network. 2021–01–22
12. Nanjing Daily. The alumni economy is a flower, and Nanjing is a fertile soil. Nanjing Daily. 2021–01–30
13. Phoenix Network. Chinese University Nuggets "Alumni Economy". Phoenix Network. 2021–01–22
14. Xu, W., et al.: The cognitive features of programming language and natural language. IFIP TC12 ICIS (2018)
15. Zou, X., et al.: The strategy of constructing an interdisciplinary knowledge center. ICNC-FSKD (2019)
16. Zou, X., et al.: New approach of big data and education: any term must be in the characters chessboard as a super matrix. In: Proceedings of the 2019 International Conference on Big Data and Education – ICBDE'19 (2019)
17. Zou, X., et al.: Smart system studied: new approaches to natural language understanding. In: Proceedings of the 2019 International Conference on Artificial Intelligence and Computer Science (2019)
18. Zou, S., et al.: How to do knowledge module finishing. IFIP TC12 ICIS (2018)
19. Luo, X., et al.: The cognitive features of interface language and user language. IFIP TC12 ICIS (2018)
20. Zou, X.: The formal understanding models. In: ICCSIP (2018)

Factor Space: Cognitive Computation and Systems for Generalized Genes

Peizhuang Wang[1], Xiaohui Zou[2]($\boxtimes$), Fanhui Zeng[1], Sizong Guo[1], Yong Shi[3], and Jing He[4]

[1] Institute of Intelligent Engineering and Math, Liaoning Technology University, Fuxin, China
[2] Searle Research Center, No. 100, Renshan Road, Hengqin 519000, Guangdong, China
zouxiaohui@puk.org.cn
[3] Chinese Academy of Sciences University, Beijing, China
[4] Oxford University, Oxford, UK

Abstract. The purpose is to generalize and expand the vector space to factor analysis, and then simplify the knowledge graph to the factor graph, focusing on the formal expression of concept, which is essentially an attempt to construct the mathematical foundation of the three categories of information, intelligence and data science & technology. The method is: first, it is clear that the vector space can only be used for mathematical analysis of quantitative real variables, and it should be generalized and extended to the qualitative causal analysis with the factor space, and that could be regarded as a special factor space; then, it is pointed out that the advantages and disadvantages of the knowledge graph, which should be simplified to the factor graph; finally, the formal expression of the factor genealogy focusing on concepts, it is essentially an attempt to construct the mathematical basis of the three major categories of science and technology of information, intelligence and data that that is different from material science and technology. The result is the discovery of the mathematics of factor as a meta-word, namely factor space. Its significance lies in: its mission is to provide the smart mathematical tools as cognitive computing system.

Keywords: Cognitive Computation System · Generalized Genes · Actor Space

1 Introduction

The purpose of this paper is to generalize and expand the vector space to factor analysis, and then simplify the knowledge graph to the factor graph, focusing on the formal expression of the factor genealogy of the concept, which is essentially an attempt to construct the mathematical foundation of the three major categories of information, intelligence and data science & technology.

Background & Reviewing the past and learning the new: Mendel deeply felt the complexity of biological attributes, and proposed the concept of genes. Each gene commands a string of biological attributes. Therefore, the factor is a generalized gene. We know that the formalization of biological genes, the formalization of chemical elements, their molecules and chemical equations, and physical symbols, equations and their dimensional systems all share a common mathematical basis. However, so far, although there are three categories of information, intelligence and data science and technology have developed by leaps and bounds, its mathematical foundation is still in the development stage of various schools of thought as: Topos category, factor space, formal concept analysis, rough set, extension, mediation logic, econometric logic, lattice-valued logic, probability logic, soft set, pleated set, thinking mathematics, philosophical mathematics, attribute theory, factor neural network, granular computing, cloud models, quotient spaces, non-optimal theory, phase theory, three-branch theory, structural meta-theory, intuitionistic fuzzy sets, hesitant fuzzy sets, set pair analysis, circular logarithms, etc., these branches are explicitly artificial intelligence services, with the characteristics of mathematical research in information, intelligence, data and other aspects, have their unique and irreplaceable avant-garde in some aspects, showing vigorous vitality. [1–25] *(the book in Chinese contains 1–44).*

With so many explorers moving forward, why worry about the development of mathematics in terms of information, intelligence, and data? What Factor Space does is to provide a common platform for all these avant-garde new mathematical theories, bringing together the forces of all parties. Factors are the elements of causal analysis and the foundation of new theories; factors are the formal expressions of various perspectives of the cognitive subject when observing things, and are the logical starting point for cognitive computing and systematic analysis of information, intelligence, data and other sciences; factors is a generalized gene, and the factor space is a universal framework for the description of thinking and things. It deserves and can take on the mission of paving the way for this new kind of mathematics.

2 Method

The method is: first, it is clear that the vector space can only be used for mathematical analysis of quantitative real variables, and it should be generalized and extended to the qualitative causal analysis with the factor space, and the vector space could be regarded as a special factor space; then, it is pointed out that the advantages and disadvantages of the knowledge graph, which should be simplified to the factor graph; finally, the formal expression of the factor genealogy focusing on concepts, it is essentially an attempt to construct the mathematical basis of the three major categories of science and technology of information, intelligence and data that that is, different from material science and technology. Since the above three aspects or three steps overlap in the actual advancement process, our following introduction is based on the narrative method of reviewing and summarizing the actual exploration method of the factor space theory, that is, the outline description method of each innovation point. The foregoing summary illustrates the induction and refinement of the internal logic of the method of "Factor Space: the Cognitive Computing System of Generalized Genes". It is for readers' reference only.

3 Overview of Method Points

From factor to meta-knowledge expression (breakthrough of 0); factor is the element of causal analysis, the key point of human brain cognitive calculation (focus); factor is a generalized variable (expanded from vector space to factor space); a factor can be truly recognized only when the phase domain is given; from the two classic problems of knowledge representation, namely, distinguishing objects and generating concepts, we can see the characteristics of the mathematical theory and method of factor space.

3.1 Knowing Meta Expressions for Explicit Factors

The first response of the human brain to things is to answer "what is this?". The nerve center transmits the object information to the memory unit, finds the storage location of the object, or establishes a new file, or uses the old file for comparison and judgment, and responds quickly. This is the most basic link of thinking activities, and it should have a basic mathematical expression. What is this meta-expression?

Previously, this meta-expression was e is p, where e was an object or entity (both things and events), and p was information about object e. This way of expression confuses the boundaries between information and objects, and fails to highlight the purpose of knowing the subject. An equal sign cannot be drawn between objects and information. From objects to information, there must be a purposeful graphping of the human brain. Factors are mathematically defined as this graphping:

Definition 1.1. A factor is a graphping $f: D \rightarrow I(f)$, where D is called the domain of definition or domain of f, and $I(f)$ is called the phase domain or information domain of f.

For example, factor f = color, D = 5 cars parked in front of the building = $\{d_1, d_2, d_3, d_4, d_5\}$, $I(f)$ = {red, white, black}. f is the graphping that turns the car into a car color, such as $f(d_1)$ = red, $f(d_2)$ = white, $f(d_3)$ = red, $f(d_4)$ = black, and $f(d_5)$ = white.

In general, a factor is a graph that turns an object into information, and information is the phase of matter under the factor graph. Information can be described mathematically only by factors, so we change the object e to its phase $f(e)$ under the factor f, and get the knowledge element expression:

$$f(e) = p \tag{1}$$

For example, e is a parked car, f represents color, p represents red, the above formula means: "The color of this car is red".

The knowledge element expression introduces a new mathematical notation at the source of perception, which is the factor.

3.2 Factors: Elements of Causal Analysis

The development of human intelligence does not come from conditioned reflexes. Animals have conditioned reflexes, but they do not have human intelligence. The human brain has the ability of causal analysis, and factors are the elements of causal analysis. The essence of machine intelligence is to rely on humans to transform the mechanism of causal analysis into a program and transplant it to the machine. The program relies on means such as looping and substitution, which can teach the machine to fish and make the machine automatically catch fish. However, machines can automatically reason, answer questions, and adjust people's emotions, but they can never automatically choose a perspective (if there is, it would be an unprecedented breakthrough), and cannot generate factors like humans. Factors depend on human input, and the machine will be smarter as many factors are input to the machine.

The core idea of causal analysis is not to search for the cause in isolation and statically from the attribute or state level, but to search for the most influential factor on the result from a deeper level. Only when this group of factors is found, can the best cause be found. From finding causes to finding factors is a sublimation of human brain cognition, and it is also the ideological core of causal science.

3.3 Factors: Generalized Variables

Factors are elements of causal analysis because of their variability. The phase domain of a factor is a string of words such as attributes, utility or desire, and the phase domain $I(f)$ is not a random combination of phases.

For example, the phase field of color can only include red, yellow, blue and other colors, but cannot be mixed with words such as "circle" and "big", the phase field must be a neat array permitted by the human brain; this permission comes from the human brain, the natural judgment of the child; this judgment is an instinct that preschoolers can display; this instinct is to turn an array into a comparable set of items, and the things that are compared are the factors.

Therefore, the extraction factor is an instinct of the human brain. Of course, this instinct can be continuously improved through training. Smart and wise or wisdom come from people's grasp and experience of factors.

The factor is in disguise, and it changes in value in its own phase domain. Factors are generalized variables.

Factors can embed the qualitative phase domain into the quantitative phase domain of Euclidean space and convert them into ordinary variables. The premise is to order the phase domain according to a certain goal.

For example, occupational domain = {worker, farmer, soldier, business owner, employee, teacher, doctor, lawyer, official,...}, there is no order among these occupations. However, when the college candidates apply for volunteers, they must sort their future careers. Wages are one way of ranking, social needs are another way, hobbies are another way, and comprehensive weighting is another way. When $I(f)$ becomes a total or partially ordered set, the qualitative phase field can be embedded in a real interval or a multidimensional hyper-rectangle. This real space can be selected as [0,1] or [0,1]^n, at this time, all phases are some degree of satisfaction to the target. And the degree of satisfaction can be transformed into a certain logical truth value, which is the platform built by the factor space for the display of the pan-logical theory.

The phase domain embedded in real space is discrete. The binary phase field $I(f) = \{0, 1\}$, or the three-valued phase field $I(f) = \{1, 2, 3\}$ or $\{1, 0, +1\}$, can be divided into more level. Discrete-valued phase domains are called trellis or brackets.

There are several special names for factors: (1) Bipolar names, such as "beautiful and ugly"; (2) Followed by a question mark, such as "beautiful?"; (3) Add "is there" or "whether" in front, such as "is it beautiful"; (4) Add the word "attributes" after it, such as "beauty".

Since factors are generalized variables, the essence of traditional mathematics can be transplanted into the theory of factor space.

3.4 Difference Between Factor and Attribute

Factors and attributes should not be confused. Attributes can ask right and wrong, such as: "Is this flower purple?", while factors cannot ask right and wrong, such as: "Is this flower a color?". Attributes are static words that describe passively; factors are dynamic words that actively pull thoughts.

This distinction is not absolute but relative, and a phase word in one context can be transformed into a factor word in another, or vice versa.

For example, taking the "evaluation standard" as the factor f for examining a piece of clothing, $I(f) = \{$color, style, price$\}$, where "color", "style" and "price" are all constant values, their status is a candidate for evaluation criteria, but once a phase is selected, it becomes a derived evaluation factor.

For example, to evaluate the price of a commodity, "price" becomes a factor and has a phase domain $I($price$) = \{$cheap, moderate, expensive$\}$. Since the same word has different status, we must distinguish whether the word is a factor or a phase at any time. When we are confused, please write the phase domain. Given the phase domain, we can truly understand a factor.

3.5 Two Classic Problems of Knowledge Representation

Two classic problems of knowledge representation are distinguishing objects and concept generation.

How does the factor space distinguish objects?
The definition of the factors: the similarities and differences of things.

Any two things have different points, but also have something in common; things in the world, apart from factors, there is no difference, only the use of factors that can distinguish the similarities and differences of things.

Definition 3.5.1. Given two things d_i and d_j, if the factor f is meaningful to both of them, and $f(d_i) = f(d_j)$, then d_i and d_j are said to be the same under the factor f; if $f(d_i) f(d_j)$, then d_i and d_j are said to be different under the factor f, or the factor f can distinguish d_i and d_j.

The Basic Problem of Object Distinction.
Given a finite set of objects $D = \{d_1,..., d_m\}$ and a simple factor space $(D, f_1,..., f_n)$ on it, a fundamental question is: can it be guaranteed that the objects in D can have two Two distinctions? How can this distinction be achieved?

Proposition 3.5.1. A sufficient and necessary condition for the conjunction factor f_g to distinguish d_i from d_j is that at least one of f and g can distinguish d_i from d_j.

How does the factor space generate concepts?

Computers have been able to automatically reason for a long time, but they have not been able to automatically generate concepts. Reasoning can only make circles in a given concept system, and cannot introduce new concepts. There is no real intelligence without new concepts. R. Wille first gave a strict definition of a concept in formal concept analysis (Wille 1982): given a set of objects E, $A(E)$ represents the set of common attributes of all objects in E, given a set of attributes A, $E(A)$ represents the shared object set of all attributes in A, $= (A, E)$ is called a concept if $A(E) = A$ and $E(A) = E$. At this time, it is said that A and E satisfy the involution and are called the intension and extension of the concept respectively. This involution can also be called Galois transformation. His algorithm is to find all the concepts in a formal background table, and construct the concept lattice according to the implication relationship. His contribution is great, but his algorithm is complex, with NP entanglement, the factor space is improved on the basis of his theory, the algorithm is simplified, the danger of exponential explosion is avoided, and the redundant concepts in the concept lattice are reduced, easy for people to understand.

Atomic Concept

The factor space makes an important simplification of the formal concept analysis:

Definition 3.5.2 (atomic concept). Given a simple factor space $(D, f_1,..., f_m)$, let R be the background relation on it, then for any $a = (a_1,...,a_n) \in R$, Denote its preimage class as $[a] = \{d \mid d \in D, f_1(d_1) = a_1,..., f_n(d_n) = a_n\}$, and say $= (a,[a])$ is an atomic concept.

Proposition 3.5.2. The atomic concept must satisfy the involution condition given by Wille.

According to this proposition, as long as there is a background relation R, all atomic concepts can be written directly from R without calculation. If you have 10 atomic concepts, you can further write 210 concepts. The automatic generation of concepts is not the fear of having too few concepts, but the fear of having too many. We just need to generate a special class of concepts:

Definition 3.5.3. A concept that satisfies involution and whose intension can be expressed in conjunction paradigm is called a basic concept.

Conjunctive normal form is a term in mathematical logic that refers to a logical expression that can be written in the form of disjunction and then conjunction: $a = (a_{11} \vee \cdots \vee a_{1n(1)}) \wedge \cdots \wedge (a_{m1} \vee \cdots \vee a_{mn (m)})$. In the information space I, its geometric image is a hyperrectangle, but each side is not necessarily connected. Every atomic concept is a fundamental concept.

4 Key Methods with Examples

From the knowledge graph, to the factor graph, and then to the factor pedigree, not only will the presentation of precise search results combined with human-computers be direct, intuitive and easy to understand, but also the knowledge base supported by the backend will be more reliable, because can be both quantitative and qualitative, so that the advantages of human and machine can be better complemented.

4.1 Factors and Knowledge Graphs

A Brief Review of Knowledge Graphs

People are born with the same brain structure, but later they are very different. That is because the human brain shapes itself in the process of processing information. The human brain is the war department for people to make intelligent decisions, and it directs the big and small battles to transform the objective world; but behind all the goals of extroverted behavior, there is the same introverted goal: to shape one's own brain structure—knowledge base. A good command must have a good knowledge base, and the two complement each other. The unified intelligence theory particularly emphasizes the goal and destination of intelligence development. Without the semantic database, artificial intelligence becomes a tree without roots. Semantic database is the goal and destination of artificial intelligence. The relational database has set a good precedent for us, and it has given birth to some intelligent algorithms of data mining. It is a pity that the relational database has not made a correct full- information interpretation of the data semantics, and cannot correctly describe the generation mechanism of intelligence. It is a passive data storage query rather than an active intelligence war department.

Mathematical Definition of Knowledge Graph. The problem of the inherent goal and destination of artificial intelligence has been paid attention to in the West. The Agent they proposed is somewhat related to what we call Tianku, but there is no specific connotation, so it is difficult to compare. however, the knowledge graph proposed by the West in recent years has a specific connotation, which is similar to what we call Tianku, but because of the different paradigms, we cannot agree. For comparison, this chapter starts with the knowledge graph.

The mathematical definition of the familiar knowledge graph is as follows:

Definition 4.1.1. A knowledge graph is a directed graph $(E; F)$, where E is a set of nodes, F is a set of directed edges; F is a binary relationship from E to E, that is, for any $f F$, there are $(e,e')EE$, let e,e' be the two nodes connected by f, each node and edge have fixed names. To give a broader definition:

Definition 4.1.2 (Wang Peizhuang et al. 2021a). A knowledge graph is a directed graph $(E, E'; F)$, where E and E' are the front node set and back node set respectively, F is the set of directed edges; A binary relationship from E to E', that is, for any $f F$, there are $e E$ and $e' E'$, so that e and e' are the front and back nodes of f respectively, and each front and back node and edge have fixed name. When $E = E'$, Definition 1.4 becomes Definition 1.3, which generalizes Definition 1.3. Every method of knowledge representation must express facts, no matter what differences exist, it must conform to the SPO expression form of subject, predicate and object in language, regard subject and object as before and after nodes, and regard predicate as directed edge, each knowledge representation can be represented as a knowledge graph, so the knowledge graph is a general form of knowledge representation, which is no exception for the knowledge base built by us.

Contribution of Knowledge Graph. The original intention of the SPARQL language is to break the monopoly of websites and establish a decentralized and transparent database on the website. This is of historical significance, and it has paved the way for the development of the blockchain. But in the West, it has been misled in the wrong direction of bitcoin, doing great harm to the ethics of intelligence. The practical effect of knowledge graphs lies in natural language understanding. text data is the storehouse of human knowledge and the most valuable thing in intelligence. Artificial intelligence should first transfer textual information from books, documents, and letters to machines, and then let machines simulate the human brain to understand and transform the world. But the opposite is true, text data is much more difficult to process than non-text data. It was not until around 2012 that the name "Knowledge Graph" was called on Google, and several giant companies that mastered Internet resources competed to use this technology to develop new search engines. The Internet is a channel for transmitting letters. the knowledge graph first processes text data. Its emergence has accelerated the research process of natural language understanding, partially realizing the digitization of books, documents, and letters, enabling natural language understanding from data-driven word frequency. Statistical methods move towards a research approach driven by the combination of knowledge and data. The knowledge graph model has achieved and will continue to achieve significant results in bridging the ambiguity gap between synonyms and antonyms.

Questions About the Development Direction of Knowledge Graph. Where is the dividing line between knowledge graph and traditional knowledge representation? It lies in the separation of their languages: the query of relational databases depends on the SQL language (Structured Query Language); the query of knowledge graphs depends on the SPARQL language (Simple Protocol And RDF Query Language). SQL is called the table library language, SPARQL is called the library language, and the name of the graph database comes from this.

Under this boundary, people only call knowledge graphs programmed in the SPARQL language as knowledge graphs. The previous relational database tables were excluded from the category of knowledge graphs because the language used was SQL instead of SPARQL, but this exclusion is irrational. Yes, it does not meet the mathematical definition of knowledge graph. The relational database table (that is, the information system table) takes the object as the row and the attribute name as the column, and the item s_{ij} represents the attribute value of the i-th object to the j-th attribute. Each entry in the table is a graph primitive with an attribute named edge from the object node to the attribute value node. A relational database table provides a set of graph primitives. There are as many grids as there are graph primitives. They form a set. According to the definition of knowledge graph, a relational database table is a knowledge graph.

4.2   Factor Pedigree and Factor Space Vine

Knowledge graph is a general form of concise and efficient knowledge representation, and the factor space should be adopted. In this section, we will discuss a new type of knowledge graph, called the factor knowledge graph. In order not to be confused with the current "knowledge graph", we use another name, called factor lineage. The essence of the factor lineage is the factor space vine, which will be the ontology structure of ours.

Knowledge Growth Expression

Concept is the basic product of intelligent creation, each concept is a semantic unit, and it is a combination of form and utility information driven by goals. The first function of a concept is to identify things. The efficacy of identification requires continuous refinement of concepts, and sub-concepts are divided from superordinate concepts, which is the most basic mode of intelligent creation. In mathematics, a concept is a 2-tuple $\alpha = (\alpha, [\alpha])$, α where is the description of the concept α, called the intension of α, and $[\alpha]$ is the set of all objects that satisfy the description of the intension, called the extension of α.

When a baby is born, there is only zero concept, the connotation is zero description, and the extension is the chaos of the whole universe. Human knowledge has evolved from zero concept through step-by-step conceptual granulation division and refinement. Each time it is split, the concept mass shrinks, the connotation description sentences increase, and a superordinate concept is split into several subordinate concepts, which is the growth of knowledge.

Knowledge Growth Expression

$$f : a \rightarrow \{a_1, a_2, \cdots, a_k\} \tag{2}$$

In the formula, α is the extension of the superordinate concept, and $\alpha_1, \alpha_2, \cdots, \alpha_k$ is α the extension of a group of subordinate concepts divided.

The word "knowledge increase" emphasizes the growth of knowledge, and formula (2) will also become the basis for quantitative calculation of knowledge.

Each knowledge-increasing expression uses a dividing factor f to decompose the extension of a superordinate concept into the extension of several sub-concepts.

The Definition of Factor Pedigree. Move the factors in front of each additive expression to the arrow to name a directed edge. This edge is the upper-level concept as the front node, and the lower-level concept is the rear node, which forms a graph primitive. Since there is more than one back node, such a graph primitive is called a multi-branch graph primitive, and the knowledge graph formed by multiple graphs is called a hyper-knowledge graph. Thanks to Mr. Chenguang LU's suggestion, the factor space should not only represent the graph primitives for concept division, but also include graph primitives for expressing relationships and associations. There may be only one back node. *For example*, the factor $f = $ *"local"*, f: "Jiangsu" and "Nanjing" corresponding graph primitives have only one back node. At this time, the factor f is not a conceptual division from Jiangsu to various cities and counties, but a contraction of attention, from Jiangsu Province to Nanjing City. Objects and concepts have different levels of granularity. If the graph primitives use objects as the front and back nodes, the attention can be transferred from one object to another, and the factors at this time will evolve into an association or relationship. Factors can generalize these knowledge representations.

The factor spectrum tree formed by dividing sentences is shown in Fig. 1, a diamond is added to the edge of each graph primitive to mark the name of the factor to highlight the status of the factor. At the same time, we also want to show that the factor acts as a program discriminator.

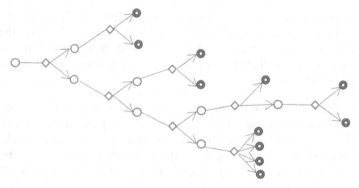

Fig. 1. The genealogy of factors formed by concept division sentences

Definition 4.2.1. A set of graph primitives whose edges are factors is called a factor lineage. The factor pedigree is the knowledge graph defined by mathematics, but not the "knowledge graph" understood by the world. The factor pedigrees identified by a single target are all factor pedigrees. The genealogy of factors identified by the multi-target is not tree-like but forest-like.

Table 1. Factor ancestry table.

	F&R	life?	animal?	spine?	breast-feeding?	plant high?	organic?	arts and sciences?
F&R		*	*	*	*	*	*	*
life?			*	*	*	*	*	
animal?				*	*	*		
spine?					*			
breast-feeding?								
plant high ?								
organic ?								
Arts& sciences ?								

In Table 1, put an asterisk "*" in the grid of 1, and leave the grid of 0 blank. In order to draw the factor pedigree, $[C]$ needs to be transformed into a parent-child matrix. Screen each non-zero element in the transitive closure $[C]$: set $c_{ij} = 1$, if there is k, so that $c_{ik} = 1$ and $c_{kj} = 1$, it means that there is still at least Across the k factor, there is no direct father-son relationship.

Ancestry Matrix of Factors. A factor pedigree is primarily a pedigree that divides concepts by factors. In the process of dividing concepts, factors also shape their own image structure. factors are restricted by the domain, outside the domain, the factors will lose their meaning. Therefore, there is a relationship between the living and being born between the factors: without the division of the "virtual and real" factors, there is no concept of "matter", and without the extension of matter, there is no definition domain of the "life" of the factors, "life" loses the soil for survival. Therefore, the factor "virtual and real" gives birth to the factor "vitality".

Embedding Structure of Factor Pedigrees. Factor pedigrees can be expanded by means of embeddings. If the ancestor node of a factor graph is an end node of another factor graph, then we can transplant the entire previous graph to this end node to form a larger graph. This process is called embedding. The inverse process of embedding is called closing; a node that can be embedded and closed is called a window. This is an indispensable feature of the current website.

Example 1.2. In Fig. 1, the concept "science" is an end node. Now take it as the ancestor concept, and introduce two increasing expressions:

"science structure": science {mathematics, physics, chemistry};
"Mathematical Structures": Mathematics {Geometry, Algebra, Analysis}.

Factors and Derived Factors. A factor pedigree is a pedigree connected by factors. Factors are innumerable, in order to reduce the number of factors, it is necessary to propose the concept of root factors and derived factors.

Basic Factor Set. There are too many levels of derivation, and it is still difficult to write. For this reason, we further propose the concept of basic factor set.

After fixing the target/utility, the form factor can be written out with a root factor x for all derived factors.

Definition 4.2.2. The root factor x is called "form". Given a basic target/utility factor y, let its factor spectrum tree be $[\alpha]^y$, then for α *of* any subconcept β ($\alpha \subset \beta$) of the pair has a derived factor $[\beta]\,x$ of the root factor x, which takes $[\beta]$ as the domain of definition, However, it has a different phase domain from x, which needs to be specified separately. In this way, the set of derived factors generated by the root factor x also constitutes a factor spectrum tree, denoted as $[\alpha]\,x^y$, which is called the derived factor spectrum tree of x with respect to $[\alpha]^y$.

Buds of Factor Genealogy and Factor Space Vines. A factor genealogy tree is a tree structure formed under a certain goal. Relative to a starting concept, each concept node has a clear generation, which can be called by the concept of the descendant of the generations. Different targets will produce different tree structures and become forests. Just like a family tree, a cousin can become an uncle, and there is a phenomenon of mixed generations. This is not terrible, and the family tree is not so messed up. Forest-like factor lineages can occur where multiple graph primitives share the same pre-node. With such a node as a window, opening is a factor space.

Definition 4.2.3. The pre-node shared by the edges of more than two factor graph primitives is called a bud of the factor lineage, and a factor lineage that only retains the bud is called a factor space vine.

In application, the real bud must have actual combat value, and it must be the combat platform required by intelligent incubation. The factor space is the arsenal of intelligent incubation, which will be introduced later.

4.3 Factor Coding

Definition of Factor Coding. Factor coding was proposed and developed by Wang Peizhuang, Meng Xiangfu, Cui Tiejun and other scholars successively. The connotation of the concept is described by the connotation elements $f_1, ..., f_n$ of the concept, and the descriptive sentence "$f_i(e) = u_i$" is abbreviated as $f^i(i)$, where the foot code (i) indicates that u_i is in $I(f_i)$ are the first few phases.

Definition 4.3.1.

$$\alpha^{\#} = f_{(1)}^1 \cdots f_{(n)}^n \tag{3}$$

$$\beta_{\alpha}^{\#} = f_{(1)}^1 \cdots f_{(n)}^n \tag{4}$$

Factor coding is so important, how can we achieve factor coding for all concepts in the world? Although there are countless concepts of human beings, they are all differentiated one by one from the higher concepts. The knowledge-increasing expression (2) gives the division process of the upper and lower concepts, which is also the way to generate factor codes.

Axiom 1.1. The connotation of the subordinate concept is equal to the conjunction of the connotation of the superordinate concept and the relative connotation of the subordinate concept about the superordinate concept. This leads to an important principle:

Factor Encoding Principle. The factor encoding of the subordinate concept is equal to the factor encoding of the superordinate concept plus the relative encoding of:

$$\beta^{\#} = \alpha^{\#} + \beta_{\alpha}^{\#} \tag{5}$$

Here, the plus sign indicates the concatenation of the codon sequences. Relative encoding can increase the length of the encoding by chain addition:

$$\beta_3{}^{\#}{}_{\beta 1} = \beta_3{}^{\#}{}_{\beta 2} + \beta_2{}^{\#}{}_{\beta 1} \tag{6}$$

It is easy to prove that the chain addition of relative codes satisfies the associative law:

$$\beta_4{}^{\#}{}_{\beta 3} + (\beta_3{}^{\#}{}_{\beta 2} + \beta_2{}^{\#}{}_{\beta 1}) = (\beta_4{}^{\#}{}_{\beta 3} + \beta_3{}^{\#}{}_{\beta 2}) + \beta_2{}^{\#}{}_{\beta 1} \tag{7}$$

The principle of factor encoding tells us that the factor encoding of a concept is uniquely determined by the relative encoding of the upper and lower positions. For a certain concept tree, no matter how long, the factor encoding is determined.

The genealogy of factors formed by different goals breaks the generational relationship of concepts, and the same concept will have different codes. But it doesn't hurt. Just as the library designs different indexes for readers, it can be more convenient for readers. Whether or not a concept may have multiple encodings, such an encoding is usable as long as no two different concepts result in the same encoding. The structuring of a system is ultimately the genealogy of concepts, and the integration and merging of complex systems is ultimately the integration and merging of system concepts. Different from general coding, the factor is the direct codeword for writing the connotation of the concept. Only it can truly compose the structure of the system and realize the integration and integration of the system. It is the key to the unification of complex structures. Factor coding is the standardization basis that needs to be solved urgently in the digital national project.

Implementation of Factor Encoding. We have theoretically solved the generation mechanism of factor encoding, and the rest is only the specific implementation problem, and the key problem is to control the codeword within a not too large range. Although there are many factors, it has already been said that the factor x = "form" can be used as a root factor to hang on a factor pedigree and derive all related form factors. The original concept is the zero concept, its inner description is empty, and its extension is the universe. There is more than one factor defined in the universe, and who to choose is related to the target factor. Assuming that our goal is to seek knowledge, we can choose the root factor $x =$ *"form"*, with the phase field $I(x) = \{real, imaginary\} = \{1, 0\}$.

There are many benefits of factor coding, but its realization requires a price: for each factor f, whether it is a root factor or a derived factor, its phase domain $I(f)$ must be defined in detail,e.g.

$I(height) = \{low, medium, high\}$
$= \{<1.6, [1.6,1.75], > 1.75\}$ *(m) (Southern man)*
$= \{<1.5, [1.5,1.65], > 1.65\}$ *(m) (southern women)*
$= \{<1.65, [1.65, 1.8], > 1.8\}$ *(m) (Northern man)*
$= \{<1.6, [1.6,1.7], > 1.7\}$ *(m) (Northern women)*
$= \{<1.9, [1.9,2.1], > 2.1\}$ *(m) (men's basketball)*

These instructions are to be stored as attachments in the library table where the factor first occurs. When necessary, the database should create a factor dictionary, a dictionary of factors across projects and industries should be compiled.

Principles of Factor Coding Application and Implementation

Try to use relative factor coding (Forth Level). Be sure to code relative factors for the superordinate concept.

Hierarchical Accountability. The starting factor spectrum tree the commander is responsible for establishing relative factor codes for i for each node concept other than i. However, there is no right or obligation to adjust the factor coding of the starting point i. The coding adjustment of the starting point is the responsibility of the upper-level commander.

Build a National Game of Chess from the Bottom Up. As long as the grassroots units do the above work, the superstructure of the factor pedigree can be successfully constructed. Our goal is to build a cross-department, cross-system, cross-industry factor pedigree and multi-objective Factor coding series. Finally, it is necessary to establish a factor dictionary, which is a national project.

Ensure Security. To distinguish friends from friends and foes, do a good job of factor coding encryption.

5 Result, Discussion and Conclusion

The result is the discovery of the mathematics of factor as a metaword, namely factor space, namely: cognitive computation and system of generalized genes. Although the induction and refinement of the internal logic of this method in this paper is preliminary, the cognitive breakthrough obtained is indeed unique.

Some people say that a relational database table is a table, and it is not drawn as a diagram. Isn't this a demarcation? The answer is: Graph theory in mathematics does need to study some abstract problems related to vision, however, knowledge graph is defined as a directed graph, requiring nodes and edges to have fixed names, which makes knowledge graph and graph theory different.

Its significance lies in: its historical mission is to provide the corresponding mathematical tools for the implementation of the intelligent generation mechanism and the intelligent incubation of all walks of life. The factor space that extends from biological genes to generalized genes is the cognitive computing and system of generalized genes.

Artificial intelligence needs a paradigm revolution. As far as mathematics is concerned, factors are the perspective from which the subject of knowledge extracts information about things, the distinguishing word between information science and material science, a gene in a broad sense, and an element of causal analysis. Factor space is the mathematical theoretical basis of information science, intelligence science and data science.

The set of graph primitives with factors as edges is called a factor lineage. A factor genealogy is a knowledge graph composed of multiple graph primitives. The knowledge structure is essentially the conceptual structure, and the factor pedigree is the ontology of the knowledge domain and the core of AI knowledge representation.

References

1. Bao, Y., Wang, Y.: Factor space: the new science of causal relationship. Ann. Data Sci. **9**(3), 555–570 (2022)
2. Bruckstein, A.M., Donoho, D.L., Elad, M.: From sparse solutions of systems of equations to sparse modeling of signals and images. SIAM Rev. **51**, 34–81 (2009)
3. Chen, D., Huang, J., Jackson, T.J.: Vegetation water content estimation for corn and soybeans using spectral indices derived from Modis near- and short-wave infrared bands. Remote Sens. Environ. **98**(2–3), 225–236 (2005)
4. Cheng, Q.F., et al.: The logistic regression from the viewpoint of the factor space theory. Int. J. Comput. Commun. Control **2**(4), 492–502 (2017)
5. Cui, T.J., Wang, P.Z., Li, S.S.: The function structure analysis theory based on the factor space and space fault tree. Clust. Comput. **20**(2), 1387–1398 (2017)
6. Cui, T.J., Li, S.S.: Deep learning of system reliability under multi-factor influence based on space fault tree. Neural Comput. Appl. **31**(9), 4761–4776 (2019)

7. Cui, T.J., Li, S.S.: Study on the Relationship between System Reliability and Influencing (2019b)
8. Factors under big data and multi-factors. Clust. Comput. **22**(1), 10275–10297
9. Cui, T.J., Wang, P.Z., Li, S.S.: Research on uncertainty of system function state from factors-data-cognition. Ann. Data Sci. 1–17 (2021)
10. Cui, T., Wang, P.Z., Li, S.: Research on uncertainty of system function state from factors-data-cognition. Ann. Data Sci. **9**(3), 593–609 (2022). https://doi.org/10.1007/s40745-021-00368-3
11. Dantzig, G.B.: Linear Programming and Extensions. Princeton University Press, Priceton (1963)
12. Dantzig, G.B.: Linear programming. Oper. Res. **50**(1), 42–47 (2002)
13. Das, S.K.: A fuzzy multi objective inventory model with production cost and set-up-cost dependent on population. Ann. Data Sci. **9**(3), 627–643 (2022)
14. Ganter, B., Stumme, G., Wille, R.: Formal Concept Analysis, Theory and Applications. Springer, Heidelberg (2005). https://www.jucs.org/jucs_10_8/formal_concept_analysis_theory/managing.html
15. Guo, J., Liu, H., Wan, R., Sun, H.: Factorial fuzzy sets theory. Ann. Data Sci. **9**(3), 571–592 (2022)
16. Gu, Y., Ma, D., Cui, J., Li, Z., Chen, Y.: Variable-weighted ensemble forecasting of short-term power load based on factor space theory. Ann. Data Sci. **9**(3), 485–501 (2022)
17. He, P.: Fuzzy non-optimal system theory and methods – the study of limiting factors in the optimum systems. In: First Joint IFSA-UC and EURO-WG Workshop on Progress of Fuzzy Sets in Europe, 25–27 November 1986, Warsaw, Poland (1986)
18. He, P.: Crime pattern discovery and fuzzy information analysis based on optimal intuition decision making. Adv. Soft Comput. Springer. **54**, 426–439 (2008)
19. He, P.: Design of interactive learning system based on intuition concept space. J. Comput. **5**(3), 535–536 (2010)
20. Huete, A., Didan, K., Miura, T., et al.: Overview of the radiometric and biophysical performance of the modis vegetation indices. Remote Sens. Environ. **83**(1–2), 195–213 (2002)
21. Jiang, B.: Research on factor space engineering and application of evidence factor mining in evidence-based reconstruction. Ann. Data Sci. **9**(3), 503–537 (2022). https://doi.org/10.1007/s40745-022-00388-7
22. Karmarkar, N.: A new polynomial time algorithm for linear programming. Combinatorica **4**(4), 373–395 (1984)
23. Kendel, A., Peng, X.T., Cao, Z.Q., et al.: Representation of concepts by factor spaces. Cybernet Syst. **21**(1), 43–37 (1990)
24. Klee, V., Minty, G.J.: How Good is the Simplex Method. Academic Press, New York (1972)
25. Kong, Q., He, J., Wang, P.Z.: Factor space: a new idea for artificial intelligence based on causal reasoning. In: 2020 IEEE/WIC/ACM International Joint Conference on Web Intelligence and Intelligent Agent Technology (WI-IAT), pp. 592–599. IEEE, Beijing (2020)

How to Improve the Quality of Academic Conversations with the Help of Human-Computer Interaction System

Shaobin Su[1], Xiaohui Zou[2(✉)] (iD), and Yezhen Su[3]

[1] Market Supervision and Administration Bureau, Yingkou, Liaoning, China
[2] Searle Research Center, No. 100, Renshan Road, Hengqin 519000, Guangdong, China
zouxiaohui@puk.org.cn
[3] Harbin University of Science and Technology (Graduate in Linguistics), Harbin, China

Abstract. It aims to explore how to extract a set of popular academic discussion models based on professional and amateur dialogues from the perspectives of interpersonal communication, human-computer interaction and computer batch processing. The method is as follows: 1. An integrated interdisciplinary theoretical model of the expert himself is unambiguously expressed; 2. The stubborn or stubborn people encountered several major confusions in the cross-border knowledge of ancient and modern Chinese and foreign in the process of trying to make the so-called common-sense paraphrase. 3. The two sides break through many difficulties, that is, a series of cognitive ambiguities or possible misunderstandings, and try to use the intelligent system to study clear word charts to achieve specific cognitive calculation and system analysis. The result is: with the help of several original knowledge maps, the higher education goals of a first-class research university such as advanced knowledge exploration and high-level talent training are applied to the process of socialized cognition, and more satisfactory positive results have been obtained. Its significance is that it not only opens up an online and offline way to popularize intellectual fusion, but also creates a standardized way to explore the active learning and thinking of the people and seek academic advancement.

Keywords: Cognitive Computing · Human-Computer Interaction System · Philosophical Common-Sense View · Scientific System Engineering Method · System Analysis

1 First Section

1.1 A Subsection Sample

It aims to explore how to extract a set of popular academic discussion models based on professional and amateur dialogues from the perspectives of interpersonal communication, human-computer interaction [1, 2] and batch processing. [3].

If it is said that Wang Wenshengyi is a major advantage of Chinese characters (and also a disadvantage, because the ambiguity or polysemy of language and words, as well

F. Sun et al. (Eds.): ICCCS 2022, CCIS 1732, pp. 211–223, 2023.
https://doi.org/10.1007/978-981-99-2789-0_18

as synonyms, etc. are common, it is easy to cause the Chinese saying that "Husband says husband is right, wife says wife is right" or "the blind man touches the elephant" as described by the Indian sayings), then it can be further found that looking at pictures and talking is a common way that all countries in the world are familiar with. As a result, a bold hypothesis was born: how to make the cognitive computing and system of computer artificial intelligence to be able to read the text and read pictures and speak? This is a problem or one of the important research backgrounds that the human-computer dialogue and further human-computer interaction and collaboration try to solve or to discuss in this paper [4–6].

2   First Section

The method is as follows: 1. An integrated interdisciplinary theoretical model of the expert himself is unambiguously expressed; 2. The stubborn or stubborn people encountered several major confusions in the cross-border knowledge of ancient and modern Chinese and foreign in the process of trying to make the so-called common-sense paraphrase. 3. The two sides break through many difficulties, that is, a series of cognitive ambiguities or possible misunderstandings, and try to use the intelligent system to study clear word charts to achieve specific cognitive calculation and system analysis. [7–14].

2.1   Fundamentals of Theoretical Smart System Studied

An integrated interdisciplinary theoretical model of the expert himself is unambiguously expressed (See Table 1 and Figs. 1, 2, 3).

Table 1.   List of knowledge fields to which the figures in this article belong.

Heading	Theoretical Smart System Studied in English	Rongzhixue in Chinese
See Fig. 1	World View of Theoretical Smart System	理论融智学的世界观
See Fig. 2	Historical View of Theoretical Rongzhixue	理论融智学的历史观
See Fig. 3	Methodology of Theoretical Smart System	理论融智学的方法论
See Figs. 4, 5, 6	The point of view of the commonsense turn of philosophy used here	

Information identities discovered by Xiaohui Zou in Rongzhixue theory.

$$[I_k + I_u = I_d] \tag{1}$$

$$[I_d = I_k + I_u] \tag{2}$$

Information equation discovered by Xiaohui Zou in Rongzhixue theory.

$$[I_u = I_d - I_k] \tag{3}$$

The information equation and its solution, namely the information identity, are the basic relational expressions that can be integrated in the microscopic, mesoscopic and macroscopic aspects under the constraints of the formal information database (series data table). However, the interpersonal communication process in this paper is limited to macroscopic discussions, while the microscopic information processing, the data processing and artificial intelligence technologies are only involved in the background human-computer interaction process.

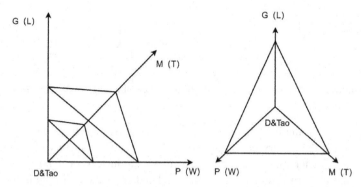

Fig. 1. The Tao function and the pyramid constructed by Xiaohui Zou in Rongzhixue theory.

It can be seen from Fig. 1 that the triangular pyramid drawn by Zou divides the macro-ontology originally summarized by seven Chinese characters, that is, the top-level classification of human knowledge, the basic category system, which must be understood and expressed by those who are familiar with Chinese characters and their thinking habits. (Rongzhixue triangular pyramid/information tetrahedron include semantic triangle), through the application of Cartesian coordinates (ie attribute coordinates), three types of thinking and information tetrahedron (triangular pyramid) are obtained qualitatively and quantitatively formalized expression, and image thinking (X-axis), abstract thinking (Y-axis), and intuitive thinking (Z-axis), which are respectively unified with the physical object world P (W), the grammatical symbolic language G (L), and the mental thought M (T) to: O The point refers to the all-encompassing D&Tao function, that is, the logical and mathematical philosophical framework and the corresponding scientific and technological method framework (operable, computable, and reasonable). This is the popular expression of the worldview of theoretical fusion/accommodating which is called Rongzhixue.

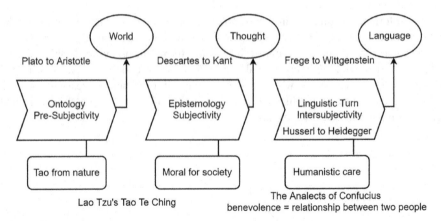

Fig. 2. Philosophy that integrates ancient and modern, Chinese and foreign, in Rongzhixue.

It can be seen from Fig. 2 that the historical view of theoretical fusion of intelligence drawn by Zou is the history of human thought, especially the history of philosophical thought (its characteristics are: it not only covers the history of mainstream thought in ancient and modern China and abroad, but also divides the first time human cognition. The Great Leap and the Second Great Leap of Human Cognition with the three core stages of the two great transitions and their great milestones in the sublimation of basic cognition). It not only inherits and develops Laozi's basic views of Tao with nature and morality with society, as well as Confucius's humanistic concern (especially benevolence covers the relationship between two persons, that can now be regarded as the earliest typical basic example of intersubjectivity), but also takes Ancient Greece's ontology (pre-subjectivity), modern epistemology (subjectivity) and contemporary linguistic turn (inter -subjectivity), three huge milestones in the search for the origin of the world, including the three referents of world, thought, and language. Although chaotic and vague (the limitation of word thinking in Western languages in terms of concept and expression), it is further adopted by Rongzhixue after philosophy and science: the world that can distinguish between matter and reason, and the meaning of thought that can distinguish its difference between the intention part and the lexical part, and the grammar of language can be divided into texts and laws. This detailed definition method lays the foundation. Therefore, based on the category, especially the basic category, the seven Chinese characters and their Chinese way of thinking (comparison with English thinking that is different from it) are highly generalized, combined with Indo-European language, especially English words, phrases and sentences, to further highlight the language/Yan and and speech/Yu, the word/Zi and phrase/Zizu, in the recognition. In terms of knowledge calculation and indirect formal methods and the construction of intelligent systems, they have three major advantages in the right time and right place. As a result, the top-level world problem that is summarized by the three super problems of natural language understanding, expert knowledge expression and software pattern recognition, that is, the combination of human and machine to resolve ambiguity (including various misunderstandings and cognitive biases) has won the high-ranking three wisdom/smart/intelligence. The opportunity of double integration (pass/close) and win-win and peace strategy (obviously surpassing war strategy) [15–19].

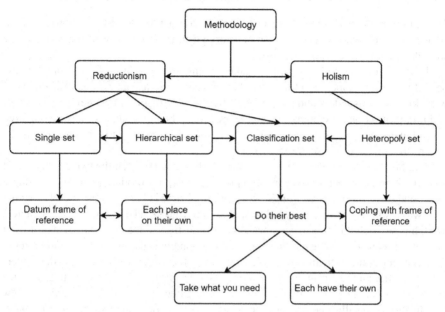

Fig. 3. Methodology of Philosophy and Fundamental Classification of Mathematics,Logic, Scientific Method in Rongzhixue theory.

It can be seen from Fig. 3 that the methodology of theoretical Rongzhixue/fusion of intelligence drawn by Zou, as well as the scientific principles of mathematical classification under its jurisdiction and its basic support, and its corresponding technical methods, not only have unique characteristics, but also have overall advantages. This is what the previous cognitive paradigms of philosophy and science lack, and it is also lacking in the execution methods of technology and art. They all require the two basic laws of logic and mathematics to constrain or regulate, and also require psychological or mental and social interaction in an individual. To enrich with group examples, it needs the support of theoretical innovation of categories, especially the 5 basic categories. Therefore, methodology and worldview, practical implementation and theoretical thinking, action guidelines and cognitive sublimation, complement each other.

Not only the logical self-consistency and mathematical rigor of the single sets and hierarchical sets and their supporting datum frame of reference and the theory of in-position are purely formal, but also the classification sets and their sub-sets of signs, attributes, characteristics, etc., echoing it, each is doing his best, getting what he needs, and getting what he deserves (different from: zero-sum game strategy). Based on them, they form a frame of reference for coping, which can deal with a heterogeneous sets (which covers a wide variety of human cognition and behavior, as well as emotion and will, consciousness, intention, and even personality or character, which not only have their corresponding formal markers, but also, and specific carrier support).

2.2 The Commonsense Turn of Philosophy Discussed Between the Coauthors

The stubborn person encountered several major confusions in the cross-border knowledge of ancient and modern Chinese and foreign in the process of trying to make the so-called common-sense paraphrase.

In the process of Su's attempt to make a common sense turn of philosophy, unexpected ambiguity appeared in the texts he relayed from various experts (For others, it is all kinds of misunderstandings, which often lead to disputes between professionals and non-professionals, experts and amateurs, amateurs and amateurs) and encountered several major confusions between ancient and modern Chinese and foreign cultures (Its generality is what this discussion is about).

The first confusion is: there is a cognitive difference between the two co-authors of this article on the understanding of "Tao produces one, one produces two, two produces three, and three produces all things". Su's point of view is based on the literal meaning (this was also the first reaction of almost everyone who understands Chinese characters and Chinese, including Zou); but since Zou has constructed the three basic categories of theoretical Rongzhixue/fusion theory (three Phenomenal-like information) and a more basic category (authentic information as defined by theoretical fusion theory) that covers three essential and real principles, meanings, and laws/rules, then go back to Lao Tzu's teachings in Tao Te Ching. When the compound category of Tao is introduced, a new insight that is usually unexpected is produced (that is, one can refer to the first basic category which is physical things, and two can refer to the second basic category which is meaning includes intent and terms, and three can refer to the third basic category which is grammatical rules in text with three referents, and then the fourth basic category, that is, everything covered by the principle/law that anything can be referred to—including the three basic categories of material, meaning, and text, that is, three types of phenomenon information. Therefore, Tao has become zeroth category).

If Su and the readers can understand this new insight, then all subsequent confusions will be solved immediately, otherwise, they will continue to struggle endlessly, go around, get stuck in the semantic quagmire, and cannot extricate themselves. Thus, 54 learning theories, 48 -ism or doctrines, 50 cognitive biases, and 100 mental models, and even a variety of other theories and methods, are endless and cannot be circumvented. Why? Readers and reviewers are invited to delve into the above-mentioned theories, doctrines, cognitive biases, thinking patterns, and various other theories and methods one by one, and repeatedly savor the following dialogues, discussions, and exchanges between the two co-authors of this article, Su and Zou, and learn from them. Explore the limits of human cognition, as well as the cognitive computing and system characteristics of computer artificial intelligence. Let's take a look at Su's ideas (picture and text - Chinese omitted here) and Zou's guide (marking, translation or explanation):

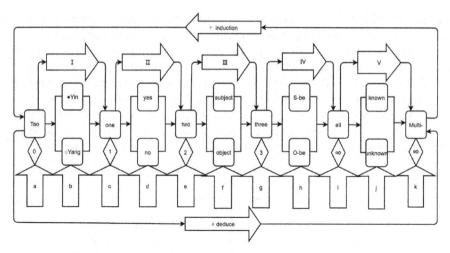

Fig. 4. The Self-justifying Ability of Taoist Ontological Thought.

It can be seen from Fig. 4 that the Taoist ontology thought described by Su has the self-justifying ability. The induction from the top to the left hand direction and the deduction from the bottom to the right hand direction (what is the basis? Su said: ∇ Induction of the unification of all methods, Δ Deduction of the Tao to produce all things); the Roman numerals above refer to "I out of nothing, as basic perspective relationship, II Basic perspective relationship between existence and non-existence, III isomorphic transition from two perspectives to three perspectives, IV isomorphic transition from three perspectives to multiple perspectives, V transition from philosophical perspective to scientific perspective"; the lowercase letters below are respectively in order refers to "a zero perspective of the multiplicity of nothing, b two perspectives of yin and yang isomorphism, c one perspective of combining two into one, d two perspectives with or without isomorphism, e two perspectives divided into two, f the same as the subject and the object, as the two perspectives of the structure, g is divided into three perspectives, h is the two perspectives of the subjective and objective existence isomorphism, i is divided into multiple perspectives, j is the two perspectives of mathematics and physics isomorphism, k is divided into many perspectives "Multiple Perspectives"; in the middle, on the one hand (above) is: yin, yes; subjective or subject; subjective existence or S-be; known; on the other hand (below): yang, no; objective or object; Objective existence or O-be; unknown; (middle left-hand direction) Tao, one, two, three, many, multi- (middle right-hand direction); corresponding to: Arabic numerals 0, 1, 2, 3, ∞, ∞.

Su said that as a non-professional in philosophy, science, and Convergence, at the early stage of self-studying philosophy, he had a requirement for philosophy to solve all his life puzzles. But after mainly studying the history of philosophy, he found that using Western philosophy can only solve some of his problems, but not all of his problems. (Note: Zou only rewrites Su's self-proclaimed first-person pronoun into a third-person pronoun; and then the Fig. 4 marked in Chinese is standardized in numbers and English, and the Chinese explanations are also simplified as much as possible, especially they are converted into texts that are easy to translate) From the beginning of 2020 to the summer of 2022, Su has continuously participated in the distance learning series of R&D Rongzhixue courses taught by Zou in the Rain Classroom of Tsinghua University (Su said that listening to Zou's lectures has brought a lot of inspiration to his thinking and writing, and Su's believes that Rongzhi -xue is very important for Su's thinking and writing. It is very helpful for Su to complete the common sense of philosophy and turn to theoretical writing, so Su hopes to jointly publish a paper. Zou used the opportunity to submit a manuscript to the first International Conference on Cognitive Computing and Systems to write a dialogue between the two sides as a cognitive aspect. Innovative attempts, we hope to give reviewers and readers some new exploration inspiration!).

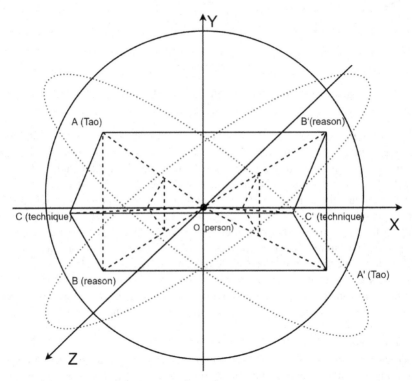

Fig. 5. Cognitive relationship between a person and the world.

As can be seen from Fig. 5, Su uses three-dimensional coordinates and geometric drawing to dynamically understand Zou's triangular pyramid model of fusion wisdom, and then draws out the geometric relationship diagram between the triangular prism and a pair of two sets of triangular pyramids. Using O point, adding a dotted line to link two (actually two groups) triangular pyramids gives people a dynamic perception of A (Tao)---O---A' (Tao), that is, the movement of the vertex of the Tao function, O as a person's central perspective, with the help of the derivative function and the transformation of the triangular pyramid, what kind of situation will it be? Readers please think! Combined with the traditional Chinese discourse habits, namely: the point of view of three parts of Taoism and magic, Su used them at both ends of the triangular prism. Among them, the technique is connected by a solid line, the Tao and the reason, the virtual and the real, respectively, in the diagonal dotted line and straight solid lines, echoing crosses, dynamically describe the vertices of the two groups of triangles in the triangular prism. It is then covered with three circles surrounded by one real and two virtual circles, echoing a person's inner world view (one of which is explained in Fig. 6 below).

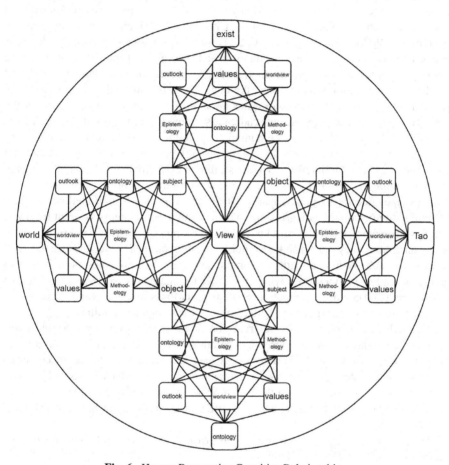

Fig. 6. Human Perspective Cognitive Relationship.

As can be seen from Fig. 6, A further regards a person's view and the O point of the Tao function as a consistent point. So, go up to existence, down to ontology, right to Tao, left to world. Going a step further, Su puts the subjectivity and objectivity, ontology, epistemology, and methodology into it. At the same time, Su also puts in the worldview, outlook on life, and values, and get the so-called intertextuality shown in Fig. 6. An intertwined web of the Human Perspective Cognitive Relationships.

To sum up, the first confusion and countless confusions are intertwined in human cognitive thinking and AI cognitive computing system.

2.3 The Two Sides Break Through Many Difficulties

The two sides break through many difficulties, that is, a series of cognitive ambiguities or possible misunderstandings, and try to use the smart system study clear word charts to achieve specific cognitive calculation and system analysis.

To sum up, it can be seen that the theory of Convergence and the eight major systems covered by it are far from being read by anyone alone (even all the series of Convergence courses taught by Xiaohui Zou in the Rain Classroom of Tsinghua University from the beginning of 2020 to the summer of 2022). Dozens of courses and hundreds of lessons) can be understood in place, let alone digested and absorbed and applied. In other words, in the era of elites, not only the learning, thinking and creation of the human brain/mind needs cognitive sublimation, but also the computer artificial intelligence needs to be re-examined or positioned. If you master the eight major formal systems, and at the same time, master the eight major academic systems (almost every discipline of the present and the past, as well as the human cognition level and system cognitive computing ability of the existing majors, can not reach the fusion level. The height, breadth and depth mentioned in intellectual studies), then the development of human beings in the near future and in the medium and long-term future will bring earth-shaking and huge changes due to the popularization of the three wisdom/smart/intelligence, double integration(accommodating/fusion), win-win and peace strategy(different from war strategy). In a word, everyone must know their own strengths, so as to avoid weaknesses; at the same time, they must also know their own shortcomings or weaknesses, and then learn from each other's strengths to complement their own weaknesses, it is best to cooperate and even synergistic with each other to create an unprecedented cognitive environment and cognitive computing. System scenarios help scholars and learners with various positions or viewpoints to better improve their cognitive abilities.

When the first International Conference on Cognitive Computing and Systems was held in the Guangdong-Macao Deep Cooperation Zone in Hengqin this year, all participants would find that the various teams gathered around the theme of cognitive computing and systems have such good complementarities. So that experts and scholars from various fields will open up their own cognitive puzzles and problems, and discuss the advantages and disadvantages of different perspectives or horizons and their complementarity.

Here, we would also like to invite all concerned to discuss the following results and their further problems and difficulties.

3 Result

The result is: with the help of several original knowledge maps, the higher education goals of a first-class research university such as advanced knowledge exploration and high-level talent training are applied to the cognitive process of socialization, and satisfactory positive results are obtained.

Below is a Knowledge Graph drawn by Su with the help of his daughter (Third author of this article) using a computer (see Fig. 7).

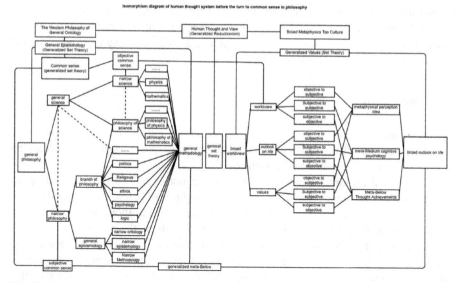

Fig. 7. Isomorphism of human thought system before the turn to common sense in philosophy.

It can be seen from Fig. 7 that, after receiving some enlightenment from the views in Rongzhixue/Convergence of Smart System Studied, Su draws a unified framework that attempts to cover both philosophy and science by combining his own philosophical commonsense-oriented viewpoints or theories. All kinds of scholars, learners and explorers in the academic world have and even today have their own so-called unified viewpoints or theories, Zou divides them into two categories: self-learning needs and academic communication needs. Whether they are recognized by others or not, they are two kinds of cognitive methods. The difference is that the former is a person's learning experience, and the latter has further developed into a road banner for later generations, or as a warning, and has the role of a cognitive beacon. This is why Zou is willing to pay attention to and support explorers in these two categories of ideological and cognitive fields. One combining the previous Fig. 4, Fig. 5, and Fig. 6 to carefully observe and analyze, especially the combination of AI cognitive computing and system modeling and simulation experiments, every serious person will feel the cognition contained in it, for that cognition, thinking and exploration and language comprehension or knowledge expression and even pattern recognition process,such findings are meaningful.

Under normal circumstances, let alone ordinary people's real life, study, work, entertainment and travel, etc., there are a series of constraints (especially further cognitive problems or difficulties) and it is difficult to realize their potential, which is a variety of special characters (those who are actively thinking and learning) also have a hard time awakening or having an epiphany. Therefore, the two authors mentioned in this article have systematically listened to and lectured for the past three years, thus, thinking deeply, learning with an open mind, and actively and earnestly focusing on long-term, persistent, and consistent focus on cognitive problems or confusion, all kinds of the folk geniuses of all countries generally need to be familiar with the cognitive computing capabilities of mainstream and non-mainstream explorers. Therefore, as far as possible, it can not only deal with all kinds of problems in an emergency, but also systematically improve people's abilities in all aspects, especially cognitive ability, and its supporting execution ability and even industrialization ability.

4 Conclusion

Its significance is that it not only opens up an online and offline way to popularize intellectual fusion, but also creates a standardized way to explore people's active learning and thinking to seek academic advancement.

Just think: if all kinds of geniuses or geeks cannot be helped in time (just right), then this will be the biggest loss for a country with a large population, a large education and a large higher education. As the saying goes: the world of those who win people's hearts/minds. Zou further said: Those who get talents will get the market.

By the way, the following is a reminder that readers will first try to understand the meaning of the text and read the pictures through the several pictures in this article (in fact, they are full of pictures and texts), and then read the text description of this article. In this way, not only the characteristics of interpersonal communication, but also the characteristics of human-computer interaction can be compared, and even the characteristics of AI cognitive computing and system analysis can be inferred. In this way, when communicating during the meeting, each of us may gain more [20–24].

References

1. Tieu, L., Schlenker, P., Chemla, E.: Linguistic inferences without words. Proc. Natl. Acad. Sci. 116(20), 9796–9801 (2019)
2. Kandler, A., Steele, J.: Modeling language shift. Proc. Natl. Acad. Sci. 114, 4851–4853 (2017)
3. Winder, D.R.: Instructional uses of the computer: batch-processing Fortran IV programs. Am. J. Phys. 38(3), 375–376 (1970)
4. Molich, R., Nielsen, J.: Improving a human-computer dialogue. Commun. ACM 33, 338–348 (1990)
5. Motger, Q., Franch, X., Marco, J.: Software-based dialogue systems: survey, taxonomy, and challenges. ACM Comput. Surv. 55(5), 1–42 (2022)
6. Sweers, T., et al.: Negation Detection in Dutch Spoken Human-Computer Conversations (2022)
7. Zou, X., Shunpeng, Z.: Basic law of information: the fundamental theory of generalized bilingual processing. ISIS (2015): n. pag
8. Zou, S., Xiaohui, Z.: Understanding: how to resolve ambiguity. In: IFIP TC12 ICIS (2017)
9. Zou, S., et al.: How to do knowledge module finishing. In: IFIP TC12 ICIS (2018)
10. Zou, S., et al.: How to understand the fundamental laws of information. In: ICCSIP (2018)
11. Zou, X.: The formal understanding models. In: ICCSIP (2018)
12. Zou, X., et al.: The strategy of constructing an interdisciplinary knowledge center. In: ICNC-FSKD (2019)
13. Zou, X., et al.: New approach of big data and education: any term must be in the characters chessboard as a super matrix. In: Proceedings of the 2019 International Conference on Big Data and Education – ICBDE 2019 (2019): n. pag

14. Zou, X., et al.: Cognitive calculation studied for smart system to lead water resources management. In: ICCSIP (2020)
15. Castellucci, G.A., Kovach, C.K., Howard, M.A., Greenlee, J.D.W., Long, M.A.: A speech planning network for interactive language use. Nature **602**(7895), 117–122 (2022)
16. Silver, D., et al.: Mastering the game of Go without human knowledge. Nature **550**(7676), 354–359 (2017)
17. Gibney, E.: Google AI algorithm masters ancient game of Go. Nature **529**, 445–446 (2016)
18. Rouse Ball, W.W.: Mathematical recreations and essays. Nature **72**, 364–365 (1905)
19. Butler, D.: Web usage data outline map of knowledge. Nature **458**, 135 (2009)
20. Mehta, R.K., et al.: Human-centered intelligent training for emergency responders. AI Mag. **43**(1), 83–92 (2022)
21. Sinayev, A., Peters, E.: Cognitive reflection vs. calculation in decision making. Front. Psychol. **6**, (2015)
22. Mayer, R.E.: Cognitive Theory of Multimedia Learning. The Cambridge Handbook of Multimedia Learning (2021): n. pag
23. Bandura, A.: Social foundations of thought and action: a social cognitive theory (1985)
24. Bandura, A.: Social cognitive theory: an agentic perspective. Annu. Rev. Psychol. **52**, 1–26 (2001)

Robot and Autonomous Vehicle

Offline Causal Imitation Learning with Latent Confounders

Siyang Huang[1], Yan Zeng[2], Ruichu Cai[1(✉)], Zhifeng Hao[3], and Fuchun Sun[2]

[1] Guangdong University of Technology, Guangzhou 510006, Guangdong, China
`cairuichu@gdut.edu.cn`
[2] Tsinghua University, Beijing 100084, China
[3] Shantou University, Shantou 515063, Guangdong, China

Abstract. Learning an imitating policy offline from the expert's demonstrations is prone to be a significant yet challenging problem. Despite great success, most methods assume that the data are uncorrupted with no latent confounders. However, such unobserved confounders could appear in many real-world applications, resulting in sub-optimal policies. Thus, in this paper, we propose an integrated two-stage algorithm to conduct the task of offline causal imitation learning, allowing the existence of latent confouders. In Stage 1, we aim at determining whether these latent variables are present or not, embracing a causal discovery method based on the conditional independence tests. In Stage 2, we adopt behavioral cloning or a variant of instrumental variable regression method for both the confounded and unconfounded cases, to eliminate the possible confounding influences. Experiments on the robotic arm control task verified the efficacy performances in both confounded and unconfounded situations.

Keywords: Causal imitation learning · Latent confounders · Instrumental variables

1 Introduction

The heart for children to learn regularities of the world and adapt to new environments lies at the ability of Imitation Learning (IL). By imitating the behaviors from the adults, they are enable to accomplish specific tasks and improve their perceptions. Inspired by this, researchers attempted to formalize the problem of learning to perform a task from expert demonstrations as IL, and developed substantive IL algorithms [15,17,37]. Such algorithms have been widely applied in various domains, including autonomous driving [7–9,25], robotic manipulation [30,36], games [2,12], etc., in that they are easy to implement and could be of no need in interactions with the environment during training [1,37].

Current IL methods can be roughly divided into two categories: Inverse Reinforcement Learning (IRL) and Behavioral Cloning (BC). IRL aims at optimizing a policy with RL algorithms after learning a reward function from expert's

---

S. Huang and Y. Zeng—Equal contribution.

F. Sun et al. (Eds.): ICCCS 2022, CCIS 1732, pp. 227–236, 2023.
https://doi.org/10.1007/978-981-99-2789-0_19

demonstrations; while BC attempts to learn a policy function from expert's covariates to actions. Despite the success of both methods, they heavily rely on the assumption that state distributions of the expert's are consistent with those of the imitator. That is, they may suffer from the distributional shift problem. When this problem occurs, applying directly IL algorithms to recover a policy would fail to gain a satisfactory performance [8,9].

One possible solution to such a problem is to consider latent confounders in recordings of the expert. Latent confounders represent those unobserved common causes of at least two observed variables in the environment, which are widespread in many real-world applications [9,31]. Examples include the weather condition for the flying aerial vehicles [31]; the wealth or poverty of patients for the medical treatment [4]; the information of auditory other than that of camera or lidar for the autonomous cars [19], etc. Latent confounders could appear in most applications where the goal is to learn policies. If ignoring these confounders, even micmicing the perfect behaviors from the demonstrator, could only attain a sub-optimal policy.

An alternative approach to circumvent this issue is to take the best of the Instrumental Variables (IVs). IVs are beneficial with their characteristics in that they can be leveraged to eliminate the spurious correlations caused by the confounders. There existed some endeavors with IVs about learning a policy [20,21,31]. [20] focused on the online learning framework while [21] studied the offline reinforcement learning based on the confounded Markov decision process. Here, we consider the offline and confounded imitation learning setting, same as [31]. The difference from the work [31] lies at the confounding detection tools: they compare the results of BC and their proposals to tell whether the confounder is present, which implies they begin detection after running their policy learning methods.

Thus, in this paper, to deal with the latent confounder issue, we propose a two-stage algorithm in the offline IL setting to learn an imitating policy. In particular, in the first stage, we embrace a method based on the independence tests, to detect the existence of confounders in the demonstrations before the policy learning procedure. It employs the confounder's characteristics that break up the asymmetry between the states or actions. In the second stage, we provide methods upon behavioral cloning or variants of IV regression techniques for both the confounded and unconfounded cases. Whether the confounders exist or not, we empirically show that our algorithm could achieve satisfactory performances compared with the state-of-art methods.

2   Related Work

Behavioral Cloning. Behavioral cloning [3,32] is one of the main approaches to deal with imitation learning problems. The main idea behind BC is to learn a policy for an MDP given expert demonstrations. This approach is powerful because it is able to imitate the demonstrator immediately without interacting with the environment. Therefore, BC has been widely used in various applications. For example, it has been successfully used in the field of control to manipulate robots to perform complex, multi-step, real-world tasks [23,26]. Our work is most similar to BC that purely considers operating on offline data.

Causal Imitation Learning. One challenge of IL is that expert trajectories may be influenced by latent confounders. It is difficult for an imitator to recover the performance of an expert policy in a confounding environment. Causal imitation learning(CIL) aims to eliminate the confounding of observed expert trajectories. Zhang et al. [35] address the problem of unobserved confounders that influence expert's actions and outcomes under a given causal graph, providing general learnability conditions. Recently, Swamy et al. [31] consider the situation where expert demonstrations may be corrupted by temporally correlated unobserved confounders, i.e., temporally correlated noise(TCN). They use IV regression to deconfound the data, but without considering pre-identifying the presence of unobserved confounders. We follow the setting with unobserved TCN and improve this problem by proposing a two-stage algorithm.

Causal Discovery with Latent Confounders. Existing methods to do causal discovery with latent confounders are flourishing, e.g., FCI [28,29], lvLiNGAM [16,27], MLCLiNGAM [6], LSTC [5], MD-LiNA [34], etc. To adapt to time series, many algorithms e.g., tsFCI [13], SVAR-FCI [22], TS-LiNGAM [18], TiMINo [24], etc. emerge. However, these methods may suffer from the low recall problem, which motivates Gerhardus et al. [14] to derive their Latent Peter-Clark Momentary Conditional Independence method (abbreviated as LPCMCI). They intuitively focused on searching the correct causal parents into the conditional sets so as to yield high-recall performances. We engage LPCMCI to determine the existence of latent confounders for demonstration trajectories in Stage 1 of our proposed algorithm.

3 Problem Formalization

In this section, we give our problem formalization with some definitions. We focus on a Markov Decision Process (MDP) setting, which is parameterized by $\langle S, A, T, r \rangle$. S is the state space, A is the action space, $T : S \times A \to S$ represents the transition function, while $r : S \times A \to [-1, 1]$ is the reward function. Let $\pi \in \Pi$ denote a policy. Since in IL, we have no access to the reward but the demonstration trajectories $\mathcal{D} = \{\tau_i\}_{i=1}^{N}$ from the expert policy π_E, where the ith trajectory $\tau_i = (s_t, s_{t+1}, a_{t+1})_{t=0,..,T}$ contains a sequence of the state s_t at time t, and the next state s_{t+1} and action a_{t+1}. N is the number of trajectories while T is the horizon.

For the generalization of our framework, we allow the non-existence and existence of latent counfounders. In the non-existence cases where there is no confounding, our model is identical as the traditional IL model, as demonstrated in Fig. 1(a). Following Fig. 1(a), the expert's demonstrations are assumed to be generated by,

$$s_t = T(s_{t-1}, a_{t-1}); \tag{1}$$

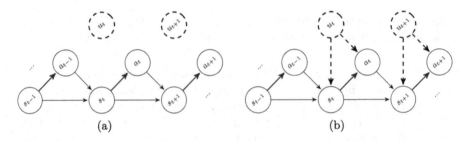

Fig. 1. The causal graphs (a) without latent counfouders and (b) with latent confounders, where s_t, a_t, u_t stand for the state, action and latent confounder at the timestep t, respectively. Note that in (a), s_t and a_t are both independent with u_t; while in (b), s_{t-1} can be served as an IV to deconfound the influences of u_t on a_t.

$$a_t = \pi_E(s_t). \tag{2}$$

In the latter case with latent confounders, we assume that the trajectories from the expert are generated by, as shown in Fig. 1(b),

$$s_t = \mathcal{T}(s_{t-1}, a_{t-1}, u_t); \tag{3}$$

$$a_t = \pi_E(s_t, u_t), \tag{4}$$

where the confounder u_t at timestep t influences the current state s_t as well as the action a_t. u_t is assumed to have zero means. In the very established causal graph, we see that the past state s_{t-1} works as an IV, in that it enjoys the following three characteristics: i) $s_{t-1} \perp\!\!\!\perp u_t$, i.e., s_{t-1} is independent of u_t; ii) $s_{t-1} \not\perp\!\!\!\perp s_t$, i.e., s_{t-1} is dependent with s_t; and iii) $s_{t-1} \perp\!\!\!\perp a_t \mid u_t, s_t$, i.e., s_{t-1} is conditionally independent of a_t given u_t and s_t. Intuitively, the IV's independence with u_t renders it offer some sources of randomness for the action a_t to break up the confounding effects of u_t.

Here our goal is to recover an imitating policy, which eliminates the estimation bias possibly induced from the confounders, and further matches the underlying data distributions.

4   Method

We now present a two-stage approach for offline causal IL, where in Stage 1, we aim at determining the presence of the latent confounder for each trajectory while in Stage 2, we perform a variant of IV regression to entail an imitating policy. In the following, we give corresponding descriptions for each stage.

In nature, due to the generation mechanism and complicated dynamics of MDP, an influential latent confounding on the current state and action could affect future states and actions as well. Hence, for brevity, once we detect the existence of latent confounders, we regard such a trajectory as confounded.

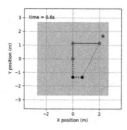 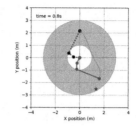

Fig. 2. (Left) Prismatic 2-link robot (called Bumblebee system) following a target trajectory (green dotted line) to a target position (green star). (Right) Rotational 3-link robot (called Butterfly system) moving towards a target position (green star). The gray area corresponds to the robot's reachable workspace, the black dashed line indicates the robot's initial position, and the orange line marks the trace of the previous position of the robot's tip. (Color figure online)

To do so, in Stage 1, we adopt a causal discovery method for time series data, called LPCMCI (Latent Peter-Clark Momentary Conditional Independence), to test latent confounders [14]. LPCMCI is a constraint-based method that mines temporally conditional independence relationships to uncover the lagged and contemporaneous causal structures between variables of interests, allowing the presence of latent confounders. It essentially takes advantages of the confounder's characteristics and some orientation rules to facilitate the test, with no assumptions on types of variables, or functional relationships. More details could be referred to in [14].

If detecting no confounders for trajectories, in Stage 2, we adopt a behavioral cloning method; while we employ a variant of IV regression method in cases where confounders are detected. In particular, as for the behavioral cloning method, we attempt to minimize an offline IL objective [11,32],

$$\min_{\pi} \mathbb{E}_{\tau_i}[(a_{t+1} - \pi(s_{t+1}))^2]. \tag{5}$$

As for the IV regression method, we seek to tackle the objective function following [10,31],

$$\min_{\pi} \max_{f} \mathbb{E}_{\tau_i}[2(a_{t+1} - \pi(s_{t+1}))f(s_t) - f(s_t)^2], \tag{6}$$

where f is a function mapping IV states s_t to Lagrange multipliers that control implicitly the slack of differences between true and predicted actions. Applying the same game-theoretic approach to imitation learning as [10,31], we yield iteratively the approximated function f and the policy pi.

5   Experiments

To evaluate the proposed method, We perform experiments on the robotic arm control task in the ROBO track of the NeurIPS 2021 learning by doing competition which considers the relations between causality, control and RL [33].

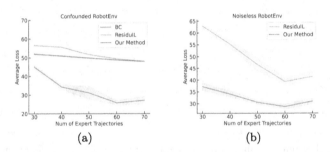

Fig. 3. (a) System losses of our method, BC and ResiduIL in the uncertain noisy robot environment. (b) System losses of our method and ResiduIL in the noiseless robot environment.

The ROBO track is a continuous task, whose goal is to learn a policy to control the tip of the robot so that it follows a given target process, as illustrated in Fig. 2. More specifically, the states of the robot include the current positions of all robotic arms and their derivatives, as well as the target position of the robot tip at the next timestep. The action is the abstract controls of the robot.

Based on the robotic arm simulator, we generate expert trajectories by the Linear-Quadratic Regulator (LQR) controller, randomly adding new Gaussian noise as well as cached noise from the previous timestep when generating the expert's action. And we control the proportion of noisy data in the dataset to be 0.5. Such cached noises are served as the latent confounders. We follow the competition's evaluation criteria to evaluate the policy, with the following loss,

$$J_i := b_i \cdot \int_0^2 ||Z^i(t) - z_*^i(t)||_2^2 dt + c_i \cdot \int_0^2 U^i(t)^T U^i(t) dt, \qquad (7)$$

where $Z^i(t)$ is the current robot tip position and $z_*^i(t)$ is the target position for each system $i \in \{1, ..., 24\}$. U_i is the control variable, and b_i and c_i are scaling constants. All results are averaged over five random seeds, and standard deviations are shown in the shaded data.

We mainly compare our method with ResiduIL [31] and Behavioral Cloning [3] in an uncertain confounded environment (with uncertain confounders). In the noise-free environment (without confounders), our method is equivalent to Behavioral Cloning, hence we only compare it with the ResiduIL.

5.1   Results of Different Environments

Here, we consider different environments, i.e., with and without latent confounders. It is worth noting that we train all three types of robot arms from the competition. As shown in Fig. 3(a), in the cases with uncertain confounders, performances of ResiduIL and BC are close, while our method performs significantly better by taking into account the detection of the presence of the latent

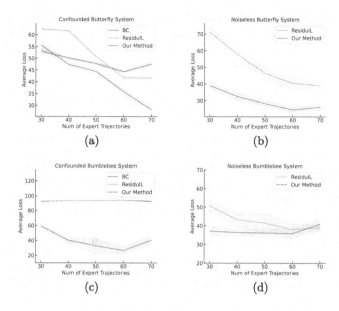

Fig. 4. Comparison results of our method with BC and ResiduIL in the (a) Confounded Butterfly System; (b) Noiseless Butterfly System; (c) Confounded Bumblebee System; and (d) Noiseless Bumblebee System.

confounder, which demonstrates the effectiveness of latent confounder identification. In the noise-free environment, from Fig. 3(b), we can see that our method obtains better results with ResiduIL.

5.2   Results of Different Robot Arms

The simulation environment provides three types of robotic arms, namely butterfly, beetle, and bumblebee. Each robotic arm has 8 subdivision types. Taking two robotic arms, butterfly and bumblebee as examples, we show the experimental results of our method in noisy and noiseless environments. As shown in Fig. 4, we can see that our method outperforms other algorithms in both robotic environments with and without uncertain confounders. Notably, the huge performance difference between our method and ResiduIL in confounded environment shows that terms of identification of latent confounder play a pivotal role in our objective.

6   Conclusions

To deal with the latent confounders issues, we propose a two-stage offline causal imitation learning algorithm to identify a policy from the expert's demonstrations. In Stage 1, we determine the existence of the confounding effects; while in Stage 2, we deploy an invariant of instrumental variable regression method to mitigate the detected confounding effects or we take the best of a behavioral cloning method for the unconfounding cases. Experiments demonstrate that our method outperforms other methods with/without latent confounders on the robotic arm control task.

References

1. Argall, B.D., Chernova, S., Veloso, M., Browning, B.: A survey of robot learning from demonstration. Robot. Auton. Syst. **57**(5), 469–483 (2009)
2. Aytar, Y., Pfaff, T., Budden, D., Paine, T., Wang, Z., De Freitas, N.: Playing hard exploration games by watching youTube. In: Advances in Neural Information Processing Systems 31 (2018)
3. Bain, M., Sammut, C.: A framework for behavioural cloning. In: Machine Intelligence 15, pp. 103–129 (1995)
4. Bareinboim, E., Forney, A., Pearl, J.: Bandits with unobserved confounders: a causal approach. In: Advances in Neural Information Processing Systems 28 (2015)
5. Cai, R., Xie, F., Glymour, C., Hao, Z., Zhang, K.: Triad constraints for learning causal structure of latent variables. In: Advances in Neural Information Processing Systems 32 (2019)
6. Chen, W., Cai, R., Zhang, K., Hao, Z.: Causal discovery in linear non-gaussian acyclic model with multiple latent confounders. In: IEEE Transactions on Neural Networks and Learning Systems (2021)
7. Codevilla, F., Müller, M., López, A., Koltun, V., Dosovitskiy, A.: End-to-end driving via conditional imitation learning. In: 2018 IEEE International Conference on Robotics and Automation (ICRA), pp. 4693–4700. IEEE (2018)
8. Codevilla, F., Santana, E., López, A.M., Gaidon, A.: Exploring the limitations of behavior cloning for autonomous driving. In: Proceedings of the IEEE/CVF International Conference on Computer Vision, pp. 9329–9338 (2019)
9. De Haan, P., Jayaraman, D., Levine, S.: Causal confusion in imitation learning. In: Advances in Neural Information Processing Systems 32 (2019)
10. Dikkala, N., Lewis, G., Mackey, L., Syrgkanis, V.: Minimax estimation of conditional moment models. In: Advances in Neural Information Processing Systems, vol. 33, pp. 12248–12262 (2020)
11. Ding, Y., Florensa, C., Abbeel, P., Phielipp, M.: Goal-conditioned imitation learning. In: Advances in Neural Information Processing Systems 32 (2019)
12. Edwards, A., Sahni, H., Schroecker, Y., Isbell, C.: Imitating latent policies from observation. In: International Conference on Machine Learning, pp. 1755–1763. PMLR (2019)
13. Entner, D., Hoyer, P.O.: On causal discovery from time series data using FCI. In: Probabilistic Graphical Models, pp. 121–128 (2010)
14. Gerhardus, A., Runge, J.: High-recall causal discovery for autocorrelated time series with latent confounders. In: Advances in Neural Information Processing Systems, vol. 33, pp. 12615–12625 (2020)
15. Ho, J., Ermon, S.: Generative adversarial imitation learning. In: Advances in Neural Information Processing Systems 29 (2016)
16. Hoyer, P.O., Shimizu, S., Kerminen, A.J., Palviainen, M.: Estimation of causal effects using linear non-gaussian causal models with hidden variables. Int. J. Approximate Reasoning **49**(2), 362–378 (2008)

17. Hussein, A., Gaber, M.M., Elyan, E., Jayne, C.: Imitation learning: a survey of learning methods. ACM Comput. Surv. (CSUR) **50**(2), 1–35 (2017)
18. Hyvärinen, A., Shimizu, S., Hoyer, P.O.: Causal modelling combining instantaneous and lagged effects: an identifiable model based on non-gaussianity. In: Proceedings of the 25th International Conference on Machine Learning, pp. 424–431 (2008)
19. Kumor, D., Zhang, J., Bareinboim, E.: Sequential causal imitation learning with unobserved confounders. In: Advances in Neural Information Processing Systems, vol. 34, pp. 14669–14680 (2021)
20. Li, J., Luo, Y., Zhang, X.: Causal reinforcement learning: an instrumental variable approach. Available at SSRN 3792824 (2021)
21. Liao, L., Fu, Z., Yang, Z., Wang, Y., Kolar, M., Wang, Z.: Instrumental variable value iteration for causal offline reinforcement learning. arXiv preprint arXiv:2102.09907 (2021)
22. Malinsky, D., Spirtes, P.: Causal structure learning from multivariate time series in settings with unmeasured confounding. In: Proceedings of 2018 ACM SIGKDD Workshop on Causal Discovery, pp. 23–47. PMLR (2018)
23. Niekum, S., Osentoski, S., Konidaris, G., Chitta, S., Marthi, B., Barto, A.G.: Learning grounded finite-state representations from unstructured demonstrations. Int. J. Robot. Res. **34**(2), 131–157 (2015)
24. Peters, J., Janzing, D., Schölkopf, B.: Causal inference on time series using restricted structural equation models. In: Advances in Neural Information Processing Systems 26 (2013)
25. Pomerleau, D.A.: Efficient training of artificial neural networks for autonomous navigation. Neural Comput. **3**(1), 88–97 (1991)
26. Ratliff, N., Bagnell, J.A., Srinivasa, S.S.: Imitation learning for locomotion and manipulation. In: 2007 7th IEEE-RAS International Conference on Humanoid Robots, pp. 392–397. IEEE (2007)
27. Salehkaleybar, S., Ghassami, A., Kiyavash, N., Zhang, K.: Learning linear non-gaussian causal models in the presence of latent variables. J. Mach. Learn. Res. **21**, 39–1 (2020)
28. Spirtes, P., Glymour, C.N., Scheines, R., Heckerman, D.: Causation, prediction, and search. MIT press (2000)
29. Spirtes, P., Meek, C., Richardson, T.: Causal inference in the presence of latent variables and selection bias. In: Proceedings of the Eleventh Conference on Uncertainty in Artificial Intelligence, pp. 499–506 (1995)
30. Sun, W., Venkatraman, A., Gordon, G.J., Boots, B., Bagnell, J.A.: Deeply aggrevated: differentiable imitation learning for sequential prediction. In: International Conference on Machine Learning, pp. 3309–3318. PMLR (2017)
31. Swamy, G., Choudhury, S., Bagnell, J.A., Wu, Z.S.: Causal imitation learning under temporally correlated noise. arXiv preprint arXiv:2202.01312 (2022)
32. Torabi, F., Warnell, G., Stone, P.: Behavioral cloning from observation. arXiv preprint arXiv:1805.01954 (2018)
33. Weichwald, S., et al.: Learning by doing: controlling a dynamical system using causality, control, and reinforcement learning. arXiv preprint arXiv:2202.06052 (2022)
34. Zeng, Y., Shimizu, S., Cai, R., Xie, F., Yamamoto, M., Hao, Z.: Causal discovery with multi-domain lingam for latent factors. In: 30th International Joint Conference on Artificial Intelligence, IJCAI 2021, pp. 2097–2103. International Joint Conferences on Artificial Intelligence (2021)
35. Zhang, J., Kumor, D., Bareinboim, E.: Causal imitation learning with unobserved confounders. Adv. Neural. Inf. Process. Syst. **33**, 12263–12274 (2020)

36. Zhang, T., et al.: Deep imitation learning for complex manipulation tasks from virtual reality teleoperation. In: 2018 IEEE International Conference on Robotics and Automation (ICRA), pp. 5628–5635. IEEE (2018)
37. Zheng, B., Verma, S., Zhou, J., Tsang, I., Chen, F.: Imitation learning: Progress, taxonomies and opportunities. arXiv preprint arXiv:2106.12177 (2021)

Efficent Gradient Propagation for Robot Control and Learning

Philipp Ruppel$^{(\boxtimes)}$ and Jianwei Zhang

Department of Informatics, Universität Hamburg, 22527 Hamburg, Germany
`ruppel@informatik.uni-hamburg.de`

Abstract. The recent wealth of discoveries in deep learning has coincided with the development of specialized automatic differentiation frameworks which can efficiently propagate gradients through repeating structures in artificial neural networks. For model-based approaches, automatic differentiation still performs relatively poorly and it is common to formulate gradients manually or to focus on low-dimensional problems. To accelerate research into model-based control of high-DOF robots such as humanoids with articulated hands and to enable hybrid approaches that combine model-based methods with deep learning, we develop a novel automatic differentiation framework that can evaluate gradients of robot models around previous candidate solutions multiple times faster than state-of-the-art methods.

Keywords: automatic differentiation · machine learning · robot control

1 Introduction

Many interesting problems in robotics and machine learning are formulated as optimization tasks. Finding solutions efficiently often depends on the ability to quickly evaluate gradients. Two different and largely incompatible approaches have been developed for model-based control and for deep learning.

Deep learning has benefited greatly from specialized automatic differentiation frameworks [1–3], which exploit the regular structure of current neural network architectures. The solvers [4] are commonly based on matrix-free first-order steepest gradient descent. Explicitly computed gradient matrices would be prohibitively large and dense.

Methods that are commonly used in robot control and motion planning rely on matrix-based higher-order approximations [5–8]. The gradient matrices are often computed explicitly using manually designed algorithms [9–13]. In principle, it is also possible to apply similar methods in a matrix-free fashion by replacing the gradient matrices with forward and reverse gradient propagation

This work was supported in part by the German Research Foundation (DFG) in project Crossmodal Learning under Grant SFB TRR 169.

and solving the inner linear equations with methods that only require matrix-vector products (equivalent to gradient propagation), such as conjugate gradient descent.

It has been a long-standing dream to directly combine deep neural networks with model-based approaches, but this would require unified optimization frameworks that can efficiently handle both robot models and neural networks. Improved automatic differentiation frameworks could also allow for higher-dimensional control problems to be solved directly (e.g. full-body control of humanoids with articulated hands).

Full gradient matrices of deep neural networks would typically be prohibitively large and dense. We therefore opt for a matrix-free approach. However, we want to support optimization methods that are also appropriate for robot control and motion planning (e.g. Gauss-Newton and sequential quadratic programming). The gradient matrix can be replaced by an automatically generated program, which computes the result of a matrix-vector product without forming the full gradient matrix. The solutions to the inner linear equations are computed via conjugate gradient descent. The original nonlinear program only has to be evaluated once for each inner linear equation while the gradient programs have to be computed repeatedly. Finding solutions efficiently requires quickly evaluating multiple gradient directions for each non-linear solution.

We identify two main challenges: nonlinearities and spatial rotations. While the gradient matrix or matrix replacement program is a linear function, it's parameterization (relation to the nonlinear program) is nonlinear. State-of-the-art automatic differentiation frameworks therefore have to evaluate non-linear operators during gradient propagation, even though the relation of output gradients to input gradients is linear. Spatial rotations introduce further problems. Small rotations or gradients can be represented efficiently via 3D vectors. Absolute orientations require higher-dimensional representations, such as quaternions or matrices. When formulating gradient terms manually, the programmer can choose appropriate representations [9–11]. State-of-the-art automatic differentiation frameworks[1–3,14], however, use the same types for values and gradients, resulting in quaternion or matrix gradients even though simple gradient vectors would suffice.

We present a novel automatic differentiation framework that can eliminate non-linear operations from the forward and reverse gradient passes by generating additional linearization and accumulation programs and automatically transform data types into simpler gradient representations. We used our framework in previous work to learn neural network policies for dexterous manipulation tasks within minutes using differentiable physics [15]. In this paper, we provide a more detailed explanation of our automatic differentiation framework and compare the performance to existing methods.

2   Related Work

Model-based approaches for robot control and motion planning typically use manually written programs to fill out gradient matrices [5,6,9–11]. These often represent local quadratic approximations of the objectives and linear approximations of the constraints [6–8,10].

Artificial neural networks are commonly trained in a matrix-free fashion via steepest gradient descent i.e. back-propagation [16]. Special software frameworks have been developed to evaluate these gradients efficiently [1–3], exploiting repeating structures and large numbers of linear operations in current artificial neural network architectures [16].

Efforts have recently been made to develop special frameworks and software libraries to enable efficient gradient propagation for model-based approaches [14,17]. These can already outperform [14] general-purpose automatic differentiation frameworks. Runtime performance can be accelerated through just-in-time compilation [14], making use of existing compiler infrastructures [18–20], which can automatically simplify certain expressions and transform the instructions into native machine code.

Automatic differentiation frameworks can use several methods to represent gradients. In the simplest case, each value is paired with a gradient and the gradients are propagated forward together with evaluating the original non-linear function [21]. This approach can be fast and easy to implement, but does not support reverse gradient propagation. By first executing the non-linear function once and recording all operations on a gradient tape, it is possible to propagate gradients in both directions [3,14,21]. An alternative approach would be to directly define the mathematical expression as a graph instead of a function, but this can be less intuitive. Some frameworks combine both models [1,2].

Current works attempt to combine model-based methods, such as physics simulators, with deep learning. Early attempts do so by training only the neural network with gradient-based methods and collecting discrete samples from model-based representations, e.g. using deep reinforcement learning [22–25]. More recent works investigate direct gradient propagation between neural networks and model-based representations [26,27].

3   Methods

3.1   Problem Representation

The optimization problem is defined as a function and first executed once to record all operations on a gradient tape. The function can be defined in C++ or Python. We provide additional methods to mark variables as free variables, objectives and constraints. Move instructions are inserted as needed to avoid ambiguities. The recorded memory addresses are rewritten and all temporary variables are packed into a single contiguous memory block by a memory allocator.

3.2   Program Simplification

We apply several program transformations to simplify the recorded instruction list. We precompute constant expressions, identify and collapse constants with equal values, replace zero constants with a zero operator, skip unnecessary move instructions, remove unused instructions and constants, and finally defragment the memory allocations again.

3.3   Type and Operator Definitions

Differentiable operators are implemented via C++ functions. Several functions can be associated with each operator to represent different modes, including the original non-linear function, a linearizer, forward and reverse gradient propagation, and an accumulation method. Data types can be associated with different gradient types via templates.

3.4   Linearization Programs

Traditional automatic differentiation frameworks have to re-evaluate non-linear operations during repeated gradient propagation even though the mapping from input gradients to output gradients is linear. Instead, we first compute a linearizer for each instance of a non-linear operator. The linearizers are combined into a linearization program, which computes linearizations for all non-linear operators. Data types of the linearizations are inferred from the function signatures of the linearizers. After evaluating the original nonlinear function once and initializing the linearizers, we can repeatedly propagate gradients back and forth, e.g. to solve a local linear or quadratic approximation via conjugate gradient descent, without having to re-evalute non-linear operations.

Using the previously generated linearization program, we can build programs for forward and reverse gradient propagation from only linear multiply and add instructions. We iterate over all instructions, allocate new memory addresses for the gradients, and insert pointers to the linearizations. For backpropagation, the process is similar, but if the same variable is used twice, we have to generate code to accumulate the incoming gradients.

Current automatic differentiation frameworks use the same data types to represent values and gradients. However, this is not always the most efficient solution. While a 3D orientation should be represented as a quaternion or matrix, the gradient can be handled more efficiently as a small 3D vector. Poses in 3D space should be represted by matrices or vector-quaternion pairs, while their derivatives can be treated as 6D vectors. We specify these substitution rules via C++ templates.

Finally, after solving a local linear or quadratic approximation of the cost landscape, the result has to be added back to the non-linear solution. To accumulate gradients with modified types, the addition operator is overloaded with the nonlinear types as the first argument and the linear type as the second argument. This also provides a convenient opportunity to enforce invariants, such as keeping orientation quaternions normalized or re-orthogonalizing transformation matrices.

4 Operators

In the following, we give exemplary descriptions of differentiable operators and their simplified linearizations, as provided by our automatic differentation framework.

4.1 Rotations

We represent differentiable rotations using a combination of 3D vectors r_n and quaternions q_n. The nonlinear program uses the quaternion representation and the gradient program uses the vector representation. We define a function $m(r_n)$ that maps from the vector representation to the quaternion representation. The orientation O_n is defined as a combination of both representations.

$$m(r_n) = \frac{r_{n,0}i}{2} + \frac{r_{n,1}j}{2} + \frac{r_{n,2}k}{2} + 1 \qquad O_n = m(r_n)\, q_n \qquad (1)$$

Concatenating two rotations with their gradients involves both the vectors and the quaternions. During the nonlinear pass $f_{nonlinear}(q_a, q_b)$, the gradient is zero and we can simply multiply the quaternions q_a and q_b. Since the gradient is premultiplied, the linearizer f_{linear} has to transform the second gradient vector r_b into the first frame and we can simply store the first quaternion q_a. The forward pass $f_{forward}$ transforms the second gradient vector r_b into the first frame and adds the rotation vectors r_a and $q_a r_b$. For the reverse pass $f_{reverse}$, the first quaternion is inverted.

$$m(r_x)\, q_x = m(r_a)\, q_a\, m(r_b)\, q_b \qquad (2)$$

$$f_{nonlinear}(q_a, q_b) = q_a\, q_b \qquad f_{linear}(q_a, q_b) = q_a \qquad (3)$$

$$f_{forward}(q_a, r_a, r_b) = r_a + q_a\, r_b \qquad f_{reverse}(q_a, r_x) = (r_x\,,\, q_a^{-1}\, r_x) \qquad (4)$$

4.2 Transformations

For poses, we also have to account for positions p_n and their derivatives v_n, in addition to orientation quaternions o_n and rotation vectors r_n. We could convert each representation into a 4-by-4 transformation matrix using a function g. When concatenating two poses, their components and their derivatives would satisfy the following equation.

$$g(p_x)\, g(v_x)\, g(r_x)\, g(q_x) = g(p_a)\, g(v_a)\, g(r_a)\, g(q_a)\, g(p_b)\, g(v_b)\, g(r_b)\, g(q_b) \qquad (5)$$

The linearizer $h_{linear}(p_a, q_a, p_b, q_b)$ stores the first orientation and precomputes the position $s_{a,b}$ of the second frame relative to the first one. The forward

Fig. 1. Robot model of a UR5 arm with a Shadow C6 hand.

$h_{forward}$ and reverse $h_{reverse}$ gradient passes transform both the translations and rotations.

$$s_{a,b} = q_a \, p_b \qquad h_{linear}(p_a, q_a, p_b, q_b) = (q_a \; , \; s_{a,b}) \qquad (6)$$

$$h_{forward}(q_a, s_{a,b}, v_a, r_a, v_b, r_b) = (v_a + q_a v_b + r_a \times s_{a,b} \; , \; r_a + q_a r_b) \qquad (7)$$

$$h_{forward}(q_a, s_{a,b}, v_x, r_x) = (v_x \; , \; s_{a,b} \times v_x + r_x \; , \; q_a^{-1} v_x \; , \; q_a^{-1} r_x) \qquad (8)$$

4.3   Angle-Axis

Robot applications often require rotating a coordinate frame by an angle α around a locally fixed axis y, e.g. to simulate revolute joints. While this can be a relatively expensive operation in the non-linear case, linearizations k with angle gradient α', axis gradient y' and rotational gradient vector r_x can be computed efficiently.

$$k_{linear}(\alpha, y) = (\alpha \; , \; y) \qquad k_{forward}(\alpha, y, \alpha', y') = (0 \; , \; \alpha \, y' + \alpha' \, y) \qquad (9)$$

$$k_{reverse}(\alpha, y, r_x) = (r_x \cdot y \; , \; \alpha \, r_x) \qquad (10)$$

5   Experiments

5.1   Correctness

We propagate gradients through sequences of 5 randomly generated spatial transformations each and compare the results to a baseline with traditional scalar automatic differentiation. The results agree with a root mean square deviation of $1 \cdot 10^{-15}$.

Table 1. Performance comparison of our automatic differentiation framework with baselines as described in Sect. 5.3 and existing methods as described in Sect. 5.2. We give average numbers of times per millisecond each method can propagate gradients from the base of the robot to the end-effector or back.

	Shadow		UR5		Shadow+UR5	
	Fwd	Rev	Fwd	Rev	Fwd	Rev
1/matrix/CasADi/interpreter	56	53	54	51	45	44
1/matrix/CasADi/JIT	60	57	60	54	55	54
1/matrix/CasADi/JIT-O3	61	57	60	58	56	55
1/matrix/CasADi/JIT-Ofast	59	55	58	57	56	55
1/quaternion/CasADi/interpreter	52	50	49	48	37	38
1/quaternion/CasADi/JIT	59	61	60	60	51	53
1/quaternion/CasADi/JIT-O3	60	61	58	58	55	53
1/quaternion/CasADi/JIT-Ofast	58	57	58	58	55	54
1/ours/baseline/baseline	42	23	111	65	32	18
1/ours/baseline/simplified	284	148	282	174	198	96
1/ours/improved/baseline	277	239	275	372	239	195
1/ours/improved/simplified	459	437	457	441	407	387
8/matrix/CasADi/interpreter	244	195	235	207	130	142
8/matrix/CasADi/JIT	327	228	321	281	220	210
8/matrix/CasADi/JIT-O3	312	268	226	254	245	244
8/matrix/CasADi/JIT-Ofast	333	275	357	297	250	227
8/quaternion/CasADi/interpreter	185	167	165	158	81	90
8/quaternion/CasADi/JIT	317	304	312	304	210	216
8/quaternion/CasADi/JIT-O3	307	324	273	284	255	269
8/quaternion/CasADi/JIT-Ofast	322	326	291	305	239	258
8/ours/baseline/baseline	47	22	152	75	34	14
8/ours/baseline/simplified	603	249	620	259	310	116
8/ours/improved/baseline	602	368	1400	1008	455	272
8/ours/improved/simplified	2283	1767	2303	1786	1524	1088
64/matrix/CasADi/interpreter	478	300	465	343	203	181
64/matrix/CasADi/JIT	682	538	708	524	372	348
64/matrix/CasADi/JIT-O3	850	638	843	613	490	449
64/matrix/CasADi/JIT-Ofast	823	621	846	593	495	440
64/quaternion/CasADi/interpreter	286	284	254	246	102	109
64/quaternion/CasADi/JIT	654	623	632	605	332	345
64/quaternion/CasADi/JIT-O3	798	696	795	737	434	501
64/quaternion/CasADi/JIT-Ofast	815	773	796	764	453	493
64/ours/baseline/baseline	33	15	138	56	24	11
64/ours/baseline/simplified	759	267	761	265	342	94
64/ours/improved/baseline	734	411	2073	1236	553	308
64/ours/improved/simplified	4550	2935	4618	2882	2319	1439

5.2   Comparison

We compare the performance of our approach to an existing automatic differentiation framework. The third-party framework CasADi [14] has been specifically designed with control and robotic applications in mind and the authors report superior performance compared to other state-of-the-art frameworks. We load URDF models of a Shadow C6 hand [28], of a UR5 arm [29], and of a C6 hand mounted to a UR5 arm (see Fig. 1), using our framework and using urdf2casadi [17], and measure the speed at which each framework can propagate gradients through kinematic chains, from the base of each robot to the end-effector (tool link for the UR5, tip of the first finger for the Shadow hand and for the combined hand-arm system). While our framework also support kinematic trees, urdf2casadi is currently limited to chains. Simulating trees using multiple parallel chains might lead to even lower performance for the third-party framework, so for a meaningful comparison, we restrict the experiments in this paper to kinematic chains. We do, however, test different numbers of parallel robots, simulating trajectory optimization across multiple time steps or learning in multiple parallel simulations. For CasADi, we repeat the experiments with and without just-in-time compilation and with different compiler options. We further repeat the experiments with matrices and with quaternions to represent spatial rotations. We perform our experiments on a single core of an Intel(R) Core(TM) i7-8700K CPU at 3.70 GHz. Our approach is multiple times faster than the previous methods. See Table 1 for results.

5.3   Ablation Study

We create additional baselines for comparison based on our own framework by disabling program simplification (Sect. 3.2) and by replacing the vector representations (Sect. 3.3–4.3) with quaternions and scalar differentiation. Each of these changes leads to a significant decrease in performance and results similar to the state-of-the-art method. Disabling both features leads to even worse performance. Results are given in Table 1.

6   Conclusions and Future Work

We developed an automatic differentiation framework that can evaluate gradients of robot models around previous candidate solutions multiple times faster than previous methods. In our other work, we could already use our approach to efficiently learn neural network policies for robotic manipulation with differentiable physics [26]. While the software is currently still under active development, we hope to release a stable version of our framework soon and that it might help to support future research on combining model-based approaches with deep learning.

References

1. Abadi, M., Agarwal, A., Barham, P.: TensorFlow: Large-scale machine learning on heterogeneous systems. (2015) http://tensorflow.org/
2. Chollet, F., et al.: Keras (2015) https://keras.io
3. Paszke, A., et al.: PyTorch: An Imperative Style. High-Performance Deep Learning Library. Curran Associates Inc., Red Hook, NY, USA (2019)
4. Kingma., D.P., Adam, J.B.: A method for stochastic optimization. In: Proceedings International Conference for Learning Representations (2014)
5. Schulman, J., et al.: Motion planning with sequential convex optimization and convex collision checking. Int. J. Robo. Res. **33**, 1251–1270 (2014)
6. Beeson, P., Ames, B.: TRAC-IK: an open-source library for improved solving of generic inverse kinematics. In: Proceedings. IEEE RAS Humanoids Conference, Seoul, Korea (2015)
7. Shanno, D.F.: Who invented the interior-point method? Documenta Mathematica. Optimization Stories, Extra Volume (2012)
8. Karmarkar, N.: A new polynomial-time algorithm for linear programming. Combinatorica **4**(4), 373–395 (1984)
9. Featherstone, R.: Rigid Body Dynamics Algorithms. Springer-Verlag, Berlin, Heidelberg (2007)
10. Smits, R.: KDL: kinematics and Dynamics Library. http://www.orocos.org/kdl
11. Coleman, D., Sucan, I., Chitta, S., Correll, N.: Reducing the barrier to entry of complex robotic software: a moveit! case study. J. Softw. Eng. Robo. (2014)
12. Todorov, E., Erez, T., Tassa, Y.: Mujoco: a physics engine for model-based control. In: 2012 IEEE/RSJ International Conference on Intelligent Robots and Systems, pp. 5026–5033, (2012)
13. Coumans, E.: Bullet physics simulation. In ACM SIGGRAPH 2015 Courses, SIGGRAPH '15, New York, USA, Assoc. Comput. Mach. (2015)
14. Andersson, J., Åkesson, J., Diehl, M.: Casadi: a symbolic package for automatic differentiation and optimal control. 87, 01 (2012)
15. Ruppel, p., Zhang, J.: Learning object manipulation with dexterous hand-arm systems from human demonstration. In: 2020 IEEE/RSJ International Conference on Intelligent Robots and Systems (IROS), pp. 5417–5424 (2020)
16. Lecun, Y., Bottou, L., Bengio, Y., Haffner, P.: Gradient-based learning applied to document recognition. Proceedings IEEE **86**(11), 2278–2324 (1998)
17. Johannessen, G., Maria, L., Arbo, H.M., Gravdahl, J.T.: Robot dynamics with URDF CasADi. In 2019 7th International Conference on Control, Mechatronics and Automation (ICCMA). IEEE (2019)
18. GNU Compiler Collection. https://gcc.gnu.org/git/gcc.git
19. The LLVM compiler infrastructure. https://github.com/llvm/llvm-project
20. Clang C language family frontend for LLVM. https://clang.llvm.org/
21. Bell, B.: Cppad: a package for c++ algorithmic differentiation. http://www.coin-or.org/CppAD
22. Open, A.I., et al.: Learning dexterous in-hand manipulation. Int. J. Robo. Res. (2018)
23. Open, A.I., et al.: Solving Rubik's cube with a robot hand. (2019)
24. Tingguang, L.I., et al.: Learning to solve a Rubik's cube with a dexterous hand. In Proceedings. IEEE International Conference on Robotics and Biomimetics (2019)
25. Rajeswaran, A., Kumar, V.: Learning complex dexterous manipulation with deep reinforcement learning and demonstrations. In: Proceedings. Robotics Science and Systems (RSS) (2018)

26. Ruppel, P., Hendrich, N., Zhang, J.: Direct policy optimization with differentiable physical consistency for dexterous manipulation. In: 2021 IEEE International Conference on Robotics and Biomimetics (ROBIO), (2021)
27. Heiden, E., et al.: Neuralsim: augmenting differentiable simulators with neural networks. In: 2021 IEEE International Conference on Robotics and Automation (ICRA), pp. 9474–9481 (2021)
28. Shadow robot common packages. https://github.com/shadow-robot/sr_common.git
29. Ros-industrial universal robots. https://github.com/ros-industrial/universal_robot.git

Robot Navigation Using Reinforcement Learning with Multi Attention Fusion in Crowd

Linxiang Li[1], Yang Chen[2], Fuchun Sun[2(✉)], and Run Guo[1]

[1] School of Electrical Engineering, Yanshan University, Qinhuangdao, China
[2] Deaprtment of Computer Science and Technology, Tsinghua University, Beijing, China
{chen_yang,fcsun}@tsinghua.edu.cn

Abstract. Due to the complex human movement and social behavior norm, robot reaching target point with collision-free in crowd safely and quickly is a challenging problem, which can be considered as human-robot interaction. In this paper, focusing on human-friendly navigation in crowd, we propose to (i) design a reinforcement learning based fusion model of Transformer network and attention module for mobile robot to achieve collision-free navigation task, (ii) adopt probability grid map as environment expression to rich robot perception information, and (iii) use an attention-based Transformer encoder to learn the human-robot interaction. A motion planning framework based on multi attention fusion mechanism is designed to learn the information of human-robot interaction and human-human interaction. Neighboring humans are assigned different importance to improve robot navigation performance. Finally, simulation experiments are conducted to demonstrate the effectiveness of the proposed method which can accurately predict human movement and make mobile robot navigate socially with collision-free.

Keywords: Robot navigation · Reinforcement learning · Multi attention fusion

1 Introduction

With the development of artificial intelligence technology, the mobile robot working environment is changing from single fixed scene to high dynamic complex scene, such as the dense crowd navigation situation as shown in Fig. 1. Human has the innate ability to precept the rapid changes of the environment, however, for robot, it is still a complex issue to encode the high dynamic environment to establish human motion prediction model accurately.

Fuchun Sun is currently a Full Professor with the Department of Computer Science and Technology, Tsinghua University, Beijing, China. His current research interests include cross-modal learning, active perception, and precise operation and teleoperation.

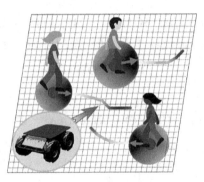

Fig. 1. Human-robot and human-human interactions for navigation in crowds.

The traditional navigation technology is mainly based on the occupied grid map, regarding the dynamic obstacles as static at a certain time for collision avoidance causes the unnatural or short-sighted robot actions [3,5,18], especially when the number of dynamic obstacles are large and dense. They are necessary to model the environment quickly, predicting the behavior of dynamic obstacles exactly, and reaching the target point to complete the task safely and timely in literature [14].

Some recent works use reinforcement learning to obtain the interaction strategy between robot and humans. These strategies use heuristics method to model human-robot interaction, some are not even considered. They also make some assumptions, such as linear human motion, to simplify the human motion prediction problem, and some works leverage the current state rather than predicting human motion state to design robot navigation strategy.

The method of dynamic crowd motion prediction by Long-Short Term Memory (LSTM) where the hidden states enable interactive learning of context information has attracted more and more researchers'concerns [10]. Based on self-attention mechanism, Transformer does not depend on the order of input data, which can learn the correlation between candidates accurately [21].

The bidirectional encoder representations from Transformers proposed by the Google for generating word vectors has greatly improved the effect of Natural language processing. The most important part of BERT is Transformer, which abandons the traditional convolutional neural network and Rerrent Neural Network (RNN). The entire network structure is designed based on attention mechanism consists of self-attention and feed forward neural network [2].

In this paper, we propose a novel RL-based collision-free motion planning framework with environment encoding and relation learning in crowd [16]. Probability Grid Map (PGM) is adopted as environment expression. Transformer as a self-attention module is employed to encode surroundings and learn the relationship between obstacles. The adaptive reward function is quoted for guiding the robot to target as far as possible. The main contributions of this paper include:

1) A novel a RL-based attention fusion motion planning framework in crowd is proposed.

2) PGM is adopted as environment expression to rich the states of perception.

3) Transformer encoder is employed to learn the information about PGM.

The rest of this paper is organized as follows. Section 2 introduces the related work on motion planning. Section 3 describes the problem statement and the proposed method. Section 4 provides the experiment results. Section 5 concludes this paper.

2   Related Work

For the problem of robot navigation in crowd, researchers mainly focused on the crowd navigation and used traditional algorithms to perform tasks. Learning human-robot and human-human interaction are the key parts of social navigation. With the development of intelligent algorithms, navigation tasks based on reinforcement learning have become an interesting research field.

Crowd Navigation. The mobile robot navigation in crowd is challenging due to the complex way of human intention and social interaction which determine human actions. In traditional methods, researchers mainly characterize the crowd based on distance and some hand-crafted rules [1]. This method has been proved effective under rigorous assumptions [9]. Previous work is done using a rule-based algorithm have been taken to describe the interaction between agents. Helbing *et al*, called this interaction between humans as "social forces" [13]. Under reciprocal assumptions, RVO [3] and ORCA [4] solve this problem by modeling mathematical description of interaction process. Interacting Gaussian process models the trajectory of each pedestrian in the crowd based on Gaussian model, an interaction potential term is proposed to keep independent [20]. In this work, we study an attention fusion navigation model to solve the crowd navigation problem.

Relational Learning. During crowd navigation, the relationship between robot and obstacles (or humans), as well as the interaction among humans are important information, which is a sequence prediction problem about time. RNN [7], such as LSTM networks, is a classic frame design for sequence prediction problem. Vemula *et al*, [22] proposed to use attention mechanism to gather the information from a series of RNNs into a spatial-temporal graph. However, conventional RNN ignores the interactions between humans. To tackle this problem, in recent years, a large number of studies have pointed to the impact of human-human interaction. Gupta *et al*, [12] added a "pooling module" after the generator to extract the interactive information between pedestrians. Xu *et al*, [25] used LSTMs to design double encoders motion encoder" and "location encoder") to learn the temporal and spatial location information in crowd. In order to fully perceive the environment and learn the relational information between them, we introduce PGM to represent the environment around the agent and Transformer encoder network is used to learn the relationship.

Reinforcement Learning. The types of RL algorithms consist of value-based (such as PPO [17]), policy-based (such as DQN [19]) and actor-critic (such as A3C [15]) are used widely. Due to off-policy DQN has the characteristics of simple structure, small calculation, fast training speed and easy convergence, it has been employed for navigation tasks. The core of RL-based algorithm lies in the design of reward function, which largely determines the effect of the experiment. Besides, effective representations of the environment are also helpful for training convergence. Through modeling the environment by CNN, Finn *et al*, [11] obtained a state transition function by collecting and learning historical frames to move forward to the goal position as fast as possible. LSTM-RL [24], CADRL [3], SARL [6], and AEMCARL [23] are all the prevalent RL crowd navigation models recently. In our work, the dynamic environment and complex social behaviors are learned by RL to make the robot complete tasks with collision-free.

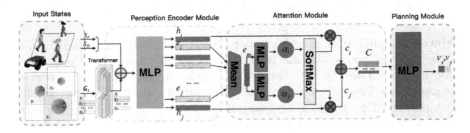

Fig. 2. Overview of our method for multi attention fusion navigation framework, containing 3 modules: perception encoder module, attention module, planning module. The joint states of robot-human and the probability grid map are used as inputs. The probability grid map first passes through the Transformer encoder network, and then is combined with the joint states to the perception encoder module composed of MLP. Through the fusion of the attention module, planning module generates motion instructions [6].

3   Method

Our task is to make the robot reach the destination as soon as possible in the dynamic crowd with collision-free. In this section, we present a novel RL-based collision-free motion planning framework includes perception encoder module, attention module and planning module in crowd, as shown in Fig. 2. A Transformer encoder is adopted to extract the relation in the possibility grid map. A simple and effective reward function is designed to guide RL algorithm to learn an optimal strategy.

3.1   Problem Definition

In this paper, we simplify a task that a robot navigates towards a goal in crowd of n humans. The robot state $r_t \in \mathbb{R}^{D_r}$ (D_r is feature size of robot) and the

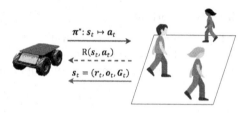

Fig. 3. RL principle based on robot-human navigation in crowd.

states of obstacles $o_t \in \mathbb{R}^{N*D_o}$ (D_o and N are the dimension and number of obstacles, respectively) are denoted as the observation of the environment at time t [6], $G_t = \{g_i, g_j, \dots, g_n\} \in \mathbb{R}^{l \times l \times n}$ is set of probability grid map around robot (or human) in a square range about $l \times l$ (m^2) at time t [23].

$$r_t = \left[p_x^r, p_y^r, v_x^r, v_y^r, v_{pref}^r, g_x, g_y\right], o_t = \left[p_x^o, p_y^o, v_x^o, v_y^o, v_{pref}^o\right] \tag{1}$$

$\boldsymbol{P^r} = \left[p_x^r, p_y^r\right]$ and $\boldsymbol{V^r} = \left[v_x^r, v_y^r\right]$ denote robot's position and velocity along the X-Y axis, respectively. v_{pref}^r is the preferred velocity in the navigation task of the goal, $\boldsymbol{G} = [g_x, g_y]$. The state of obstacle o_t includes position, $\boldsymbol{P^o} = \left[p_x^o, p_y^o\right]$, velocity, $\boldsymbol{V^o} = \left[v_x^o, v_y^o\right]$, and preferred velocity, v_{pref}^o. $s_t = (r_t, o_t, G_t) \in \mathbb{R}^{N*D_o+D_r+l^2}$ is the robot's input joint states of our network. The objective of RL in crowd navigation is to find an optimal motion strategy, $\boldsymbol{\pi^*}: \boldsymbol{s_t} \mapsto \boldsymbol{a_t}, (\boldsymbol{a_t} = [v_x, v_y]$ is action taken at time t), which through repeated exploratory training to guide the robot to reach the target point as soon as possible under the condition of safety, as shown in Fig. 3.

$$\pi^*(s_t) = argmaxR(s_t, a_t) + \gamma^{\Delta t \cdot v^{pref}} \int_{s_{t+\Delta t}} P(s_t, a_t, s_{t+\Delta t})V^*(s_{t+\Delta t})ds_{t+\Delta t}, \tag{2}$$

$$V^*(s_t) = \sum_{k=t}^{T} \gamma^{k \cdot v^{pref}} R^k(s_k, \pi^*(s_k)), \tag{3}$$

where $R(s_t, a_t)$ is the reward received at time t, γ is the discount factor (default $\gamma = 0.9$ in this paper), $\gamma \in [0,1]$, and V^* is the optimal state-action value obtained, $P(s_t, a_t, s_{t+\Delta t})$ is the probability of transitioning from s_t to $s_{t+\Delta t}$ when a_t is implemented at time t [23].

This paper focuses on the following problems:

1) How to express the environment as complete as possible without adding too much computation.
2) How to fully learn the knowledge in the environment, such as robot-human interaction, human-human interaction.

3.2 Perception Encoder Module

Probability Grid Map. In order to more accurately represent the environment, considering the states of agent (robot) and obstacles (humans), we add environment perception - probability grid map in this paper, as shown in Fig. 4.

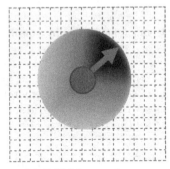

(a) PGM for one human   (b) PGM for robot in one frame

Fig. 4. Probability grid map

In order to reduce the computation, we grid the square area of $l \times l(m^2)$ (default $l = 4$ in this paper) around the robot, which denoted as $p_{grid}^t(x^t, y^t)$ to represent the occupancy probability at the position (x^t, y^t) in grid map at time t, can be compute as fellow [23]:

$$p_{\text{grid}}^t \left(x^t, y^t \right) = \sum_{i=1}^{N} p^x \left(x^t - x_i^o \right) * p^y \left(y^t - y_i^o \right) * p^\theta \left(\theta^t - \theta_i^o \right), \tag{4}$$

$$p^x = N\left(0, \delta_x\right), p^y = N\left(0, \delta_y\right), p^\theta = N\left(0, \delta_\theta\right), \tag{5}$$

$$\theta_i^o = \tan^{-1}\left(\frac{v_i^y}{v_i^x}\right), \theta^t = \tan^{-1}\left(\frac{y^t - y_i^o}{x^t - x_i^o}\right), \tag{6}$$

where N is the number of obstacles in the square range of $l \times l(m^2)$, and p^x, p^y, p^θ is the hyperparameter obeying Gauss distribution ,which all the mean are 0 and the variance are δ^x, δ^y, δ^θ. (x_i^o, y_i^o), (v_i^x, v_i^y), θ_i^o are the position, velocity and heading angle of obstacle i, respectively. θ^t is the angle between (x^t, y^t) and (x_i^o, y_i^o) in the x-axis. $p_{grid}^t(x^t, y^t)$ is the cumulative probability at each grid. As shown in Fig. 4(b), for obstacle i, we only consider a circle range where center is (x_i^o, y_i^o), radius is $\sqrt{(v_i^x)^2 + (v_i^y)^2} * \Delta t$, where Δt is the time prediction length. Due to the environment state added, our reward function ate consisted of [6,23], which penalizes collisions or uncomfortable distances, as follow:

$$R(s_t, a_t) = \begin{cases} -0.25, & collision \\ 1, & arrival \\ \alpha \cdot \left(-0.1 + \frac{d_t}{2}\right) + (1 - \alpha) \cdot \beta \cdot P_{collision}, & otherwise \end{cases}, \tag{7}$$

$$P_{collision} = \sum_{(x,y) \in \phi_{agent}} p_{grid}(x,y), \tag{8}$$

where α, β are the hyperparameters(in this paper, $\alpha = 0.5$, $\beta = -0.5$), d_t is the minimum distance between robot and humans, $P_{collision}$ is the sum of probability in the coverage ϕ_{agent} of the agent as shown in Fig. 4(b).

Transformer Encoder Network. Extracting important information from the environment can help agent complete the navigation without collision, we introduce the Transformer network based on self-attention mechanism as the perception encoder. The self-attention of Transformer module is given by [21]:

$$\begin{cases} \boldsymbol{Q} = MLP(\boldsymbol{W}^q \boldsymbol{G}_t + b^q), \\ \boldsymbol{K} = MLP(\boldsymbol{W}^k \boldsymbol{G}_t + b^k), \\ \boldsymbol{V} = MLP(\boldsymbol{W}^v \boldsymbol{G}_t + b^v), \\ \boldsymbol{Attention}(\boldsymbol{Q}, \boldsymbol{K}, \boldsymbol{V}) = SoftMax(\frac{\boldsymbol{Q}\boldsymbol{K}^T}{\sqrt{d_k}})\boldsymbol{V}, \end{cases} \tag{9}$$

where $\boldsymbol{G}_t$ is grid map at time t, $\boldsymbol{W}^q, \boldsymbol{W}^k, \boldsymbol{W}^v$ are the weights of $MLP(\cdot)$, which denote multi layer perception with baises b^q,b^k,b^v, respectively. d_k is the dimension of the feature, $\boldsymbol{Q}$, $\boldsymbol{K}$, $\boldsymbol{V}$ are three feature embedding layers. $\boldsymbol{Attention}(\boldsymbol{Q}, \boldsymbol{K}, \boldsymbol{V})$ is output of the module.

$\boldsymbol{Attention}(\boldsymbol{Q}, \boldsymbol{K}, \boldsymbol{V})$ and $\boldsymbol{s}_t$ are embeded into a fixed length vector $\boldsymbol{e}_i$ by $MLP(\cdot)$ with $ReLU$, which hidden units is (150,100) [23]. And then the embedding vector $\boldsymbol{e}_i$ is fed to a subsequent $MLP(\cdot)\cdot$ that hidden units is (100,50) to obtain the pairwise interaction feature between the robot and person i: (Note that $\boldsymbol{W}_e$ and $\boldsymbol{W}_h$ are the different weights.)

$$\boldsymbol{e}_i = ReLU(MLP(\boldsymbol{s}_t, \boldsymbol{Attention}(\boldsymbol{Q}, \boldsymbol{K}, \boldsymbol{V}), \boldsymbol{W}_e)), \tag{10}$$

$$\boldsymbol{h}_i = ReLU(MLP(\boldsymbol{e}_i, \boldsymbol{W}_h)). \tag{11}$$

3.3   Attention Module

The number of human are dynamically changing in robot perception range, so we need to map the variable input to a fixed length. In a fixed range, different distances, speeds and directions have different effects on the agent. Recently, a large number of studies have explored the different human-robot impact [8]. The interaction embedding $\boldsymbol{e}_i$ is transformed into an attention score α_i as follow [6]:

$$\boldsymbol{e}_m = \frac{1}{n} \sum_{k=1}^{n} (\boldsymbol{e}_k), \tag{12}$$

$$\alpha_i = ReLU(MLP(\boldsymbol{e}_i, \boldsymbol{e}_m, \boldsymbol{W}_\alpha)), \tag{13}$$

where $\boldsymbol{e}_m$ is a fixed-length embedding vector obtained by mean pooling, $\boldsymbol{W}_\alpha$ is weight of $MLP(\cdot)$ with (100,100) hidden units and $ReLU$.

Given vector $\boldsymbol{h}_i$ and the corresponding attention score α_i for each neighbor i in a fixed range, the representation of the scene is a weighted linear combination as fellow [6]:

$$c = \sum_{i=1}^{n} SoftMax(\alpha_i) \cdot \boldsymbol{h}_i. \tag{14}$$

3.4 Planning Module

For our experiment, the output action of robot $\boldsymbol{a}_t = [v_x, v_y]$ is obtained through a simple $MLP(\cdot)$ where hidden units is (150,100,100) with $ReLU$ [6].

$$\boldsymbol{a}_t = ReLU(MLP(\boldsymbol{s}_t, \boldsymbol{c}, \boldsymbol{W}_p)) \tag{15}$$

where $\boldsymbol{s}_t$ is the joint state, $\boldsymbol{c}$ is the output of Attention module, $\boldsymbol{W}_p$ is weight.

4　Experiments

Three state-of-the-art methods LSTM-RL [24], CADRL [3], SARL [6] are implemented as a baseline to compare with our method. We took 12 h to train the module by 10000 episodes with 1 agent (robot) and 5 obstacles (humans). Success rate, collision rate, navigation time and reward value are shown in the Fig. 5.

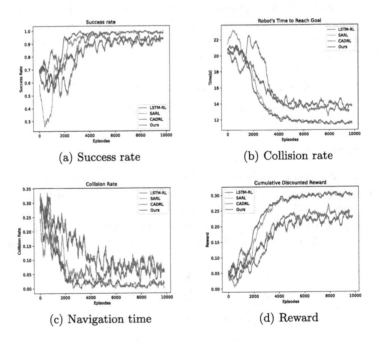

(a) Success rate　　　　　　　　(b) Collision rate

(c) Navigation time　　　　　　　(d) Reward

Fig. 5. Learning curves in training. Blue, orange, green and red represent LSTM-RL, SARL, CADRL, Ours, respectively. (Color figure online)

Due to the multi attention fusion module, our (Red) method improves the success rate, cumulative discounted reward continuously in the early stage, outperforms all baseline methods. As for collision rate and navigation time, our (Red) method also show excellent performance. Although the indexes of SARL (Orange) are generally consistent with ours in the later stage, it is obvious that the advantages of ours are reflected in the first 3000 episodes significantly.

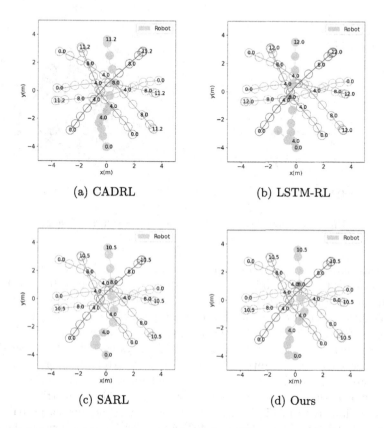

Fig. 6. Trajectory and time comparison in an invisible test case. Circles are the positions of robots at the labeled time, where yellow is robot, and other colors represent humans. (Color figure online)

Our goal is to reach the destination as fast as possible without collision, so the navigation time and the length of trajectory are important indices. We further compare the navigation time and the length of trajectory of different methods when robots and humans are invisible to each other, as shown in Fig. 6. The robot moves to the intersection of the crowd at 4.0s, and different algorithms tend to different navigation strategies. CADRL goes straight through the crowd, LSTM-RL first judges the wrong direction of movement, resulting in navigation path growing. Moreover, when encountering a large number of humans, it directly adopts the "freezing-point" strategy to wait, which eventually led to maximum navigation time. The navigation time of SARL and ours are same, SARL experiences several obvious direction changes. In contrast, our motion path is more comfortable and smoother.

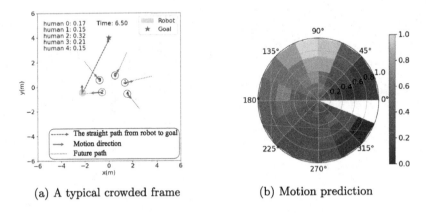

(a) A typical crowded frame   (b) Motion prediction

Fig. 7. Motion prediction based on multi attention fusion.

In this paper, we use Transformer encoder to learn the interaction between humans, and attention module to assign different importance to each human. Figure 7 shows the attention score of our algorithm for a typical crowd frame in the navigation. Because human #2 is moving towards the robot and the distance between robot and human #2 is nearest, giving the maximum attention score 0.32. The current motion state of human #3 shows that it will appear on the path of the robot in the future, so giving a large attention score 0.21. Other humans are far away from the robot and tend to be more farther, so they are given low attention score. Figure 7(b) shows the probability distribution of the robot's motion trend in Fig. 7(a). It can be seen that the motion probability of the robot at 90° is highest, but (0°, 45°) and (290°, 315°) is lowest, the result is consistent with the analysis in Fig. 6.

5   Conclusion

In this work, a RL-based fusion model consisting of Transformer encoder and attention module is proposed to solve the problem of the robot navigation in crowd safely and quickly. Probability grid map is used to enrich environment perception and improve navigation efficiency. In order to extract the interactive information between human-human and human-robot, we use Transformer encoder for relationship learning and fuse the attention score to judge the motion direction. Through 10000 episodes of training, the experiment results shows that our method outperforms obvious advantages in the early stage of training, maintains a high success rate and reward in the later, and also possesses strong judgment function in motion prediction.

Acknowledgements. This work is supported in part by National Key Research and Development Program of China with No. 2021ZD0113801 and in part by Beijing Science and Technology Plan Project No. Z191100008019008.

References

1. Alahi, A., Ramanathan, V., Fei-Fei, L.: Socially-aware large-scale crowd forecasting. In: Proceedings of the IEEE Conference on Computer Vision and Pattern Recognition pp. 2203–2210 (2014)
2. Bahdanau, D., Cho, K., Bengio, Y.: Neural machine translation by jointly learning to align and translate. arXiv preprint arXiv:1409.0473 (2014)
3. Van den Berg, J., Lin, M., Manocha, D.: Reciprocal velocity obstacles for real-time multi-agent navigation. In: 2008 IEEE International Conference on Robotics and Automation, pp. 1928–1935. IEEE (2008)
4. Berg, J.V.D., Guy, S.J., Lin, M., Manocha, D.: Reciprocal n-body collision avoidance. In: Robotics Research, pp. 3–19. Springer, Heidelberg (2011). https://doi.org/10.1007/978-3-642-19457-3_1
5. Borenstein, J., Koren, Y.: Real-time obstacle avoidance for fast mobile robots. IEEE Trans. Syst. Man Cybern. **19**(5), 1179–1187 (1989)
6. Chen, C., Liu, Y., Kreiss, S., Alahi, A.: Crowd-robot interaction: crowd-aware robot navigation with attention-based deep reinforcement learning. In: 2019 International Conference on Robotics and Automation (ICRA), pp. 6015–6022. IEEE (2019)
7. Cho, K., et al.: Learning phrase representations using rnn encoder-decoder for statistical machine translation. arXiv preprint arXiv:1406.1078 (2014)
8. Conneau, A., Kiela, D., Schwenk, H., Barrault, L., Bordes, A.: Supervised learning of universal sentence representations from natural language inference data. arXiv preprint arXiv:1705.02364 (2017)
9. Coscia, P., Castaldo, F., Palmieri, F.A., Alahi, A., Savarese, S., Ballan, L.: Long-term path prediction in urban scenarios using circular distributions. Image Vision Comput. **69**, 81–91 (2018)
10. Everett, M., Chen, Y.F., How, J.P.: Motion planning among dynamic, decision-making agents with deep reinforcement learning. In: 2018 IEEE/RSJ International Conference on Intelligent Robots and Systems (IROS), pp. 3052–3059. IEEE (2018)
11. Finn, C., Levine, S.: Deep visual foresight for planning robot motion. In: 2017 IEEE International Conference on Robotics and Automation (ICRA), pp. 2786–2793. IEEE (2017)
12. Gupta, A., Johnson, J., Fei-Fei, L., Savarese, S., Alahi, A.: Social gan: socially acceptable trajectories with generative adversarial networks. In: Proceedings of the IEEE Conference on Computer Vision and Pattern Recognition, pp. 2255–2264 (2018)
13. Helbing, D., Molnar, P.: Social force model for pedestrian dynamics. Phys. Rev. E **51**(5), 4282 (1995)
14. Kretzschmar, H., Spies, M., Sprunk, C., Burgard, W.: Socially compliant mobile robot navigation via inverse reinforcement learning. Int. J. Rob. Res. **35**(11), 1289–1307 (2016)

15. Mnih, V., et al.: Asynchronous methods for deep reinforcement learning. In: International Conference on Machine Learning, pp. 1928–1937. PMLR (2016)
16. Mnih, V., et al.: Human-level control through deep reinforcement learning. Nature **518**(7540), 529–533 (2015)
17. Schulman, J., Wolski, F., Dhariwal, P., Radford, A., Klimov, O.: Proximal policy optimization algorithms. arXiv preprint arXiv:1707.06347 (2017)
18. Snape, J., Van Den Berg, J., Guy, S.J., Manocha, D.: The hybrid reciprocal velocity obstacle. IEEE Trans. Rob. **27**(4), 696–706 (2011)
19. Sorokin, I., Seleznev, A., Pavlov, M., Fedorov, A., Ignateva, A.: Deep attention recurrent q-network. arXiv preprint arXiv:1512.01693 (2015)
20. Trautman, P., Krause, A.: Unfreezing the robot: navigation in dense, interacting crowds. In: 2010 IEEE/RSJ International Conference on Intelligent Robots and Systems, pp. 797–803. IEEE (2010)
21. Vaswani, A., et al.: Attention is all you need. Adv. Neural Inf. Process. Syst. **30**, 1–11 (2017)
22. Vemula, A., Muelling, K., Oh, J.: Social attention: modeling attention in human crowds. In: 2018 IEEE International Conference on Robotics and Automation (ICRA), pp. 4601–4607. IEEE (2018)
23. Wang, S., Gao, R., Han, R., Chen, S., Li, C., Hao, Q.: Adaptive environment modeling based reinforcement learning for collision avoidance in complex scenes. arXiv preprint arXiv:2203.07709 (2022)
24. Wang, X., Girshick, R., Gupta, A., He, K.: Non-local neural networks. In: Proceedings of the IEEE Conference on Computer Vision and Pattern Recognition, pp. 7794–7803 (2018)
25. Xu, Y., Piao, Z., Gao, S.: Encoding crowd interaction with deep neural network for pedestrian trajectory prediction. In: Proceedings of the IEEE Conference on Computer Vision and Pattern Recognition, pp. 5275–5284 (2018)

Three-Dimensional Force Sensor Based on Deep Learning

Qingling Duan[1,2,3], Qi Zhang[4,5], Dong Luo[1], Ruofan Yang[1,3], Chi Zhu[1], Zhiyuan Liu[1], and Yongsheng Ou[1(✉)]

[1] Shenzhen Institute of Advanced Technology, Chinese Academy of Sciences, Shenzhen 518055, Guangdong, People's Republic of China
ys.ou@siat.ac.cn
[2] Key Laboratory of Human-Machine Intelligence-Synergy Systems, Shenzhen Institutes of Advanced Technology, Chinese Academy of Sciences, Shenzhen, China
[3] Shenzhen College of Advanced Technology, University of Chinese Academy of Sciences, Shenzhen, China
[4] Guangdong Provincial Key Laboratory of Robotics and Intelligent System, Shenzhen Institutes of Advanced Technology, Chinese Academy of Sciences, Shenzhen, China
[5] Guangdong-Hong Kong-Macao Joint Laboratory of Human-Machine Intelligence-Synergy Systems (#2019B121205007), Shenzhen, China

Abstract. Human skin can accurately sense subtle changes of both normal and shear forces. However, tactile sensors applied to robots are challenging in decoupling 3D forces due to the inability to develop adaptive models for complex soft materials. Therefore, a new soft tactile sensor has been designed in this paper to detect shear and normal forces, including a soft probe and image acquisition device. First, to capture the deformation of the sensor, colored silicone squares were embedded in the soft probe. Capcamera movement of the colored squares under external forces. The image dataset collected at different 3D forces is then input into a deep learning model. Finally, a custom miniature image device is acquired and embedded in the soft probe to miniaturize the sensor. Computing results obtained from experimental datasets show that the proposed method can accurately decouple the 3D forces. Robots can grap vulnerable objects with sensors prepared at the robot's tip. The tactile sensors studied in this paper are expected to be applied in robotics fields such as adaptive grasping, dexterous manipulation and human-computer interaction.

Keywords: Tactile sensing · Robot technology · Deep learning

1 Introduction

Human skin contains four mechanoreceptors (SA-I, SA-II, RA-I, RA-II), allowing humans to perceive subtle changes in force during contact with objects accurately [1,2]. Moreover, force perception is a natural appeal to barriers to visual

---

Q. Duan, Q. Zhang and D. Luo—These authors contributed equally to this work.

© The Author(s), under exclusive license to Springer Nature Singapore Pte Ltd. 2023
F. Sun et al. (Eds.): ICCCS 2022, CCIS 1732, pp. 259–268, 2023.
https://doi.org/10.1007/978-981-99-2789-0_22

perception [3]. However, when a robot has visual perception impairment, such as insufficient light supply and occlusion, in complex tactile contact task scenarios such as robot dexterous operation, good tactile feedback (such as contact force) can provide rich proprioception, resulting in more reliable operation and control strategies [4]. Therefore, designing soft force sensors like human skin is critical to the robot field, which can promote robot development.

The traditional soft mechanical sensor generally detects the force signal through the change of capacitor [5,6], resistance [7,8] and other electrical signals caused by the deformation under the action of external force. However, with the development of vision algorithms, visual and tactile sensors have emerged, which model the contact colloidal deformation information captured by the camera as tactile information such as force signals through visual algorithms. With the advantages of high spatial resolution, low cost and rich tactile information, it has gradually become the hot spot direction of tactile sensors [9].

The design of vision-tactile sensors comprises the colloidal contact layer, light source structure and camera imaging system. In MIT Gelsight's visual-tactile sensor [10], they introduced labeled points on the reflective film inside the soft elastomer to capture the displacement of labelled points under 3 D forces and established the mapping relationship between the labelled point displacement and 3 D forces through finite element analysis to realize 3 D force detection in the soft environment. The TacTip series optic tactile sensor [11] was proposed by the University of Bristol Nathan et al. The TacTip sensor mimics the human fingertip touch-body receptor structure by embedding an array-distributed pin in the colloidal layer, thereby conducting the deformation information on the sensor surface using the camera system to observe the movement of the pin array. Meta's Digit sensor, the [12], optimizes the structure of the sensor to integrate it into the fingertips for robotic operations. In addition, Sui et al. of Tsinghua University have developed the Tac3D tactile sensor [13], which contains an optical path system refracted by four light mirrors and can achieve a virtual binocular imaging effect with a monocular camera. Cui et al. of the Institute of Automation, Chinese Academy of Sciences, have proposed the GelStereo visual and tactile sensing series [14] based on binocular vision, which can obtain binocular tactile images simultaneously and then recover the contact depth information through the stereo matching algorithm.

A soft 3 D force sensor is presented in this paper that can accurately decouple the 3 D force. It is achieved by soft silicone, camera, light source, etc., as shown in Fig. 1 (a). The 3 D force information was converted into the image information of the silicon surface, as shown in Fig. 1 (c), and a deep learning method was used to decouple the 3 D forces. The proposed force sensor has a small shape, high decoupling accuracy and high flexibility, which is suitable for various operations of robot fingertips.

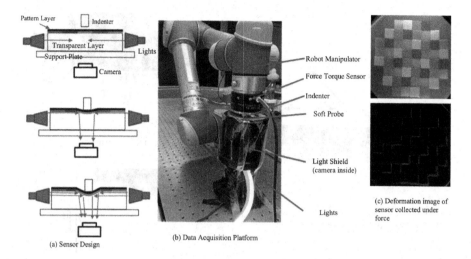

Fig. 1. The principle design of 3d force sensor, data acquisition platform, and the image data collected.

2 Sensing Principle and Sensor Algorithm

2.1 Deformation Mode of the Soft Probe

The main idea of decoupling the 3D force is to record the deformation of the sensor through the camera and calculate the 3D force applied to the sensor through the deformation image taken. As shown in Fig. 1(a), place the camera at the bottom of the sensor to capture the silicone deformation. Sorta-clear^{TM}12 silicone gel was used to prepare the sensor. As shown in Fig. 2, to obtain quantifiable features, the top of the transparent silicone is covered with a pattern layer made of the arrangement and combination of a plurality of small colored silicone squares of the same softness. The top of the pattern layer is covered with a black silicone layer called the bottom layer. The soft probe consists of a transparent layer, a pattern layer, and a base layer. A transparent bottom plate is placed at the bottom of the soft probe, serving as a support plate for the soft probe so that the soft probe can move relative to the bottom. Light from the LED light source is reflected into the camera through the soft probe. When the soft probe is stationary, i.e., when no force is applied, the light captured by the camera comes from the pattern layer of the soft probe. The following is the analysis and quantification of mechanical deformation and resulting pattern changes in the presence of shear and normal forces. Suppose the silicone is a cuboid with a height of H and a bottom length of L.

Shear Deformation. When the upper surface of the soft probe moves in parallel relative to the bottom, the resulting angular shape variable is proportional to the

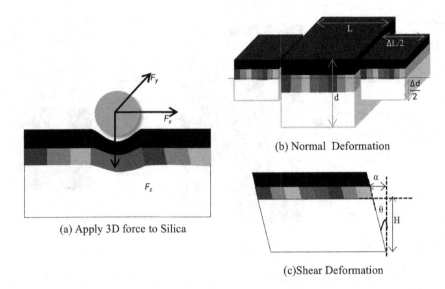

(a) Apply 3D force to Silica

(b) Normal Deformation

(c)Shear Deformation

Fig. 2. 3D Force Applied to the Soft Probe Cause the Normal Deformation and Shear Deformation of the Surface

shear force applied to the surface, $F = G\theta$, where G stands for shear coefficient, and angular variable θ can be calculated by measuring shear deformation α:

$$\theta = tan^{-1}\alpha/H. \tag{1}$$

Small color squares in the pattern layer will move to different degrees according to the direction and size of the shear force.

Normal Deformation. When the cylindrical soft probe stretches along its horizontal axis, according to the poisson effect, the thickness change of the cross-section Δd is :

$$\Delta d = dv\Delta L/L, \tag{2}$$

where d is the original thickness, v is the poisson ratio, L is the length before stretching, and ΔL is the change of length. As a result, the periphery of the pattern layer extends outward.

2.2   Deep Learning Model Adopted to Decouple the 3D Forces

The pixel values of the corresponding image taken by the camera will also change due to the deformation of the color image layer. It is not easy to map the force deformation to the inductive surface. Encouraged by the success of deep learning in tactile perception, this paper presents a convolutional neural network model. Specifically, a multi-output CNN model is used to extract features of high-resolution deformation images and, eventually, get three-dimensional forces. The accuracy of the prediction depends on the sensor material and preparation.

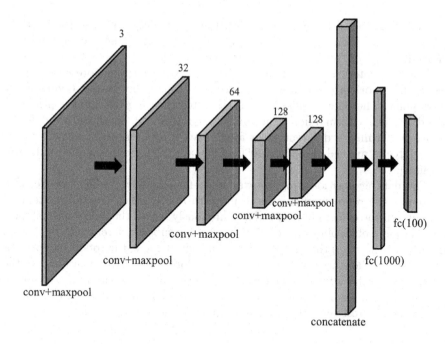

Fig. 3. The Decoupling 3D Force Network Structure

The decoupling 3D force network structure is shown in Fig. 3, divided into feature extraction and regression. The feature extraction block is composed of 4 convolutional layers and pooling layers. The convolution kernel size is 5×5 for the first and second layers and 3×3 for the last two layers, the number of channels is 32, 64, 128, 128, respectively, and the step size is 2. Max pooling size is 2×2, and the step size is 2. Finally, the features obtained by the feature extraction are connected and passed to the regression layer. The regression layer consists of two fully connected layers, and the number of channels is 1000, 100. The final output is a three-dimensional force. The means square error (MSE) was used as a loss function to carry out backpropagation during training. The neural network was optimized using a stochastic gradient descent optimizer(SGD), with a batch size of 64 and a learning rate of 10^{-5}. The GPU server Tesla p40 was trained for 120,000 iterations.

3   Experiment and Results

3.1   Automatic Data Acquisition Platform

As shown in Fig. 1(b), the main body of the collection platform uses an industrial robot(UNIVERSAL ROBOTS, UR5e Robot), which is used as a force application device and is equipped with a force/torque sensor(ROBOTIC, FT 300-S Force Torque Sensor) to record force data in real-time. A 20 mm diameter indenter is mounted at the end of the force/torque sensor used to act on the surface of the soft probe. Communication with the mechanical arm uses the TCP/IP protocol. A position servo of 125 hz was used to control the arm-end-indenter applied to the soft probe at the same speed.

The fabricated sensor is fixed to the optical platform. Before each experiment, the robotic arm's end was moved to the top of the sensor, serving as the start point. The data acquisition process is to reach a given eight depths at a speed of 2 mm/s, form different normal forces, under each depth to 1000 different positions, forming a different shear force, back to the starting point after each shear force is applied. Finally, 24000 (1000 × 8 ×3) group sampled data was obtained.

The data acquisition program was written in Labview, which was realized to control the robot manipulator to move to the specified position, automatically obtain the camera image, and synchronously record the 3-dimensional force information. Soft probes vary over time, bringing subtle differences to the image. In order to eliminate the bias, after applying shear force, the robot manipulator will return to the start point to collect the no-force data, each experiment of the normal force (shear force) image minus the original image (no force) to obtain the difference map. The differential-treated dataset is divided into the train set (70%) and the test set (30%) for the deep learning model.

Table 1. Sensor Performance

Size		$22\,\text{mm} \times 22\,\text{mm} \times 22\,\text{mm}$
Shore Hardness		12A(soft)
Measuring Range	F_x	$\pm 40\,\text{N}$
	F_y	$\pm 40\,\text{N}$
	F_z	$0\text{--}70\,\text{N}$
3D Force Measurement Error (RMSE)	F_x	$0.34\,\text{N}$
	F_y	$0.43\,\text{N}$
	F_z	$0.68\,\text{N}$
Bandwidth		$30\,\text{Hz}$

3.2 Sensor Performance

The final response of the manufactured sensor is obtained from the color image, which depends on multiple factors. The size of the sensor depends mainly on the camera, and using a more miniature camera will reduce the size of the sensor. Collect the data into a deep learning model for training and got the root-mean-square error (RMSE) of the test set for F_x, F_y and F_z were 0.34 N, 0.43 N, and 0.68 N. Running under a computer with a graphics processing unit (NVIDIA GeForce GTX 3080 Ti). The sensor bandwidth is around 30 hz. The measuring range of the sensor is shown in Table 1. Figure 4 shows that the force measured by the designed sensor is close to the ground truth after the calibration.

Table 2. Comparison of state-of-the-art three-dimensional force sensor and force decoupling methods.

	Gelsight [10]	Digit [12]	Kakani V [18]	Tac3D [13]	GelStereo [14]	Ours
Decoupling force method	Array marker points, finite element modeling	Not reported	Binocular vision, deep learning (vgg-16)	Finite element modeling	Finite element analysis	Deep learning
Normal force (shear force) measurement	Yes	Not reported	Yes	Yes	Yes	Yes
Size	Bulky	20 mm * 27 mm * 18 mm	Bulky	Bulky	Bulky	22 mm * 22 mm * 22 mm
Sensor Construction/ Features	Bulky, poor model adaptability	Compact, cheap, undeveloped force measurement function	Bulky	Bulky, poor model adaptability	Poor model adaptability	Soft, Strong, Compact, minimum size

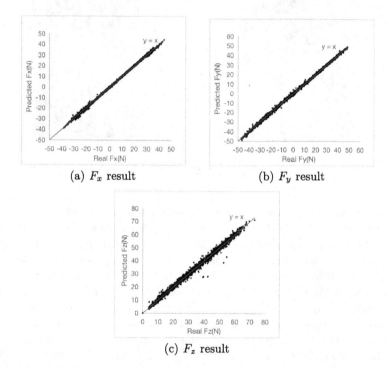

(a) F_x result   (b) F_y result

(c) F_z result

Fig. 4. The comparison of the force measured by the designed sensor and the ground truth in x, y and z direction. The red lines represent the ideal result, and the black dots represent measured result. (Color figure online)

Conventional capacitive [5,6] and resistive [7,8] tactile sensors can achieve one-dimensional force (pull force or pressure) measurement, but this method is challenging to achieve in three-dimensional force detection. Because in the detection process, the plane direction of the shear force (Fx, Fy) and the vertical direction of the normal force (Fz) will cause the deformation of the sensor simultaneously, making the generated signals interfere [15]. Through the method

(a) The Sensors are Fixed on the Gripper   (b)The Robot Arm Wears

    at the End of the Manipulator   Sensors to Grab Fragile Objects

Fig. 5. Color Image under Normal Signal Direction Shear Force

of structural innovation [16,17], there are still the problems of complex decoupling process, easy interference and low decoupling accuracy.

Visual-tactile sensors such as gelsight [10]. They introduced the marker point on the reflective film inside the soft elastomer to capture the marker point displacement under the 3 D force. They established the mapping relationship between the marker point displacement and the 3 D force through the finite element analysis to realize the 3 D force detection in the soft environment. Also, using finite element modeling are the Tac3D [13] and GelStereo [14]. However, the established three-dimensional force model is relatively small due to the need to set more preconditions and simplify the problem in finite element modelling. Meta's Digit sensor [12] optimizes the structure of the sensor to integrate it into the fingertips for robotic operations, but force measurements are not reported. Kakani et al. [18] improved the VGG-16 deep neural network to measure the contact position, the contact area, and the contact force distribution of the binocular tactile images. Still, the sensor structure is bulky and difficult to use. As shown in Table 2, the designed sensors are based on optical principles, which not only can accurately measure the 3 D force compared to the existing optical tactile sensors but also are the smallest size of the current optical tactile sensors and are very soft.

Using the sensors studied in this paper, the grasping of fragile objects can be achieved. As shown in the Fig. 5, attaching the sensor to the clip claw at the end of the robotic arm can receive and control the magnitude of the force while grasping the fragile object so that the fragile object will neither fall nor be crushed. Thus, the force control problem in the process of grasping the fragile object is solved.

4 Conclusion

To detect 3 D forces under soft conditions, this paper designs a structurally innovative sensor, ranging from data acquisition to data analysis and processing to hardware implementation. This study can be used for force control in robot compliant control, which is expected to solve the problems in the robot field and promote the intellectual development of robots. In the future, it is hoped to realize the update and iterative development of soft 3D force sensors and hope to realize mechanical arm force control equipment to solve different mechanical arm control problems.

Acknowledgements. This work was financially supported by the Key-Area Research and Development Program of Guangdong Province (2019B090915002), National Natural Science Foundation of China (Grants No. U1813208, 62173319,62063006), Guangdong Basic and Applied Basic Research Foundation (2020B1515120054) and Shenzhen Fundamental Research Program (JCYJ20200109115610172).

References

1. Abraira, V., Ginty, D.: The sensory neurons of touch. Neuron **79**, 618–639 (2013)
2. Johnson, K.O.: The roles and functions of cutaneousbegingroup(); mechanoreceptors. Curr. Opin. Neurobiol. **11**, 455–461 (2001)
3. Ranasinghe, A., Sornkarn, N., Dasgupta, P., Althoefer, K., Penders, J., Nanayakkara, T.: Salient feature of haptic-based guidance of people in low visibility environments using hard reins. IEEE Trans. Cybern. **46**(2), 568–579 (2015)
4. Li, Q., Kroemer, O., Su, Z., Veiga, F.F., Kaboli, M., Ritter, H.J.: A review of tactile information: perception and action through touch. IEEE Trans. Rob. **36**(6), 1619–1634 (2020)
5. Shen, Z., Zhu, X., Majidi, C., Gu, G.: Cutaneous ionogel mechanoreceptors for soft machines, physiological sensing, and amputee prostheses. Adv. Mater. **33**, 09 (2021)
6. Fu, M., Zhang, J., Jin, Y., Zhao, Y., Huang, S., Guo, C.: A highly sensitive, reliable, and high-temperature-resistant flexible pressure sensor based on ceramic nanofibers. Adv. Sci. **7**, 09 (2020)
7. Ji, B., Zhou, Q., Wu, J., Yibo, G., Wen, W., Zhou, B.: Synergistic optimization toward the sensitivity and linearity of flexible pressure sensor via double conductive layer and porous microdome array. ACS Appl. Mater. Interfac. **12**, 31021–31035 (2020)
8. Zhou, Y., et al.: Significant stretchability enhancement of a crack-based strain sensor combined with high sensitivity and superior durability for motion monitoring. ACS Appl. Mater. Interfaces **11**, 7405–7414 (2019)
9. Abad, A.C., Ranasinghe, A.: Visuotactile sensors with emphasis on gelsight sensor: a review. IEEE Sens. J. **20**(14), 7628–7638 (2020)

10. Technology, Y.: Tactile measurement with a gelsight sensor (2015)
11. Ward-Cherrier, B., et al.: The tactip family: soft optical tactile sensors with 3D-printed biomimetic morphologies. Soft Rob. **5**(2), 216–227 (2018)
12. Lambeta, M., et al.: Digit: a novel design for a low-cost compact high-resolution tactile sensor with application to in-hand manipulation. IEEE Rob. Autom. Lett. **5**(3), 3838–3845 (2020)
13. Zhang, L., Wang, Y., Jiang, Y.: Tac3d: a novel vision-based tactile sensor for measuring forces distribution and estimating friction coefficient distribution. arXiv preprint arXiv:2202.06211 (2022)
14. Cui, S., Wang, R., Hu, J., Wei, J., Wang, S., Lou, Z.: In-hand object localization using a novel high-resolution visuotactile sensor. IEEE Trans. Ind. Electron. **69**(6), 6015–6025 (2021)
15. Pang, C., et al.: A flexible and highly sensitive strain-gauge sensor using reversible interlocking of nanofibres. Nat. Mater. **11**, 795–801 (2012)
16. Song, Y., Wang, F., Zhang, Z.: Decoupling research of a novel three-dimensional force flexible tactile sensor based on an improved bp algorithm. Micromachines **9**, 236 (2018)
17. Zhu, Y., Jiang, S., Xiao, Y., Yu, J., Sun, L., Zhang, W.: A flexible three-dimensional force sensor based on pi piezoresistive film. J. Mater. Sci. Mater. Electron. **29**, 12 (2018)
18. Kakani, V., Cui, X., Ma, M., Kim, H.: Vision-based tactile sensor mechanism for the estimation of contact position and force distribution using deep learning. Sensors **21**(5), 1920 (2021)

Gated Attention Unit: An Attention-Based Recurrent Neural Network in an Intelligent 3C Assembly Framework

Hao Wang[1], Rumo Wang[1], Linxiang Li[2], Bo Zhang[3], and Fuchun Sun[4](✉)

[1] Beijing Institute of Petrochemical Technology, Beijing, China
[2] Yanshan University, Hebei, China
[3] Beijing Information Science and Technology University, Beijing, China
[4] Department of Computer Science and Technology, Tsinghua University, Beijing, China
fcsun@tsinghua.edu.cn

Abstract. 3C industry is the general name of computer, communication and consumer electric industry. Driven by China's policy and market, the demand for 3C products has further expanded. At the same time, the development of artificial intelligence has pushed 3C intelligent control and manufacturing to a new level. This paper proposes an intelligent assembly method for smart phone parts. Firstly, we proposed a Gated Attention Unit (GAU), which can classify the pressure data collected by pressure sensors during pressing smartphone part. In addition, we design and implement a smartphone assembly framework with a dexterous end-effector, which can control the parts through the cooperation of visual module and pressure module. Experiments were carried out on the pressure dataset and showed that the method we proposed can effectively classify the pressure date. The classification accuracy is 93.75% with only 0.0487 million parameters, which is the most suitable networks for smartphone assembly.

Keywords: 3C assembly · Recurrent neural network · Dexterous end-effector

1 Introduction

In recent years, 3C industry [13], mainly computer, communication and consumer electronics products, has maintained rapid growth. 3C assembly intellectualization has become a major trend. New technologies represented by robots and artificial intelligence have accelerated the integration with manufacturing

---

F. Sun—His current research interests include cross-modal learning, active perception, and precise operation and teleoperation.

F. Sun et al. (Eds.): ICCCS 2022, CCIS 1732, pp. 269–279, 2023.
https://doi.org/10.1007/978-981-99-2789-0_23

industry, which has promoted the development of intelligent manufacturing. The automation ratio of electronic manufacturing equipment is on the rise [20]. At the same time, 3C products update faster, the future demand for further expansion [19]. Based on the trend of great development, more efficient production methods are urgently needed.

Automatization. The drawbacks of automatic assembly are gradually emerging. 3C industry is a labor-intensive industry, its automation level is about 25%–35%. The rest of the tasks need to be completed manually. Gradually, the current assembly environment is showing drawbacks: high labor costs and moderate productivity. Specifically, the rapid updating and replacement of 3C products make their manufacturing process and assembly methods continuously upgrade, which increasingly requires the quality of workers. Particularly for mobile phone and computer assemblies, workers are required to have experience in assembling fine parts [15]. Secondly, to cope with the huge demand for 3C products, it is difficult for human to cooperate with automated production lines, and the assembly speed needs to be improved [23]. The transformation from the mode of combining automation and manual work to intelligent assembly has become a difficult problem to be solved.

Intellectualization. The development of artificial intelligence provides a new idea for 3C assembly [16,24]. The 3C industry is facing enormous challenges and difficulties. The key point is to use intelligent industrial robots instead of automation and simple and tedious manual actions. Artificial Intelligence (AI) combined with multimodal information can solve many difficult problems in assembly process [17,18], thus greatly improving production efficiency [8]. In the face of more complex and variable assembly scenarios, the following basic issues must be considered: (1) Accuracy. (2) Adaptability. (3) Lightweight. After meeting the basic requirements, the intelligent scheme should also have the following potential: (1) Transferability [22]. (2) Learnability. (3) Imitativeness [12].

Smartphone Assembly. Smartphone assembly also needs to have the basic ability and potential mentioned above [5]. We study the accuracy and adaptability of basic assembly problems. There are two reasons that affect the accuracy and adaptability of smartphone assembly: internal reasons and external reasons. Special-shaped parts lead to internal causes. The irregular shape of the parts makes it difficult to control and operate. Also, some special-shaped parts are the combination of the rigid part and the non-rigid part, which puts forward new requirements for the dexterous end-effector. Inevitable mechanical errors lead to external causes. In the mode of combination of automation and manual, the errors that can not be eliminated can be compensated by workers' senses and experience, which puts forward new requirements for intelligent algorithms. A reasonable assembly framework can minimize or even eliminate the uncertainty caused by internal and external factors.

Recurrent Neural Network. We use the recurrent neural network to classify the pressure data generated in the process of pressing parts. We will introduce the data source and overall framework in Sect. 2 and Sect. 3.2, respectively. RNN is a sequential model that began in the 1980s s and has been a brilliant performer in natural language processing, such as machine translation [14], speech recognition [2] and generating image descriptions, etc. [25]. The recurrent neural network remembers the previous information and applies it to the calculation of the current output, that is, the nodes between hidden layers are no longer disconnected but connected. The input of the hidden layer includes not only the output of the input layer but also the output of the hidden layer from the previous moment. With the growth of data, the traditional RNN is prone to gradient diffusion. In response to this situation, researchers have introduced the concept of gate, from which the long short-term memory have been born [9]. This recurrent neural network combines long-term memory with short-term memory. Gate Recurrent Unit (GRU), an improvement based on LSTM, has an update gate and reset gate, which can effectively reduce the number of LSTM parameters and improve performance [6]. The LSTM and GRU is shown in Fig. 1. Subsequently, in order to get a better loop model, the multi-layer or bidirectional RNNs have been developed continuously, which can effectively solve all kinds of problems in the field of natural language processing [7]. A better result has been obtained due to the introduction of gates, but the a more complex structure leads to a long training time [11,21]. At the same time, for the task of smartphone assembly, it needs a lightweight structure rather than blindly pursuing the classification accuracy [1]. The RNN with attention unit has the less parameters and even performs better than the more complex network [3].

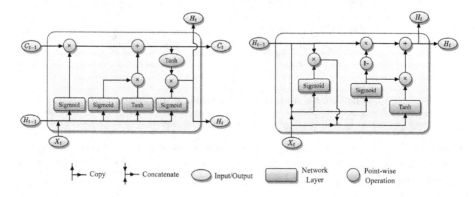

Fig. 1. Two classical RNN network architectures: LSTM (left) and GRU (right)

To solve the above problems, we have carried out the following work: (1) We propose a gated attention unit, which could accurately classify the pressure data. (2) We design a general framework for smartphone assembly, and implemented a dexterous end-effector for assembly tasks. (3) We verify the method through pressure data and show the overall framework.

The structure of this paper is as follows. Section 2 introduces the assembly task and problem formulation. Section 3 presents the implementation for the proposed method and the overall framework. Section 4 discusses the experiments which were conducted with the purpose of verifying the proposed method. Section 5 presents the conclusion of this paper and the proposals for future work.

2   Problem Formulation

We take the flexible printed circuit (FPC) of a smartphone as an example. Specifications for this smartphone and its parts are not detailed. The key steps to achieve assembly are: (1) Detect the FPC in material area. (2) Suction the FPC by the end-effector. (3) Classify the pressure data generated by pressure sensors during the process of pressuring the part. (4) Adjust the position of the part. (5) Assemble. The following are the details.

Object Detection. There are several FPCs in the material area. The positions of the FPCs are randomly placed on the material tray to simulate the randomness caused by transportation or disturbance. There are several male connectors in FPCs needed to be pressed onto the female connectors, as it is shown in Fig. 2. After assemble all the FPCs, the new FPCs will be added to the material tray to ensure the normal assembly process.

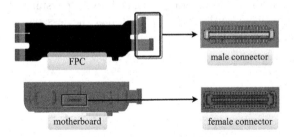

Fig. 2. FPC and its male and female connectors

Control. There are two suction nozzles on the end-effector. After obtaining the coordinates calculated by computer vision, the suction nozzle on the dexterous end-effector suction the FPC.

Classification of Deviation Direction. Due to the inevitable error of the computer vision algorithm, it is necessary to judge the deviation direction of the part. The calibration stand is equipped with a pressure sensor. The pressing block presses the male connector onto the pressure sensor and get the pressure data. As it is shown in Fig. 3, Fig. 3-(a) shows the 44 force bearing points on the pressure sensor. Figure 3-(b) shows that the male connector is on the center of the sensor. Figure 3-(c), 3-(d), 3-(e), 3-(f) show the failure status.

Adjustment. After getting the deviation direction, the robot arm will continuously adjust to reach a position within an acceptable error.

Assembly. When reaching the suitable position, the end-effector can assemble the parts.

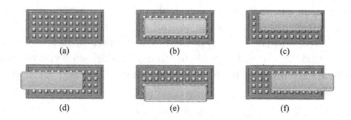

Fig. 3. Pressure sensor and several buckling states

3 Methodology

3.1 Gated Attention Unit

The recurrent neural network with GAU can classify the time series data collected by pressure sensors. The overall framework of GAU is shown in Fig. 4. Compared with simple sequential tasks, the following three factors affect the classification effect. (1) The buckle side of the male connector is full of pins, which makes it uneven. (2) The material is non rigid. When suction nozzles suction the FPC, the center of gravity changes caused by deformation. (3) The pressure data sampled each time is different due to the compliance control. As a

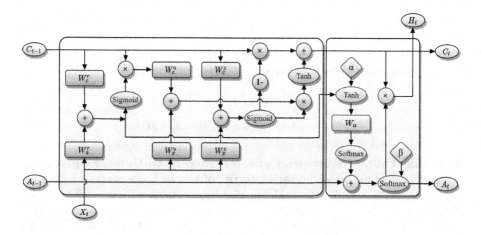

Fig. 4. The gated attention unit we proposed

result of the three aspects, the analysis of pressure data can not only rely on the pressure value in the steady state, but also focus on the process of compliance control. The method we proposed is based on GRU, and the attention unit is added to the GRU. X_t denotes the inputs. C_{t-1} represents the previous cell state and A_{t-1} represents the previous attention state. At time t, X_t, C_{t-1}, and A_{t-1} are fed to the gated attention unit. At the meanwhile, the outputs of unit consists of three parts, hidden state H_t, new cell state C_t, and new attention state A_t. Different from the meaning shown in Fig. 1, we separate the network layer from the activation function. In the overall structure, the rectangle represents the network layer, and the circle represents the operation. In the attention unit, there are two variables α and β, and the diamond represents the variable. There are two activation functions in the GAU, Softmax [4,10] and Tanh, as shown in formula (1)(2).

$$Softmax(x_i) = \frac{e^{x_i}}{\sum_{k=1}^{K} e^{x_n}} \tag{1}$$

$$tanh(x) = \frac{e^x - e^{-x}}{e^x + e^{-x}} \tag{2}$$

Gate Unit. There are two gates in gate recurrent unit, reset gate and update gate, as is shown on the left side in Fig. 4. There are four intermediate tensors, g_1, g_2, g_3, and g_4, as shown in formula (3) (6). d denotes the dimensionality of the various parameter matrices W. There are two types of d, input dimension d_x and hidden layer dimension d_h. In the reset gate, the previous cell state and the previous attention state are fed to W_c^r (a $d_x * d_h$ matrix) and W_x^r (a $d_x * d_h$ matrix), respectively. Then a Sigmoid is used for a transformation. In the update gate, the output of the reset gate g_2 add the output of W_x^z (a $d_x * d_h$ matrix) to get g_3. In the output part, the previous cell state and the previous attention state are fed to W_c^z (a $d_x * d_h$ matrix) and W_x^z (a $d_x * d_h$ matrix), respectively. Then a Sigmoid is used to get g_4. The new cell state is c_t.

$$g_1 = W_c^r * c_{t-1} + W_x^r * x_t \tag{3}$$

$$g_2 = Sigmoid(g_1) * c_{t-1} \tag{4}$$

$$g_3 = W_c^u * g_2 + W_x^u * x_t \tag{5}$$

$$g_4 = Sigmoid(W_c^z * c_{t-1} + W_x^z * x_t) \tag{6}$$

$$c_t = c_{t-1} * (1 - g_4) + Tanh(g_3 * g_4) \tag{7}$$

Attention Unit. The attention unit, is behind the GRU, could separate and transform cell state, which makes a integrity of memory, as is shown on the right side in Fig. 4. First, adjust the tanh function by a variable α. Next, g_1 is fed to W_a after the *Tanh* function. Then a *Softmax* function is used for a transformation, where the variable /beta is used to adjust the Softmax. The output of attention state a_t. The ability of the reset door to determine which cell states are valuable coincides with the idea of the attention unit, so the reset door state can be used for the attention unit.

$$a_1 = Softmax(W_a * Tanh(\alpha * (g_1 + x_t))) \tag{8}$$

$$a_t = Softmax_\beta(a_1 + a_{t-1}) = \frac{e^{\beta*(a_1+a_{t-1})}}{\sum e^{\beta*(a_1+a_{t-1})}} \tag{9}$$

3.2 Overall Assembly Framework

For the assembly tasks, we have designed a general assembly framework. The overall framework is shown in Fig. 5. The entire assembly process takes place in three areas: material area, calibration stand and assembly area. The assembly process is generally divided into three parts. Firstly, detect the FPCs in the material area. Secondly, classify the pressure data and adjust continuously according to the classification results. Thirdly, buckle the FPCs to complete the assembly task. Each step carries one or more intelligent algorithm. In the material area, a lightweight object detection network is used to detect the FPCs on the material plate. On the calibration stand, the GAU is used to classify the pressure data generated by pressure sensor, a compliant control network to adjust the force, and a reinforcement learning model to adjust the position of FPCs continuously. In the assembly area, the compliance control network is also needed. In order to run the intelligent assembly framework, we design and implement a dexterous end-effector, as it is shown in Fig. 6. There are two sensory modalities on the end-effector, vision and haptics, which provide and collect data for the neural network. There is a fixed camera on the left and a pressing block under the end-effector. In the middle is an expandable and detachable connector, which can be expanded or changed according to the needs of different assembly tasks. In addition, there are two air valve and two suction nozzles under the effector for suctioning the FPCs.

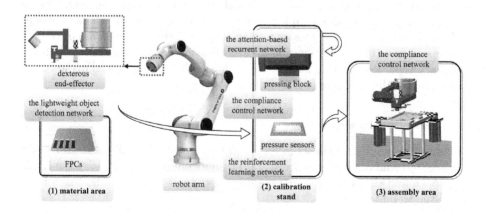

Fig. 5. The overall assembly framework

4   Experiments

4.1   Preparation and Setting

The experimental environment, experimental data and experimental parameters are set as follows.

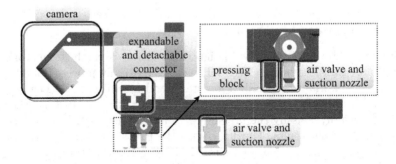

Fig. 6. Dexterous end-effector for smartphone assembly

Software. The codes were compiled in Windows 10 with Python 3.7. The deep learning framework were PyTorch 1.9.0 and TorchVision 0.10.0. The design application was SoildWorks 2018.

Hardware. The codes were run in personal computer with CPU AMD R7 4800H, GPU NVIDIA GeForce RTX 2060 laptop and 16G memory. The robot arm was HansRobot Elfin. The smartphone was provided by a company.

Dataset and Experiment Setting. The pressing data was divided into nine types, central and eight neighborhoods. There are 100 samples in each type, a total of 900 samples. 600 samples were chosen as training set, 150 samples as validation set and 150 samples as testing set. The learning rate was set to 0.01 and epoch was 100. The evaluation for classification included the accuracy top1, accuracy top5, loss and model size. The other optimization algorithms were not considered such as feature engineering, etc.

4.2   GAU on Pressure Data

The training results are shown in Fig. 7. The Fig. 7-(a), 7-(b) illustrate the trend for acc top1 and acc top5, respectively. The horizontal axis represents the training epoch and the vertical represents the acc top1 or acc top5. The Fig. 8 illustrate the trend for loss.

In addition, the parameter size of the network model also needs to be considered on devices with limited storage. Figure 9 shows the number of parameters for different networks. All the results are shown in Table 1.

Table 1. The final results for different networks

	Bi-Layer GRU	Bi-Layer LSTM	GRU	LSTM	GAU (ours)
Acc Top1 (%)	93.53	92.97	88.28	87.50	**93.75**
Acc Top5 (%)	100	100	100	99.22	**100**
Loss	0.1874	0.1890	**0.1867**	0.2556	0.2568
Paras (M)	0.0566	0.0814	**0.0392**	0.0516	0.0487

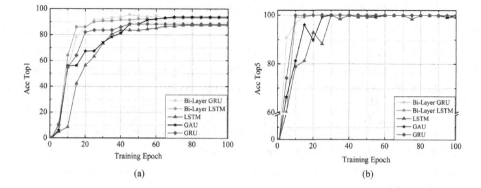

Fig. 7. The trend for acc top1 and acc top5 based on different networks

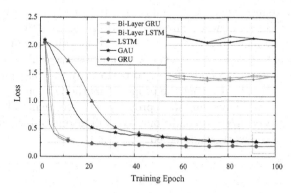

Fig. 8. The trend for loss based on different networks

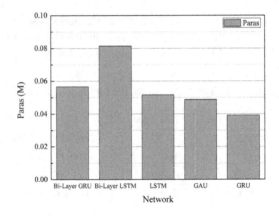

Fig. 9. The number of parameters for different networks

The experimental results show that the proposed method achieves 93.75% classification accuracy on the pressure dataset, and the number of parameters is only 0.0487 Million. The accuracy of single-layer GAU can exceed that of double-layer GRU, which proves that our model is more suitable for running on devices with limited storage space. In addition, although our method has a slow convergence speed before the 50th epoch, it can achieve the best effect in the 60th epoch, which is similar to other networks.

5   Conclusion

In order to achieve intelligent 3C assembly, we propose a gated attention unit and an efficient assembly framework. Gated Attention Unit separates cell states and extracts important memories through the attention unit to achieve the classification task. In addition, we design a general framework for smartphone assembly and implement a dexterous end-effector to perform various assembly tasks. The experiments carried out show that the proposed method can achieve precise classification. Future work may focus on improving the convergence ability under the different random initialization conditions. We will also detail the other modules of the assembly framework we designed.

References

1. Agarwal, P., Alam, M.: A lightweight deep learning model for human activity recognition on edge devices. Procedia Comput. Sci. **167**, 2364–2373 (2020)
2. Baevski, A., Hsu, W.N., Conneau, A., Auli, M.: Unsupervised speech recognition. Adv. Neural Inf. Process. Syst. **34**, 27826–27839 (2021)
3. Basiri, M.E., Nemati, S., Abdar, M., Cambria, E., Acharya, U.R.: Abcdm: an attention-based bidirectional cnn-rnn deep model for sentiment analysis. Future Gener. Comput. Syst. **115**, 279–294 (2021)

4. Brown, P.F., Della Pietra, V.J., Desouza, P.V., Lai, J.C., Mercer, R.L.: Class-based n-gram models of natural language. Comput. Linguist. **18**(4), 467–480 (1992)
5. Chin, K.S., Ratnam, M.M., Mandava, R.: Force-guided robot in automated assembly of mobile phone. Assembly Automation (2003)
6. Cho, K., et al.: Learning phrase representations using rnn encoder-decoder for statistical machine translation. arXiv preprint arXiv:1406.1078 (2014)
7. Dhyani, M., Kumar, R.: An intelligent chatbot using deep learning with bidirectional rnn and attention model. Mater. Today: Proc. **34**, 817–824 (2021)
8. Gao, S., Qiu, T., Wang, G., Huang, A., Yu, J.: Printing characters recognition of chip resistors based on the combination of image segmentation and artificial neural network. In: 2021 16th International Conference on Computer Science & Education (ICCSE), pp. 643–647. IEEE (2021)
9. Gers, F.A., Schmidhuber, J., Cummins, F.: Learning to forget: continual prediction with LSTM. Neural Comput. **12**(10), 2451–2471 (2000)
10. Goodman, J.: Classes for fast maximum entropy training. In: 2001 IEEE International Conference on Acoustics, Speech, and Signal Processing. Proceedings (Cat. No. 01CH37221), vol. 1, pp. 561–564. IEEE (2001)
11. Huddar, M.G., Sannakki, S.S., Rajpurohit, V.S.: Attention-based multi-modal sentiment analysis and emotion detection in conversation using rnn (2021)
12. Hussein, A., Gaber, M.M., Elyan, E., Jayne, C.: Imitation learning: a survey of learning methods. ACM Comput. Surv. (CSUR) **50**(2), 1–35 (2017)
13. Lin, Y., Zhou, L., Shi, Y., Ma, S.: 3C framework for modular supply networks in the Chinese automotive industry. Int. J. Logist. Manag. (2009)
14. Lopez, A.: Statistical machine translation. ACM Comput. Surv. (CSUR) **40**(3), 1–49 (2008)
15. Luo, D., Guan, Z., He, C., Gong, Y., Yue, L.: Data-driven cloud simulation architecture for automated flexible production lines: application in real smart factories. Int. J. Prod. Res. **60**, 1–23 (2021)
16. Ming, W., et al.: Application of convolutional neural network in defect detection of 3C products. IEEE Access **9**, 135657–135674 (2021)
17. Ngiam, J., Khosla, A., Kim, M., Nam, J., Lee, H., Ng, A.Y.: Multimodal deep learning. In: ICML (2011)
18. Radu, V., et al.: Multimodal deep learning for activity and context recognition. Proc. ACM Interact. Mobile Wearable Ubiq. Technol. **1**(4), 1–27 (2018)
19. Sun, P.: Computers, communications and consumer electronics (3C) manufacturing. In: Unleashing the Power of 5GtoB in Industries, pp. 179–205. Springer, Singapore (2021). https://doi.org/10.1007/978-981-16-5082-6_12
20. Tong, M., Lin, W., Huo, X., Jin, Z., Miao, C.: A model-free fuzzy adaptive trajectory tracking control algorithm based on dynamic surface control. Int. J. Adv. Rob. Syst. **17**(1), 1729881419894417 (2020)
21. Wang, F., Tax, D.M.: Survey on the attention based rnn model and its applications in computer vision. arXiv preprint arXiv:1601.06823 (2016)
22. Weiss, K., Khoshgoftaar, T.M., Wang, D.D.: A survey of transfer learning. J. Big Data **3**(1), 1–40 (2016)
23. Yan, D., Sha, W., Wang, D., Yang, J., Zhang, S.: Digital twin-driven variant design of a 3C electronic product assembly line. Sci. Rep. **12**(1), 1–12 (2022)
24. Yang, X., Xie, Y., Zhang, N., Wang, Y., Lou, Y.: Pose estimation algorithm of 3C parts based on virtual 3d sensor for robot assembly. In: 2021 40th Chinese Control Conference (CCC), pp. 4126–4131. IEEE (2021)
25. Zaremba, W., Sutskever, I., Vinyals, O.: Recurrent neural network regularization. arXiv preprint arXiv:1409.2329 (2014)

Skill Manipulation Method of Industrial Robot Based on Knowledge Graph for Assembly Scene

Daming Zhong[1], Shengyi Miao[1], Runqing Miao[2], Fuchun Sun[1,3], Zhenkun Wen[4], Haiming Huang[1], and Na Wang[1(✉)]

[1] College of Electronics and Information Engineering,
Shenzhen University, Shenzhen 518060, China
2110436226@email.szu.edu.cn, wangna@szu.edu.cn
[2] College of Modern Post, Beijing University of Posts and Telecommunications,
Beijing 100083, China
tsingm@bupt.edu.cn
[3] Department of Computer Science and Technology, Tsinghua University,
Beijing 100083, People's Republic of China
[4] College of Computer Science and Software,
Shenzhen University, Shenzhen 518060, China

Abstract. As the labor cost of manual assembly increases, it is urgent to use industrial robots to carry out assembly tasks. This work presents an industrial robot skill manipulation method based on knowledge graph to realize the intelligent assembly of products. Firstly, the knowledge base based on parameter expression is designed and deployed on the cloud server, so as to facilitate the real-time communication of robots when they perform tasks. In the second stage, the manipulation skills in the assembly scene are expressed in knowledge and stored in the knowledge base, and each action primitive is mapped into the bottom-level code that the robot can execute. Finally, the business system including the execution terminal and turntable is designed, and the memory stick assembly experiment is carried out to verify the effectiveness of this design.

Keywords: skill manipulation · knowledge graph · parameter expression · industrial robot · action primitive

1 Introduction

Intellectualization is the trend of robot technology development. At present, robots have been widely used in military, medical, industrial, agricultural and other fields. Robots can already be seen on many production lines, and they are used to perform some repetitive and boring tasks instead of humans. With the progress of science and technology, the execution efficiency of robots becomes a difficult problem in the face of complex scenes and tasks. In the existing robot operating platform, faced with complex task scenes, robots need to learn constantly in order to obtain the motion parameters and attribute information of the corresponding target tasks. Moreover, the acquired skills

F. Sun et al. (Eds.): ICCCS 2022, CCIS 1732, pp. 280–290, 2023.
https://doi.org/10.1007/978-981-99-2789-0_24

and action expression mechanisms are only suitable for learning individuals, and they are not common, so they can't share knowledge. When replacing an agent to perform a task, the new agent needs to learn again from 0, which poses a challenge to the robot to quickly perform the target task. The reference [1] puts forward a framework of human-machine collaborative reasoning based on semantic knowledge reasoning, which aims to understand human demonstration and closely combine with knowledge base to infer assembly-related knowledge, so that robots have the ability to perceive, understand, plan and execute manipulations. Reference [2] introduces SME robotics, which introduces robot manipulation expertise into robots and automation systems to achieve seamless man-machine cooperation and precision assembly. In view of the task planning process in the field of robot manipulation, reference [3] puts forward a knowledge framework of perception and motion (PMK) to provide information for autonomous robot manipulation. In order to meet the challenges related to robot manipulation in the real world, reference [4] suggests that to successfully complete a complex task, robots must learn knowledge effectively and keep clarity in their motion decisions.

All the above studies emphasize the importance of knowledge expression in the actual manipulation of robots. In fact, many task planning and skill learning are knowledge-oriented. Reference [3] establishes the framework of knowledge representation based on the concept of ontology, and uses meta-ontology, patterns and examples to represent knowledge, but it does not establish the relationship between entity knowledge. In order to reflect the relationship between the levels of operational knowledge, reference [4] emphasizes five types of knowledge related to operational task: object, agent, space, task and action, and characterizes these five types of knowledge from three levels: class knowledge, instance knowledge and attribute. In order to reflect the hierarchical relationship of manipulation knowledge, reference [5] characterizes manipulation knowledge from three levels: primitive, skill and task. The so-called primitive is actually the bottom-level action that robots can perform, but they are not verified in actual scenes based on knowledge base. Reference [6] plans manipulation task from the perspective of action, and proposes that action include high-level action and low-level action. The so-called high-level action is pure semantic knowledge, similar to abstract natural language, and is mainly used for task planning, while the low-level action is robot-oriented executable unit.

In this work, a robot skill manipulation method based on knowledge graph is proposed for assembly manipulation scene, our key contributions are as follows:

(1) In this work, a knowledge graph is constructed, which includes four levels: entity, task, skill and action, in which the action level represents the action primitive, which is the smallest action unit that a robot can perform. In this work, each action primitive is expressed parametrically in order to improve the accuracy of robot manipulation.
(2) The manipulation skills in assembly scenes are expressed in knowledge and stored in the knowledge base. Based on the knowledge graph, a skill manipulation framework is proposed, which aims to improve the execution efficiency of robots in the face of complex assembly manipulation tasks and unfamiliar scenes.
(3) Taking the memory stick assembly task in assembly scene as an example, the effectiveness of this design is verified.

2 Related Work

Knowledge graph is now becoming one of the core drivers of the development of artificial intelligence. In 2012, Google put forward the concept of knowledge graph for the first time [7]. Knowledge graph is essentially a semantic network of various entities and their relationships. In the past decade, artificial intelligence technology has been developing continuously, and the research on knowledge representation has also continued. In order to effectively complete the abstract task instructions, robots need complex knowledge processing methods. Tenorth M et al. put forward the robot knowledge processing system KnowRob [8], extended and designed the interface based on Know-rob, and put forward the second generation robot knowledge processing framework KnowRob2.0 [9]. Reference [10] introduces a large-scale knowledge engine-RoboBrain, which is a collaborative project. RoboBrain uses graphic structure to represent knowledge and learn and share knowledge from different sources. To enable robots to complete complex manipulation tasks independently, they must be equipped with knowledge and reasoning mechanisms. Reference [11] puts forward a service to promote knowledge representation and reasoning-Open ease, which aims to enable robots and operators to analyze and learn manipulation data. Reference [12] puts forward a knowledge-based system-RoboEarth, which includes a semantic mapping system. According to RoboEarth cloud service, ordinary robots can build semantic maps needed for daily tasks. In view of the problem of insufficient real data of robot manipulation, reference [13] proposes a cross-platform data collection platform Roboturk, which also supports manual teaching, and automatically stores the successfully demonstrated data into the cloud server to increase the amount of data.

Aiming at the knowledge base in the field of robot manipulation, the research team of National University of Defense Technology proposed an ontology knowledge base for robot task planning-RTPO [14]. Starting from the concept of ontology, the knowledge base was constructed with the ontology knowledge of task, environment and robot in three dimensions, which was used for specific task planning.

The complex manipulation task of robot has the following characteristics: 1) the task is usually composed of multiple sub task sequences; 2) Subtasks usually need to be composed of multiple action primitive sequences; 3) Action primitives are usually the lowest level actions when a robot performs a task.

3 Robot Skill Manipulation Framework

This work is knowledge-oriented, and designs a robot skill manipulation framework based on knowledge graph by introducing parametric representation of action primitives and hierarchical expression of manipulation knowledge, so as to realize the robot's autonomous manipulation and improve the robot's task execution efficiency in complex scenes. The overall scheme structure is shown in Fig. 1. This chapter introduces the overall framework of skill manipulation, and then introduces the analytical process of robot manipulation tasks, the logic of knowledge graph construction, and the parametric representation of actions.

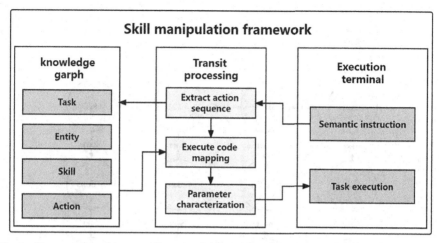

Fig. 1. The overall architecture diagram of this paper. First of all, it is necessary to establish a knowledge graph based on parameter expression, including four levels: entity, task, skill and action. In the process of carrying out the assembly manipulation task, the execution terminal acquires the task instructions of human natural language, and at the same time, it also serves as an manipulation flow display interface, which can display the robot execution process in real time. Then, according to the given semantic instruction of the manipulation task, search in the knowledge base, query the corresponding task knowledge, extract the knowledge, and obtain the serialized action primitives. Each action corresponds to a bottom-level code that the robot can execute. After parameter assignment, the robot can execute according to the sequence of actions, thus completing the specified manipulation task.

3.1 Task Analysis Process Framework

Figure 2 is the flow chart of task analysis: it mainly includes task-skill-action, and the relationships among hierarchical nodes include: contain, next, start, end, etc. Each task includes different serialization skills. A task starts with the first skill it contains and ends with the last skill it contains. Each skill includes different serialization actions. A skill starts with the first action it contains and ends with the last action it contains. So in essence, a task is a combination of a series of bottom-level actions. A task starts with the first action of the first skill and ends with the last action of the last skill. In this work, the concept of skills is put forward between tasks and actions. Firstly, it is to facilitate the task analysis, so that operators and robots can understand tasks easily. Secondly, it is to facilitate the reasoning of knowledge at the skill level. Then, we can learn skills transfer based on the knowledge graph designed in this work.

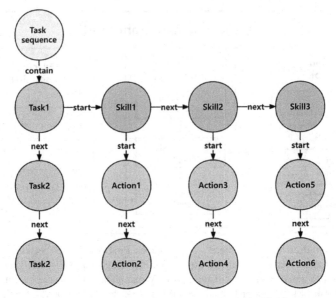

Fig. 2. The task analysis process framework. For a task, it can be divided into a series of skill combinations, such as skill 1, skill 2 and skill 3. Each skill can be combined into a series of actions in stages. For example, skill 1 includes action 1 and action 2. Skill2 include actions 3 and action 4. Every practical task can be finally summed up as a sequential combination of a series of actions.

3.2 Construction of Knowledge Graph

In view of the above robot task analysis process, this work analyzes robot manipulation skills in assembly scene, and designs a knowledge base of robot manipulation domain, which includes four levels of entity, task, skill, and action and the relationships among related levels of nodes. The overall logical framework is shown in Fig. 3. The entity layer includes the agents, robot parts, manipulation objects, end-effectors and the relationships among different manipulation objects, such as subject, object, etc., involved in the robot's task execution.

The task layer is established based on the objectives and task requirements of the robot in the actual manipulation process, including the manipulation tasks in different assemble scenes such as flexible cable installation, camera installation, memory stick installation, etc. The skill level is established based on the motion planning of the robot in the actual manipulation process. The serialized skill combination can complete a task, and the skill can be understood as a subtask or an abstract advanced action. The action layer is established based on the action primitives in the actual manipulation of the robot. Serialized action primitives can be combined to complete a skill. Each action primitive represents the smallest action unit that the robot can perform, and the so-called smallest action unit can be mapped to the bottom-level code that the robot can execute.

In order to facilitate the knowledge expression of robot manipulation skills, this work constructs the hierarchical relationship between the entity level and tasks, skills and actions into a knowledge graph. Figure 3 is a schematic diagram of the knowledge graph composed of four levels. Representing knowledge in the form of knowledge graph

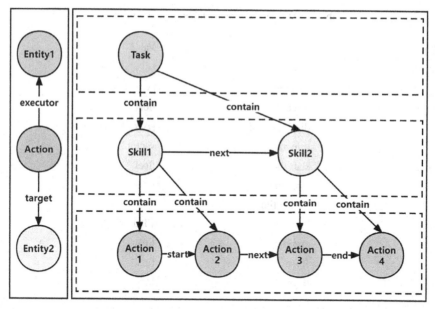

Fig. 3. Knowledge graph diagram of four levels including entity, task, skill and action. Entity 1 includes related agents and end effectors that perform actions, and entity 2 includes operational objects of actions, such as memory chips, cameras, and so on. Task includes specific manipulation task, skill includes specific manipulation skill under current task, and action includes specific action primitives under current skill.

is not only intuitive and concise, but also conducive to knowledge inquiry and knowledge reasoning.

3.3 Parametric Representation of Action

The actions described in this work are action primitives, and each action primitive can be mapped to the bottom-level code that the robot can execute. The execution of the bottom-level code is realized by the motion-related parameters of the agent itself, such as joint angle, coordinate value, force, etc. These motion-related parameters are closely related to the function configuration of the agent. Because every task will eventually be carried out to the specific action, when the robot receives the natural language instruction provided by human beings, it can quickly determine the required action primitives and corresponding execution parameters, and finally control the corresponding actuators and components according to the specific execution parameters to complete the task.

Table 1. Attribute parameter table of action primitive

Type	Key	Value
Move	position	[x, y, z]
	rotation	[x, y, z, w]
Force	force	float
	displacement	[Δ x, Δ y, Δ z]
	orientation	vector3
Time	start time	float
	stop time	float
	during time	float
Gripper	bool	{True, False}
	speed	float
	force	float
	degree	int

In this work, the daily actions of robots involved in assembly scene are classified according to categories, and the attributes are labeled to facilitate parametric expression, as shown in Table 1. Movement-like actions mainly have two attributes: position and rotation angle. The position attribute is represented by three-dimensional coordinate parameters including [x, y, z], and the rotation angle attribute is represented by four-dimensional coordinate parameters including [x, y, z, w]. Force-like actions mainly include three attributes: force, displacement and direction. The force attribute uses floating-point data to represent the force, the displacement attribute uses three-dimensional coordinates, and the direction attribute uses vector. Class actions include three attributes: start time, end time and duration, all of which are represented by floating-point data. Grabbing action includes four attributes: open, close, speed, force and degree of opening and closing. The attributes of open and close are represented by Boolean values, True indicates grabbing, False indicates loosening, and the attributes of speed and force are represented by floating-point data. The attribute of opening and closing degree is represented in the form of shaping data, ranging from 0 to 100. If it is 100, it means that the gripper is completely closed. Reference [15] mentions that the combination of binary codes 0 and 1 is used to represent actions, but the classification of actions is too abstract to be suitable for actual manipulation scenes.

In the knowledge graph designed in this work, you can query the node information and execution parameters related to specific tasks by searching and looking up tables.

4   Experiment

In this work, PC memory stick assembly task in assembly scene is taken as an example.

The designed skill manipulation platform based on knowledge graph is tested, and the memory stick assembly manipulation platform is shown in Fig. 4.

The testing of task analysis framework and knowledge graph mainly focuses on the performance of analytical manipulation tasks and the real-time communication performance of knowledge graph during actual manipulation. While the assembly experiment of skill manipulation platform aims to verify whether the platform has good assembly effect and reliability during actual manipulation.

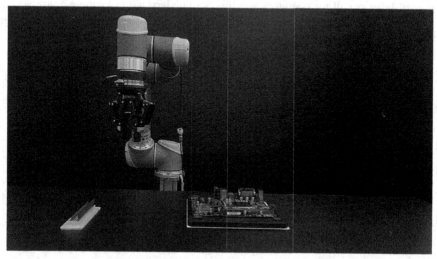

Fig. 4. Manipulation platform for memory stick assembly. Comprises an operating platform, a memory stick and a base for placing the memory stick, a computer PC motherboard equipped with the memory stick, a UR5 mechanical arm and a two-finger clamping jaw robotiq.

4.1 Analysis of Memory Stick Assembly Task

According to the framework of task analysis process, the memory stick assembly task is analyzed first. The memory stick assembly task can be decomposed into six semantic skills that human beings can understand: locate, grip, move, locate, insert and home. According to the minimum action primitives that robots can perform, the first skill locate includes two action primitives: diagonal-move and posture-adjustment, the second skill grip includes two action primitives: vertical move and close-gripper, the third skill move includes two action primitives: vertical-move and diagonal-move, the fourth skill locate includes two action primitives: vertical move and posture adjustment, the fifth skill insert includes two action primitives: vertical move and press, and the sixth skill home includes three actions: open-gripper, vertical move and diagonal-move. The whole memory stick assembly task can be completed by the combination of 13 serialization action primitives.

4.2 Demonstration of Knowledge Graph of Memory Stick Assembly Task

This work builds the knowledge graph of assembly domain based on the neo4j graph database, and represents the relationship between entities based on the logic of triples.

The demonstration of the knowledge graph of memory stick assembly task is shown in Fig. 5. In the memory stick assembly scene, the entity relationships involved mainly take action as the core. This work focuses on the action primitives and strings the direct entities involved in the action. For example, the action of pressing is performed by the UR5 robot and the target is the memory stick; The gripper closing action is performed by robotiq and the target is the memory stick. In the knowledge graph, the entity layer, task layer and skill layer are all for operators and robots to understand tasks more conveniently. The actions stored in the final action primitive layer are closely related to the execution of robots.

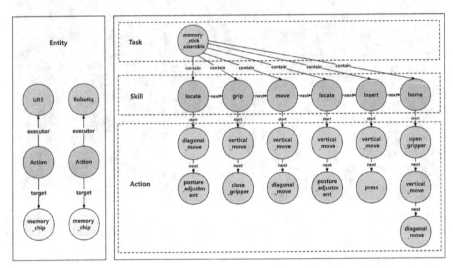

Fig. 5. Knowledge graph demonstration of memory module assembly task. The entity layer includes UR5 robot, end effector-robotiq and manipulation object memory. The core action primitive layer includes 13 action primitives, such as diagonal-move, posture adjustment, vertical move and close gripper, etc. Robots only need to execute the lowest action primitives to complete the top tasks.

4.3 Memory Stick Assembly Manipulation Experiment

After completing the memory stick assembly task analysis and building the memory module assembly knowledge graph, we carried out the actual memory stick assembly experiment verification, and the experiment was very smooth. The experimental execution process is shown in Fig. 6. After the robot obtains the human PC memory stick assembly task instruction, the robot will establish communication with the robot manipulation domain knowledge base through the middle turntable, find and obtain a series of action primitives under the current task instruction, map each action primitive obtained to the corresponding execution code, and assign specific action primitive parameters to the execution code through the vision system or other manipulation algorithms, So as to ensure the accurate execution of each action primitive and the smooth completion of the task.

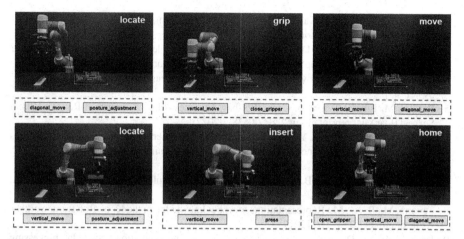

Fig. 6. The snapshot sequences of the actual memory module assemble experiment. The task includes 13 serialized action primitives. Video URL link: https://youtu.be/9Vk60uyiMN4

5 Conclusion

In this work, we carefully studied the skill manipulation methods and manipulation knowledge representation of industrial robots, and limited the research to assembly scenes. First of all, we propose a knowledge graph of robot manipulation field for assembly scene, which aims to solve the problem of low execution efficiency when robots face complex tasks and unfamiliar scenes. In order to meet the operational accuracy of the robot when performing tasks, we propose an action primitive based on parameter expression, summarize the action categories involved in the assembly manipulation scene, and mark their attributes and parameters. In order to verify the feasibility of this design, on the basis of the built knowledge base, we designed a business system based on the knowledge base, the execution terminal and the middle turntable. Taking the PC memory module assembly under the assembly scene as an example, we carried out the actual manipulation verification. Through the analysis of the experimental results, the business system we designed can well complete the manipulation instructions issued by operator. Subsequently, we will conduct experiments on assembly tasks in other assembly scenes based on this knowledge base, and conduct certain knowledge reasoning and skill transfer based on a large number of real experimental data.

Acknowledgement. This work was supported by the Major Project of the New Generation of Artificial Intelligence (No. 2018AAA0102900), the National Natural Science Foundation of China (No. 62173233), the Shenzhen Science and Technology Innovation Commission project (JCYJ20210324094401005, JCYJ20220531102809022), the Guangdong Basic and Applied Basic Research Foundation (No. 2021A1515011582).

References

1. Akkaladevi, S.C., Plasch, M., Hofmann, M., et al.: Semantic knowledge based reasoning framework for human robot collaboration[J]. Procedia CIRP **97**, 373–378 (2021)

2. Perzylo, A., Rickert, M., Kahl, B., et al.: SME robotics: Smart robots for flexible manufac-turing[J]. IEEE Robot. Autom. Mag. **26**(1), 78–90 (2019)

3. Diab, M., Akbari, A., Ud Din, M., et al.: PMK—A knowledge processing framework for autonomous robotics perception and manipulation[J]. Sensors **19**(5), 1166 (2019)

4. Liua W, Darunaa A, Chernovaa S. A Survey of Semantic Reasoning Frameworks for Robotic Systems[J]

5. Pantano M, Eiband T, Lee D. Capability-based Frameworks for Industrial Robot Skills: a Survey[J]. arXiv preprint arXiv:2203.00538, (2022)

6. Zhang, M., Tian, G., Zhang, Y., et al.: Service skill improvement for home robots: Autonomous generation of action sequence based on reinforcement learning[J]. Knowl.-Based Syst. **212**, 106605 (2021)

7. Chen, Z., Wang, Y., Zhao, B., et al.: Knowledge graph completion: A review[J]. Ieee Access **8**, 192435–192456 (2020)

8. Tenorth, M., Beetz, M.: Representations for robot knowledge in the KnowRob framework[J]. Artif. Intell. **247**, 151–169 (2017)

9. Beetz M, Beßler D, Haidu A, et al. Know rob 2.0—a 2nd generation knowledge processing framework for cognition-enabled robotic agents[C]//2018 IEEE International Conference on Robotics and Automation (ICRA). IEEE, pp.512–519. (2018)

10. Saxena A, Jain A, Sener O, et al. Robobrain: Large-scale knowledge engine for robots[J]. arXiv preprint arXiv:1412.0691. (2014)

11. Beetz M, Tenorth M, Winkler J. Open-ease–a knowledge processing service for robots and robotics[J]. AI researchers, 374. (2015)

12. Riazuelo, L., Tenorth, M., Di Marco, D., et al.: RoboEarth semantic mapping: A cloud enabled knowledge-based approach[J]. IEEE Trans. Autom. Sci. Eng. **12**(2), 432–443 (2015)

13. Mandlekar A, Zhu Y, Garg A, et al. Roboturk: A crowd sourcing platform for robotic skill learning through imitation[C]//Conference on Robot Learning. PMLR, pp.879–893. (2018)

14. Sun, X., Zhang, Y., Chen, J.: RTPO: a domain knowledge base for robot task planning[J]. Electronics **8**(10), 1105 (2019)

15. Paulius D, Eales N, Sun Y. A motion taxonomy for manipulation embedding[J]. arXiv preprint arXiv:2007.06695. (2020)

Sponge-Based Mobile Soft Robot
with Multimodal Locomotion

Zehao Liang[1] , Lingxi Ma[1] , Xiaoyi Liu[1] , and Haiming Huang[1,2,3]([⊠])

[1] College of Electronics and Information Engineering, Shenzhen University, Shenzhen, China
haimhuang@163.com
[2] Shenzhen University-HAN's Robot Intelligent Robot Joint Innovation Center, Shenzhen University, Shenzhen, China
[3] Intelligent Manufacturing Industry Research Center, Shenzhen University, Shenzhen, China

Abstract. In recent years, mobile soft robots have been widely studied because of their simple structure and friendly interaction. However, most mobile soft robots have only one locomotion mode, such as crawling or rolling, which can't adapt to unstructured environments. In this paper, we propose a novel Sponge-based mobile soft robot with multimodal locomotion. This is the first mobile soft robot with multimodal locomotion based on sponges. The proposed sponge-based mobile soft robot comprises 12 edge actuators and 8 corner actuators. By regulating the vacuum pressure in edge actuators and corner actuators, the proposed robot can crawl or roll. The proposed robot can locomote in unstructured environments by choosing a suitable locomotion mode. In this paper, the shrinkage of edge actuators is examined, and the deflection angle of two kinds of locomotion modes is compared. According to the experiment on the shrinkage of edge actuators, as the vacuum pressure increases, the shrinkage of edge actuators increases and finally reaches a steady state. The deflection angle experiment shows that crawling is more stable than rolling, and rolling requires less cycle than crawling at a certain distance.

Keywords: Soft robot · Mobile robot · Locomotion · Multimodal · Sponge

1 Introduction

The mollusk can effectively locomote in the environment. By observing and imitating the locomotion characteristics of mollusks in nature, people use soft elastic materials to design the mobile soft robot. The mobile soft robot can accomplish dangerous tasks like entering unstructured environments or working under extreme pressure. Because the mobile soft robot has strong adaptability, excellent compatibility, high safety, and flexibility, the research based on the mobile soft robot has gradually become a hot spot in soft robotics.

In recent years, the sponge has been widely used in making soft robots. Under vacuum pressure, the sponge in a cavity has a large deformation, higher driving efficiency, and less energy consumption due to elastic deformation. Therefore, the sponge has a broad application prospect in the design of soft robots [1–4].

The mobile soft robot has two common forms of locomotion: crawling and rolling. The mobile soft robots designed by Shepherd et al. [5] and Tolley et al. [6] can crawl by changing the friction between the head/tail and the ground, and crawl by using the friction difference. Steltz et al. [7] changed the center of gravity of the robot by regulating the inflation degree of the balloon and the stiffness of the jamming structure, so the robot could complete the rolling action. Compared with crawling robots, rolling robots move faster, but the large rolling action has adverse effects on tasks, such as unstable transportation of objects and complex control [8]. Compared with rolling robots, crawling robots have a large contact surface with the ground, which increases the wear and perforation of the robot. Moreover, the locomotion of the crawling robot is stable but slow.

In this paper, a sponge-based mobile soft robot with multimodal locomotion is proposed. To the best of the authors' knowledge, this is the first mobile soft robot with multimodal locomotion based on sponges. By using two different control methods, the proposed robot realizes two modes of locomotion: crawling and rolling. By analyzing the environment and choosing an appropriate locomotion mode, the proposed soft robot can efficiently accomplish the task. For the above reason, the proposed robot has higher adaptability than traditional single-mode robots.

2   Design

2.1   Design of the Sponge-Based Mobile Soft Robot

The structure of the proposed robot is shown in Fig. 1(a). It comprises 12 edge actuators and 8 corner actuators. Each corner actuator connects three edge actuators by magnetic force.

As shown in Fig. 1(b), each edge actuator is composed of a sponge, magnets, plastic film and an air tube. Two opposite surfaces of the sponge are attached by magnets. The sponge and magnets are sealed with an air tube in the elastic film. As shown in Fig. 1(c), the edge actuator contracts under vacuum pressure.

Each corner actuator is composed of a hollow iron cube, latex film and an air tube. As shown in Fig. 1(d), each surface of the hollow iron cube contains nine suction holes. As shown in Fig. 1(e), the latex film wraps the hollow iron cube flatly but not tautly. The latex film has only one opening, and the opening is connected to an air pipe. When under positive pressure, the latex film inflates, as shown in Fig. 1(f). When under negative pressure, part of the latex film shrinks into the suction holes, as shown in Fig. 1(g).

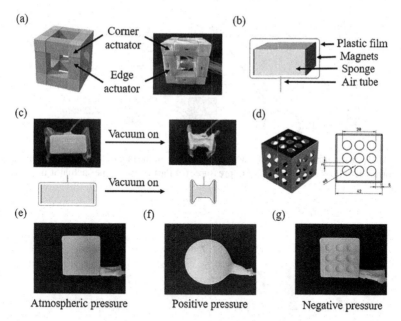

Fig. 1. Design of the proposed sponge-based mobile soft robot. (a) Structure of the proposed robot. (b) Structure of the edge actuator. (c) Schematic of the contracting of the edge actuator. (d) Structure of the hollow iron cube. (e) Structure of the corner actuator under atmospheric pressure. (f) Schematic of the inflation of the corner actuator under positive pressure. (g) Schematic of the shrinkage of the corner actuator under negative pressure.

2.2   Sucking Principle of Corner Actuators

Based on Chen's [9, 10] research on indirect vacuum adhesion, we designed the corner actuator. As shown in Fig. 2(a), the corner actuator is placed on the ground under atmospheric pressure. Suppose the initial pressure and volume between the latex film and the ground are $P_{initial}$ and $V_{initial}$. The value of $P_{initial}$ is equal to atmospheric pressure. When the air in the corner actuator is evacuated, part of the latex film will be sucked into suction holes, as shown in Fig. 2(b) and (c). In the above process, the volume between the latex film and the ground increases from $V_{initial}$ to V_{final}. Based on the ideal gas equation shown in Eq. (1), the pressure between the latex film and the ground decreases from $P_{initial}$ to P_{final}. Since P_{final} is less than atmospheric pressure, the corner actuator is pressed by a force directed to the ground. From the above, the corner actuator can fix on the ground under negative pressure.

$$P_{final} = \frac{P_{initial} V_{initial}}{V_{final}} \tag{1}$$

where P_{final} and V_{final} are the final pressure and the volume between the latex film and the ground.

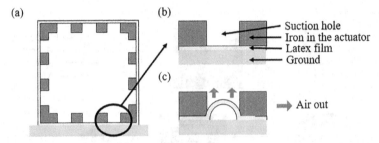

Fig. 2. Analysis diagram of the corner actuator. (a) The initial state of the corner actuator. (b) Part of the corner actuator under atmospheric pressure. (c) Part of the corner actuator under vacuum pressure.

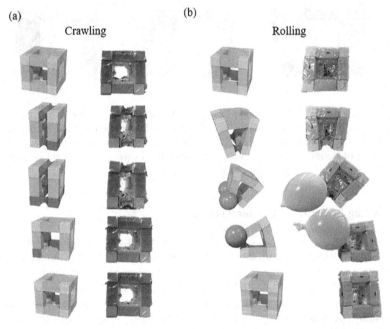

Fig. 3. Two locomotion modes of the proposed robot. The actuator marked in blue is in the negative pressure state, while the actuator marked in green is in the positive pressure state. (a) Crawling. (b) Rolling. (Color figure online)

2.3   Process of Crawling

Through crawling, the soft-bodied reptiles can locomote stably on the flat ground. By supplying negative pressure to edge actuators and corner actuators in a certain order, the proposed soft robot achieves worm-like crawling, as shown in Fig. 3(a).

Firstly, two front corner actuators are provided negative pressure, so that two corner actuators fix on the ground. Secondly, the four middle edge actuators contract under negative pressure. At this time, the tail of the proposed robot moves forward. Thirdly, the rear corner actuators fix on the ground under negative pressure. Fourthly, two front corner actuators and four edge actuators return to their initial state, so that the head of the robot rebounds ahead. Finally, two rear corner actuators return to their initial state. The process mentioned above completes one cycle of the proposed robot's crawling.

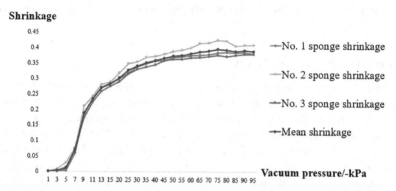

Fig. 4. The relationship between the shrinkage of the edge actuator and the vacuum pressure.

2.4 Process of Rolling

By rolling, the soft-bodied reptiles can pass through the uneven area. The rolling process of the proposed robot is shown in Fig. 3(b).

Firstly, two bottom edge actuators contract under vacuum pressure. The above contraction can raise the center of gravity of the proposed robot, making it easier for the proposed robot to roll. Secondly, rear corner actuators inflate under positive pressure, until the robot rolls. Finally, corner actuators and edge actuators return to their initial state. The process mentioned above completes one cycle of the proposed robot's rolling.

3 Experiment

3.1 Experiment on Shrinkage of Edge Actuators

To study the locomotion performance of the proposed robot, the shrinkage of edge actuators should be considered primarily. A vernier caliper (of precision 0.02 cm) is used to measure the length of the edge actuator.

In this test, three edge actuators (40 mm × 40 mm × 89 mm, 40D) were used as samples. The length of each edge actuator was measured before and after the contraction. Suppose x is the differential value of the length of the edge actuator before and after contraction, y is the length of the edge actuator before contraction, and s is the shrinkage of the actuator, as $s = x/y$. For each edge actuator, we examined the relationship between vacuum pressure and shrinkage. The vacuum pressure was divided into two ranges, namely, 0 to −15, and −15 to −95 kPa in steps of −2 and −5 kPa, respectively. As shown in Fig. 4, Three groups of test data were obtained under vacuum pressure, and the average values were calculated.

As shown in Fig. 4, when the vacuum pressure is between 0 kPa and –5 kPa, the shrinkage is close to 0 as the vacuum pressure increases. When the vacuum pressure is between –5 kPa and –30 kPa, the shrinkage increases as the vacuum pressure increases. When the air pressure is lower than –30 kPa, the shrinkage gradually reaches a steady state.

3.2 Deflection Angle Experiment

The deflection angle is an important factor to measure the locomotion stability of the proposed robot. The larger the deviation angle, the more unstable the locomotion mode will be. To compare the deflection angle of two locomotion modes of the proposed robot, an experimental platform was built, as shown in Fig. 5(a). The platform is white cardboard with a red central axis. The proposed robot is controlled to walk along the red line until it leaves the platform. Suppose L is the distance of the robot's movement, and D is the distance between the robot and the red central axis. The deviation angle can be expressed in Eq. (2), as follows:

$$\alpha = \arctan\left(\frac{D}{L}\right) \tag{2}$$

(a)

(b)

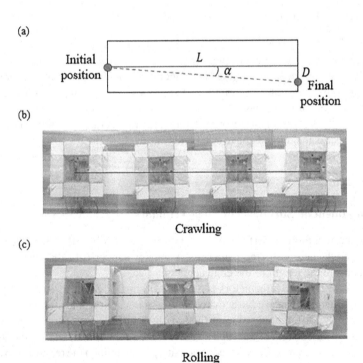

Crawling

(c)

Rolling

Fig. 5. Deflection angle experiment. (a) Experimental platform (b) The path of the robot crawls. (c) The path of the robot rolls.

where α is the deviation angle.

During the text, two kinds of locomotion deflection angles were measured: crawling and rolling. Firstly, record the initial position of the robot. Then control the robot to walk along the red line by crawling and rolling, respectively, as shown in Fig. 5(b) and (c). Take pictures to record the new position of the robot when it completes one cycle. The experiment shows that $L_{crawl} = 880$ mm, $D_{crawl} = 37$ mm, $L_{roll} = 1023$ mm, $D_{roll} = 66$ mm, $\alpha_{crawl} \approx 2.41°$, $\alpha_{roll} \approx 3.66°$. The deviation angle of crawling is smaller, which means that crawling is more stable than rolling. In addition, it takes 4 cycles to complete the whole crawling process, while rolling only takes 3 cycles. It means that rolling requires fewer cycles than crawling at a certain distance.

4 Conclusion

In this study, we proposed a novel sponge-based mobile soft robot with multimodal locomotion. This is the first mobile soft robot with multimodal locomotion based on sponges. The proposed soft robot consists of edge actuators and corner actuators. Under negative pressure, the edge actuators contract and the corner actuators suck like suckers. Under positive pressure, corner actuators inflate. By using two actuators above, the robot can locomote in two different modes: crawling or rolling. As the vacuum pressure increases, the shrinkage of edge actuators increases and finally reaches a steady state. Experiments show that crawling is more stable than rolling, and rolling requires less cycle than crawling at a certain distance.

Although the proposed robot has many advantages, there are still several problems in its practical application. One of the problems is the slow-moving speed, which is limited by the air velocity. To the best of the authors' knowledge, the air velocity is limited by the output of the air pump and the side of the air pipe. Optimizing the above parts may improve the speed and maneuverability of the proposed robot.

Acknowledgements. The authors are grateful for the support by the National Natural Science Foundation of China (No. 62173233), the Guangdong Basic and Applied Basic Research Foundation (No. 2021A1515011582), the Shenzhen Science and Technology Innovation Commission project (JCYJ20210324094401005), the 2021 National Undergraduate Training Programs for Innovation and Entrepreneurship (202110590023).

References

1. Liu, X., Liang, Z.: Soft actuator using sponge units with constrained film and layer jamming. Ind. Robot Int. J. Robot. Res. Appl. (2022)
2. Robertson, M.A., Paik, J.: Low-inertia vacuum-powered soft pneumatic actuator coil characterization and design methodology. In: 2018 IEEE International Conference on Soft Robotics (RoboSoft), pp. 431–436. IEEE (2018)
3. Yamada, Y., Nakamura, T.: Actuatable flexible large structure using a laminated foam-based soft actuator . In: 2020 IEEE/SICE International Symposium on System Integration (SII), pp. 74–79

4. Mitsuda, T., Otsuka, S.: Active bending mechanism employing granular jamming and vacuum-controlled adaptable gripper. IEEE Robotics and Automation Letters **6**(2), 3041–3048 (2021)
5. Shepherd, R.F., Ilievski, F., Choi, W., et al.: Multigait soft robot. Proc. Natl. Acad. Sci. **108**(51), 20400–20403 (2011)
6. TolleyMichael, T., ShepherdRobert, F., GallowayKevin, C., et al.: A resilient, untethered soft robot. Soft Robot. (2014)
7. Steltz, E., Mozeika, A., Rodenberg, N., et al.: JSEL: jamming skin enabled locomotion. In: 2009 IEEE/RSJ International Conference on Intelligent Robots and Systems, pp. 5672–5677. IEEE (2009)
8. Li, W.B., Zhang, W.M., Zou, H.X., et al.: A fast rolling soft robot driven by dielectric elastomer. IEEE/ASME Trans. Mechatron. **23**(4), 1630–1640 (2018)
9. Chen, R., Wu, L., Sun, Y., et al.: Variable stiffness soft pneumatic grippers augmented with active vacuum adhesion. Smart Mater. Struct. **29**(10), 105028 (2020)
10. Chen, R., Zhang, C., Sun, Y., et al.: A paper fortune teller-inspired reconfigurable soft pneumatic gripper. Smart Mater. Struct. **30**(4), 045002 (2021)

Motion Languages for Robot Manipulation

Chunfang Liu$^{(\boxtimes)}$, Xiaoyue Cao, Ailin Xue, and Xiaoli Li

Chunfang Liu, Xiaoli Li and All Other Authors Are with Faculty of Information
and Technology, Beijing University of Technology, Beijing, China
cfliu1985@126.com, lixiaolibjut@bjut.edu.cn, xueailin@emails.bjut.edu.cn

Abstract. In the cooperation between humans and robots, it is essential
for robots to translate natural human language into continuous action
sequences to complete complex collaborative tasks. In this article, a hierarchical system is established to realize the conversion from human natural language to robot motion sequence in complex tasks. The system
consists of three layers: task layer (top layer), semantic motion layer
(middle layer) and motion primitive layer (bottom layer). When humans
tell robots tasks through natural language, they first input language sentences at the task level; Then, by combining oral description with visual
cues, the sentence is translated into motion language in the middle layer,
which consists of the predicate object structure and the six-dimensional
state (position and posture) of the object. Among them, the sequence
of predicate object structure uses words to describe complex tasks at
semantic level, and the 6D state of the object mainly includes the initial
state before operation and the target state after operation. In addition,
we propose a novel search algorithm of motion sequences which integrates
our knowledge base with Deep Q-learning. Furthermore, the new knowledge base is established which is used to encode various characteristics
of motions, objects and relationships. To verify the effectiveness of this
method, we set up an actual robot experimental platform (consisting of
aubo-i5 manipulator and robotiq mechanical claw) for typical complex
operation experiments.

Keywords: Knowledge base · Search algorithm of motion sequences ·
Deep Q-learning

1 Introduction

In daily life, we hope that robots can help us complete some tasks. For example,
we tell the robot to "give me a book" through natural language. Then, the robot
can understand the task content and complete the task of "passing the book"
through its own operational skills. In order to realize this process, the following
three problems need to be solved urgently: (1) How do robots understand human
natural language? (2) How do robots master various operating skills? (3) How
is the relationship between natural language and operational skills established?

© The Author(s), under exclusive license to Springer Nature Singapore Pte Ltd. 2023
F. Sun et al. (Eds.): ICCCS 2022, CCIS 1732, pp. 299–314, 2023.
https://doi.org/10.1007/978-981-99-2789-0_26

The solution to the first problem of "understanding human natural language". First of all, robots need to have the ability to analyze grammar, including analyzing word parts of speech (verbs, nouns, prepositions, etc.) and the dependencies between words. Then, the robot plans its trajectory according to the semantic information extracted from the grammar analysis. From the point of view of robot operation, semantic information consists of the sequence of predicate object structure, specifically, it consists of two parts: operation skills and operated objects. From the object-centered point of view, semantic information refers to the changes of the initial state and target state of an object. We transform natural language into a set of predicate object structures and object states. This process completes the construction of motion language through the combination of semantic and visual input.

For the second question "mastery of robot operation skills". Imitation learning has been widely used in robot systems in recent years. The method comprises the following steps: firstly, collecting the teaching trajectory of human beings by dragging teaching, visual capture system or other means; Then, the motion characteristics of the teaching trajectory are extracted and generalized to the new sports scene. However, imitation learning focuses on the description of a single skill in a simple task. For complex tasks, it is still a great challenge for robots to choose appropriate operation skills independently according to the tasks.

The latest neurophysiological research shows that for vertebrates and invertebrates, exercise is a combination of multiple exercise primitives (MPs), and different MPs are executed in turn, resulting in more complex movements. Inspired by this, we are committed to building a robot MPs library composed of skill primitives, object primitives and relation primitives to ensure that robots can master a large number of operation skills and object attributes. Then, after sorting the MPs according to the movement language, the robot can independently complete the planning of the movement track according to the task.

The third question plays a vital role in human-computer interaction. On the one hand, at the semantic level, robot operation is described in words. On the other hand, in the robot skill operation layer, robot operation is characterized by skill features, and the skill features can be directly used to generate motion trajectories. Therefore, if the relationship between the two levels can be established, the robot can understand "the meaning of human natural language". A developed an end-to-end, language adjustment control strategy composed of high-level semantic modules and low-level controllers for operation tasks, and realized the framework of integrating language, vision and control. However, the experiment involves only a few operations (dumping, picking, etc.), and the operation words (verbs) correspond to specific motion primitives. In reality, a verb in natural language may need 3 or 4 operation skills of a robot to complete; Even if the same verb corresponds to different objects, robots may need to perform different operation skills.

At the same time, we make robots have various operating experiences by building robot skill library. In the skill library, each skill primitive is represented by Dynamic Motion Primitive (DMPs)[7], which is an algorithm of imitation learning. It can realize the generalization of point-to-point and periodic motion only by learning a single teaching trajectory.

The system block diagram of this paper is shown in Fig. 1. Aiming at the storage form of action knowledge in knowledge base, this paper puts forward the representation method of operation skill primitives, provides text description to match verbs in instruction sentences, learns the movement characteristics of operation skills based on dynamic movement primitives, and expresses the functions of operation skills in coding form. In order to make primitives executable by robots, a parametric description of operation skill primitives is proposed.

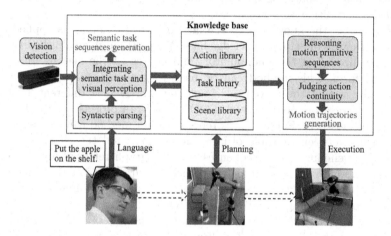

Fig. 1. The proposed framework for robot planning motion sequences based on a knowledge base

2   Related Work

In order to complete complex tasks, the robot should have a large number of MPs reserves, and can reasonably combine the MPs when receiving task instructions. On the one hand, it is necessary to build a knowledge base which connects the upper human language with the lower robot motion sequence. On the other hand, there are hidden Markov models [20], hybrid models and other hierarchical methods to generate MPs sequences. Most of these methods accomplish complex tasks through visual presentations or manual annotation of skill sequences. Different from these works, we can build a knowledge base to enable robots to obtain information of complex tasks through natural language in the process of interacting with people.

2.1   Semantic Analysis

Because the content and structure of knowledge base involve the functional description of different actions, it is very important for the reasoning of motion sequences. In 2010, Ude [18] et al. established a library containing multiple dynamic motion primitives, which was used to store the demonstration trajectory of the same action. Although this way of building database can realize the generalization of the action to the new position, it is not suitable for complex task scenes with multiple actions and multiple targets. Forte [6] introduces query points into the database, and uses query points to represent primitives belonging to each task. In the process of robot movement, sequences can be generated through the connection of primitives, but this library can't reflect the physical meaning of primitives, that is to say, robots can't generate sequences independently by analyzing the target state, but can only move according to the established primitive sequences.

In addition, the knowledge in the field of knowledge map is applied to the storage structure of action knowledge base. Yang [21] et al. established the operation tree library of the manipulator, and expressed the actions in symbolic high-level form. This method can be used to generate action sequences, but the premise is that the movement should be expressed at different levels. The functional object-oriented network FOON [13] is an object-oriented functional network proposed by Paulius. This representation method can well express the relationship between multiple objects and actions and the changes of object states, and generate action sequences through graph retrieval algorithm. The disadvantage is that the state description between objects at the same time is not general, which makes the sequence generation algorithm unable to match the names of corresponding primitives when facing different objects. Yiwen Ding [5] of Wuhan University of Science and Technology and others designed a knowledge map for robot task in 2019 for the disassembly task of man-machine collaboration. This type of knowledge base constructs knowledge with graph structure according to ontology, but the application of its skills is still limited to the patterns stored in the repository.

2.2   Action Sequence Reasoning

The design of sequence reasoning algorithm is closely related to the organization form of robot knowledge base. In literature [3], the action is decided according to the current visual scene, which will be limited by the state of the object appearing in the demonstration. Reference [1] uses the learned symbol planner for planning. In the action reasoning process of reference [16], all implicit semi-Markov models of the center of the object need to be cascaded into a large model, and the most probable action is selected according to the probability. If the operation skills are increased, the cascading model will be huge, which will bring great inconvenience to calculation and search. The reasoning in reference [15] applies the classical tree search algorithm, but the limitation of this algorithm is that all the tasks and objectives that can be reasoned must be stored in the knowledge network in advance. Later, they improved the generalization ability of the network on different kinds of objects [14] and different robots [13], but the problem of the limitation of tasks and objectives mentioned above has not been solved. In 2019, Liutikov [11] proposed an attribute grammar optimized by Markov chain Monte Carlo to complete the reasoning process. This method introduces the concept of attributes into actions, and transforms the probabilistic context-free grammar into the sequence of actions. However, at present, their task scenarios only consider picking and placing in different positions. When there are many basic skills stored in the library, in order to ensure the possible correlation between them, it may be necessary to provide a lot of prior knowledge.

3   Establishment of Skill Knowledge Base

Robot operation skill library is mainly composed of two sub-libraries: action library and scene library.

3.1   Action Library

The motion library stores motion primitives for trajectory planning. Each primitive consists of semantic names, functional codes and motion features. Specifically, the semantic name is used to retrieve the motion primitives in the library; The function code describes the action function through the change of the object state before and after the action is executed, which plays a very important role in the inference of the motion sequence of complex tasks. The motion features are extracted by DMPs calculation and used to generate the motion trajectory.

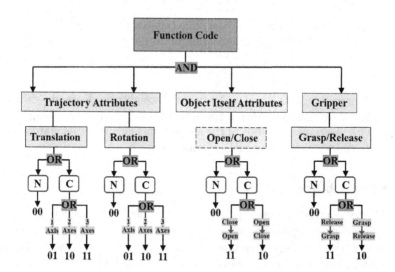

Fig. 2. The graph of the function code

Functional Codes. The function code in Fig. 2 reflects the action's effect on the change of the object's state, and adopts binary code. Where "And" means "grouping attribute codes of different branches"; "OR" means "only one branch occurs"; "N" and "C" represent "unchanged" and "changed" respectively.

The trajectory is described by manipulating the object's "translation" or "rotation". Both translation and rotation can be divided into three situations: the object moves along one-dimensional coordinate axis, two-dimensional plane and three-dimensional space; Gripper attributes include "grasp" or "release", which means that the gripper changes state between two states; Attributes of objects can be obtained from object attributes in the scene library.

Motion Features. Action is the feature extracted from human teaching trajectory by DMPs. DMPs is an imitation learning algorithm to describe the motion trajectory. It adopts the damping spring model, and according to the dynamic system properties, it can finally converge to the target point at the speed of $0\,\mathrm{m/s}$. Dynamic transformation system of damping spring system is as follows:

$$\tau \dot{z} = \alpha_z(\beta_z(g - y) - z) + f \tag{1}$$

$$\tau \dot{y} = z \tag{2}$$

where y and z denote the position and velocity respectively, τ is a time constant, α_z and β_z are positive constants. With $\beta_z = \alpha_z/4$, y converges toward the goal g. f is the forcing term expressed by

$$f(x, g) = \frac{\sum_{i=1}^{N} \psi_i(x)\omega_i}{\sum_{i=1}^{N} \psi_i(x)} x(g - y_0) \tag{3}$$

where $\psi_i(x)$ are fixed exponential basis functions, $\psi_i(x) = exp(-h_i(x - c_i)^2)$. y_0 is the initial state and the parameter g is the target state. x defines as

$$\dot{x} = -\alpha_x x \tag{4}$$

where α_x is a constant. $\omega_i(i = 1,...,\text{N})$ denoted weights are learned as the action features. We store these features in the action library. If the input language verb matches with the semantic name of the action, the feature of this matched action will be selected for generating motion trajectory.

3.2   Scene Library

Scene library contains knowledge about how to deal with objects in the scene and the positional relationship of these objects. This library enables robots to have the experience of inputting objects and remember their lease relationships.

Nowadays, the advanced robot grasping work at home and abroad mainly uses convolutional neural networks to learn and detect the grasping position of objects [2,8–10]. Based on the generated residual convolution neural network (GR-ConvNet) proposed in reference [9], this paper is used to learn the task-related object grabbing patterns, so as to facilitate the subsequent operation. As shown in Fig. 3, GR-ConvNet network consists of five convolution layers, five residual layers and convolution transpose layers. Input RGB images, and the trained model generates three images related to grasping angle, grasping width and grasping quality score, which can be used to predict 3D grasping patterns related to tasks. Accurate prediction and real-time execution are two typical advantages of this method.

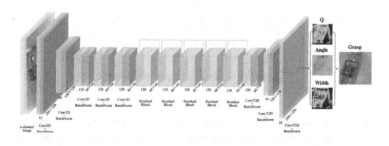

Fig. 3. Detection of grasp position and posture

4   Task Sequence Extraction

In this paper, humans assign tasks to robots through natural language. We have developed a multi-modal interaction model of visual language, which is used to extract the key information of complex tasks from natural language, and transform it into a task sequence that robots can understand by combining visual detection.

4.1   Extraction of Task Information

We define $\mathbf{v}$ as a verb indicating operation, and $\mathbf{N}$ as the set of objects involved in the operation task. Moreover, $\mathbf{N}$ includes an operated object set $\mathbf{Q}$ and a related object set $\mathbf{R}(Q \subseteq N, R \subseteq N)$. Our first key work is to extract verb-object semantic tuples (v, q) from each short sentence. This extraction can be done by StanfordCoreNLP parsing tool, and the analysis result of this tool is shown in Fig. 4.The part of speech of each word is marked in the sentence (" Put the book on the shelf.") and the dependencies between words are described by a dependency graph. The chart shows that the core verb in a sentence is the central component that dominates other components. According to the dependency of "obj", we can extract verb-object semantic tuple: (Put, Book), in which part of speech VB and NN represent verb-object respectively.

4.2   Language and Visual Perception Generate Task Sequences

In this section, we perform the task sequence generation by integrating syntactic analysis and visual perception. Firstly, a pre-trained target detection network (YOLOv5 method) is used to detect the targets in the scene, $F = \{f_0, .., f_c\}$, each of which is represented by a feature vector $f = (f^0, f^b)$, which consists of the detected object category f^0 and their bounding box $f^b \in R^4$ in the robot workspace.

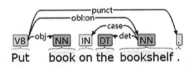

Fig. 4. Dependency graph of example

However, if the objects in set **N** don't match any objects in visual perception set **F** (see q_1 in the Fig. 5, the manipulated objects may not appear in the robot's field of vision), the robot will retrieve the relational knowledge map in the library, find the path from the objects in set **N** to set **F**, and save all 4-tuple relationships in this path.

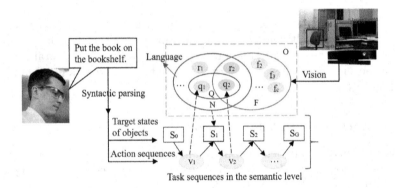

Fig. 5. Task sequences generation by fusing languages and vision

The state $S_i(i = 1,2,...,G)$ is inferred by combining the semantic target position relationship with visual perception. For example (Book, on, Bookshelf, 1) is a 4-tuple semantic target relationship. Combined with visual inspection, we convert this semantic relationship into digital target position data of the manipulated object according to the Table 1. Finally, the task sequence can be expressed as:

$$\xi = S_0(v_1, q_1)S_1(v_2, q_2)...S_{G-1}(v_G, q_G)S_G, \tag{5}$$

in which $(v_1, q_1)...(v_G, q_G)$ are denoted as the semantic predicate-object sequences and $S_0, S_1, ...S_G$ represent the initial state and the series of goal states of the manipulated objects.

Table 1. The relation descriptions of prep words

Prep	Relation
In	$P_h \subset P_t$
On	$P_h > P_t$
To	$P_h \subseteq P_t$
Of	$P_h = P_t$

5 Motion Primitive Sequences Inference for Complex Tasks

In this paper, the Deep Q Network (DQN) is used to reason the motion primitives in each semantic subtask. If the verbs in the subtasks can't match any skill primitives in the library, or if the state of the object after executing the skills (function codes) can't match the target state, we input the semantic subtasks into the DQN to infer the appropriate motion primitives. Finally, by synthesizing the motion primitives of all subtasks, the sequence of motion primitives of complex tasks is obtained.

DQN is a typical value-based reinforcement learning algorithm, which introduces deep neural network into Q-learning to approximate the action value function called Q function. Specifically, the neural network is composed of an evaluation network and a target network with the same structure but different parameters. The critic network is used to predict $Q(s_t, a_t)$ while the target network is used to predict the target Q value. Then, the agent's experience $e_t = (s_t, a_t, r_t, s_{t+1})$ at each time step t is stored in the data set $D = (e_1, ..., e_t)$, which is the so-called replay memory. Associated with the target Q value, the loss function of DQN is designed as:

$$L(\theta) = E[(R + \gamma \max_{a'} Q(s', a'; \theta') - Q(s, a; \theta))^2], \tag{6}$$

where θ' and θ are the weight parameters of target network and the critic network, respectively. These networks iteratively update by the gradient descent algorithm. During learning, the ϵ-greedy policy is utilized to choose the actions with maximum Q value.

In this work, a_t denotes the robot primitive action, s_t in DQN represents object state after executing a_t. Here, s_t refers to the change of the state in the trajectory, object itself, gripper and the position relation among objects. The previous three attributes describe by utilizing the function codes of the skills.

For simplicity, we use the state matrix to represent the object state. Input the initial and target state matrix of the object into DQN learning algorithm, and output the required skill primitive sequence. In order to reach the target state quickly, we construct the reward function from three aspects: (1) Does the new action repeat the previous action? (2) the difference between the target state and the state after skill execution; (3) the number of actions that have been taken at present. Therefore, the reward function can be expressed as:

$$r = \alpha_1 x_{same} + \alpha_2 y_{dif} + \alpha_3 n_{step}, \tag{7}$$

where $\alpha_1, \alpha_2, \alpha_3$ is the weight of each item, with values of (1, 2, 3) respectively; The value of x_{same} is -1 when it is the same as the migration action, and 1 when it is different; y_{dif} is the number of different states between the current state and the target state; n_{step} is the number of actions taken in the current round.

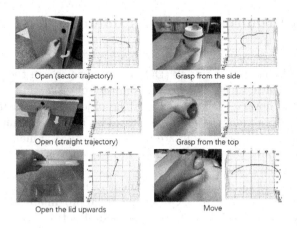

Fig. 6. The motions of Open, Grasp, Move

6   Experiments

6.1   Construction of Knowledge Base

According to the definition of operation skill and the operation task scenario considered in this paper, the set of operation skill primitives S established in this paper contains 12 basic operation skills, such as open (sector track), open (straight track), open (flip up), close (sector track), close (straight track), close (close down), move, take (from side), take (from top), release (from top), dump, etc. As shown in Table 2, the demonstration actions and their acquisition tracks are shown in Fig. 6.

6.2   Task Key Parameter Extraction

The task is "put the medicine bottle on the shelf", and the mandatory sentence is "Put medicine bottle on the shelf". The task scene, recognition results and the task sequence of obtaining its semantic layer by applying the action parameter extraction method proposed in this chapter are shown in Fig. 7.

Table 2. Common operation skill primitives set

Number	Text Description	Motion Function Coding
0	Open (Sector Track)	0 11 00 0
1	Open (Straight Track)	0 11 00 0
2	Open (Turn Up)	0 11 00 0
3	Close (Sector Track)	0 10 00 0
4	Close (Straight Track)	0 10 00 0
5	Close (Close Down)	0 10 00 0
6	Move	1 00 00 0
7	Take (From the Side)	0 00 11 0
8	Take (From Above)	0 00 11 0
9	Release (From the Side)	0 00 10 0

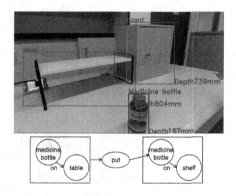

Fig. 7. Scene and task key information extraction result of task

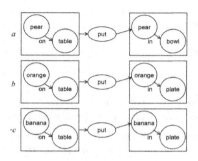

Fig. 8. Task key information extraction results of task a, b and c

Aiming at the mandatory sentences and visual scene information of task input, this part proposes a method of extracting the key information of complex tasks based on multi-mode interaction of language and vision. Firstly, according to the types of objects considered in the knowledge base, a common object data set is established and YOLO model is trained. Based on dependency parsing, the imperative sentence is parsed, and the object of interest, target relation and verb-object phrase sequence are obtained. Then, using the image collected by Kinect v2 depth camera, the information such as the category and position of the objects in the scene can be obtained by YOLO recognition. Finally, the transformation of the mandatory sentences and visual scene information to the task sequence at semantic level is realized, which prepares for the reasoning and execution of the operational skill primitive sequence that the robot can execute later.

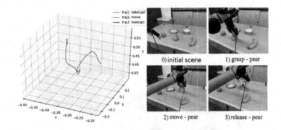

Fig. 9. Movement trajectories and execution of Task a

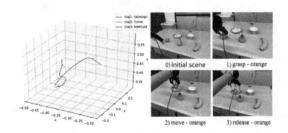

Fig. 10. Movement trajectories and execution of Task b

6.3   Complex Task Reasoning

The instruction statement of task A is "Put pears in a bowl", task B is "Put oranges on a plate" and task C is "Put bananas on a plate". The extracted mission key information is shown in Fig. 8. It can be seen that the predicates are all verbs "put", but their operating objects are different, namely "pear", "orange"

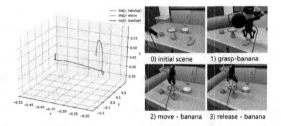

Fig. 11. Movement trajectories and execution of Task c

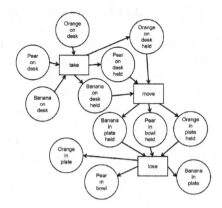

(a) Functional object-oriented network

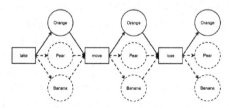

(b) The action sequence deduced by our method

Fig. 12. Functional object-oriented network of tasks a,b,c compared with method in this paper

and "banana", and their target states are also different. The action trajectories and execution diagrams of these three tasks obtained by the method in this paper are shown in Fig. 9, Fig. 10 and Fig. 11. Although the primitives corresponding to the library are the same, the target points of "take (from above)" actions of different tasks change according to the position of the operated object in the environment (shown by the red trajectory in the figure); According to the end point and target state of the last action, the trajectory of "Move" action is different in different tasks (shown by the green trajectory in the figure).

When the predicates in the analyzed operation skill primitive sequences are the same, and the objects correspond to objects in different scenes, the robot can recognize the positions of each object and select the correct object to operate. Comparing the method of this paper with the functional object-oriented network [15], the functional object-oriented network needs to first establish the network structure diagram containing the information in Fig. 12(a)) in order to complete the above tasks, and trace back the knowledge network according to the target state in the reasoning stage until it finds the same node as the current state. The method in this paper makes reasoning according to the change of state, so it is not necessary to store all possible action sequences, so the action sequences are shown in Fig. 12(b)). Compared with the functional object-oriented network, the generalization ability of the method in this paper is enhanced.

This part of the design includes 30 common physical operation tasks of the above four types, and whether the target state of the task reaches the expected state of the task is the basis for judging the completion of the task. When the initial scene of the task is the same as the task goal, the method in this paper can successfully complete 21 tasks, and the method in [17] can successfully complete 17 tasks. By analyzing the experimental results, when the operated object of the task is not detected in the initial scene, the method in this paper can search the scene to find the corresponding object, thus achieving the task goal and improving the success rate of completing the task.

7 Conclusion

Aiming at the task input in language form, this paper puts forward a method of task-critical information extraction based on multi-mode interaction of language vision: extracting task-critical information (object-target relationship diagram and verb-object phrase sequence) from human imperative sentences by using part-of-speech and dependency syntax analysis; The object detection in the current scene is realized based on YOLO v5 visual recognition method, and the initial relationship diagram between objects is obtained by analysis. Finally, combining with the key information of the task, the task sequence of semantic level is generated and verified by experiments. The experimental results show that the task sequence containing key information can be obtained from the instruction statement and vision by this method.

In order to transform the semantic level task sequence into robot executable action sequence, an operation skill primitive sequence reasoning algorithm based on deep Q network is proposed. Aiming at each subtask in the task sequence, action sequence reasoning with the achievement degree of object target state and relationship as reward function is designed in combination with reinforcement learning, which solves the problems that the operated object does not appear in the operation scene, the same semantic verb corresponds to multiple operation skills, the current object state does not meet the action execution conditions, and the action trajectory is connected. At last, an actual robot experimental platform (consisting of aubo-i5 manipulator and robotiq mechanical claw) is built, and the effectiveness of the proposed method is verified by typical complex operation experiments.

Next, we will further improve the running efficiency of reasoning algorithms, improve the interaction between robots and the environment, and make the execution of tasks more effective. At the same time, it saves the use of computer resources during the execution of the algorithm.

Acknowledgment. The work was jointly supported by Beijing Natural Science Foundation (4212933), Scientific Research Project of Beijing Educational Committee (KM202110005023) and National Natural Science Foundation of China (62273012).

References

1. Ahmadzadeh, S.R., Paikan, A., Mastrogiovanni, F., Natale, L., Kormushev, P., Caldwell, D.G.: Learning symbolic representations of actions from human demonstrations. In: 2015 IEEE International Conference on Robotics and Automation (ICRA), pp. 3801–3808 IEEE (2015)
2. Cao, H., Chen, G., Li, Z., Lin, J., Knoll, A.: Lightweight convolutional neural network with gaussian-based grasping representation for robotic grasping detection. (2021) arXiv preprint arXiv:2101.10226
3. Chella, A., Dindo, H., Infantino, I.: A cognitive framework for imitation learning. Robot. Auton. Syst. **54**(5), 403–408 (2006)
4. Defferrard, M., Bresson, X., Vandergheynst, P.: Convolutional neural networks on graphs with fast localized spectral filtering. Advances in Neural Information Processing Systems, 29(2016)
5. Ding, Y., Xu, W., Liu, Z., Zhou, Z., Pham, D.T.: Robotic task oriented knowledge graph for human-robot collaboration in disassembly. Procedia (CIRP) 83, 105–110 (2019)
6. Forte, D., Gams, A., Morimoto, J., Ude, A.: On-line motion synthesis and adaptation using a trajectory database. Robot. Autono. Syst. **60**(10), 1327–1339 (2012)
7. Ijspeert, A.J., Nakanishi, J., Hoffmann, H., Pastor, P., Schaal, S.: Dynamical movement primitives learning attractor models for motor behaviors. Neural Comput., 25(2), 328–373 (2013)
8. Knollmüller, J., Enßlin, T.A.: Bayesian reasoning with trained neural networks. Entropy. 23(6), 693 (2021)
9. Kumra, S., Joshi, S., Sahin, F.: Antipodal robotic grasping using generative residual convolutional neural network. In: 2020 IEEE/RSJ International Conference on Intelligent Robots and Systems (IROS), pp. 9626–9633 IEEE (2020)
10. Kumra, S., Kanan, C.: Robotic grasp detection using deep convolutional neural networks. In: 2017 IEEE/RSJ International Conference on Intelligent Robots and Systems (IROS), pp 769–776 IEEE (2017)
11. Lioutikov, R., Maeda, G., Veiga, F., Kersting, K., Peters, J.: Learning attribute grammars for movement primitive sequencing. Int. J. Rob. Res. **39**(1), 21–38 (2020)
12. Monti, F., Boscaini, D., Masci, J., Rodola, E., Svoboda, J., Bronstein, M.M.: Geometric deep learning on graphs and manifolds using mixture model CNNs. In: Proceedings of the IEEE Conference on Computer Vision and Pattern Recognition, pp. 5115–5124 (2017)

13. Paulius, D., Dong, K.S.P., Sun, Y.: Task planning with a weighted functional object-oriented network. In: 2021 IEEE International Conference on Robotics and Automation (ICRA), pp. 3904–3910 IEEE (2021)
14. Paulius, D., Jelodar, A.B., Sun, Y.: Functional object-oriented network: Construction expansion. In: 2018 IEEE International Conference on Robotics and Automation (ICRA), pp. 5935–5941 IEEE (2018)
15. Paulius, D., Huang, Y., Milton, R., Buchanan, W.D., Sam, J., Sun, Y.: Functional object-oriented network for manipulation learning. In: 2016 IEEE/RSJ International Conference on Intelligent Robots and Systems (IROS), pp. 2655–2662 IEEE (2016)
16. Rozo, L., et al. Learning and sequencing of object-centric manipulation skills for industrial tasks. In: 2020 IEEE/RSJ International Conference on Intelligent Robots and Systems (IROS), pp. 9072–9079. IEEE (2020)
17. Stepputtis, S., Campbell, J., Phielipp, M., Lee, S., Baral, C., Amor, H.B.: Language-conditioned imitation learning for robot manipulation tasks. Advances in Neural Information Processing Systems. 33, 139–150 (2020)
18. Ude, A., Gams, A., Asfour, T., Morimoto, J.: Task-specific generalization of discrete and periodic dynamic movement primitives. IEEE Trans. Robo. 26(5), 800–815 (2010)
19. Wang, P., Li, Z., Hou, Y., Li, W.: Action recognition based on joint trajectory maps using convolutional neural networks. In: Proceedings of the 24th ACM international conference on Multimedia, pp. 102–106 (2016)
20. Wu, D., et al: Deep dynamic neural networks for multimodal gesture segmentation and recognition. IEEE Transactions on pattern Analysis and Machine Intelligence, 38(8), 1583–1597 (2016)
21. Yang, Y., Guha, A., Fermüller, C., Aloimonos, Y.: Manipulation action tree bank: a knowledge resource for humanoids. In: 2014 IEEE-RAS International Conference on Humanoid Robots, pp. 987–992 IEEE (2014)

Adaptive RBF-Neural-Network Control with Force Observer for Teleoperation Robotic System

Zhaohui An[1,2,3] iD, Gaochen Min[1,2,3] iD, Xiangzuo Jiang[4], Xinbo Yu[1,2,3] iD,
Wei He[1,2,3(✉)] iD, and Carlos Silvestre[5,6] iD

[1] School of Intelligence Science and Technology, Institute of Artificial Intelligence,
University of Science and Technology Beijing, Beijing 100083, China
weihe@ieee.org
[2] Beijing Advanced Innovation Center for Materials Genome Engineering,
University of Science and Technology Beijing, Beijing 100083, China
[3] Key Laboratory of Intelligent Bionic Unmanned Systems, Ministry of Education, University
of Science and Technology Beijing, Beijing 100083, China
[4] Donald Bren School of Information and Computer Sciences, University of California,
Irvine, CA 92697, USA
[5] Department of Electrical and Computer Engineering, Faculty of Science and Technology,
University of Macau, Macau, China
[6] Institute for Systems and Robotics (ISR/IST), LARSyS, Instituto Superior Técnico,
Universidade de Lisboa, Lisbon, Portugal

Abstract. In this paper, an adaptive radial basis function neural network
(RBFNN) control strategy is proposed for bilateral teleoperation robotic system
with uncertainty and time delay. Meantime, the force observer method is used
to estimate the interaction force between the slave robot and environment. The
model of teleoperation robotic system is analyzed, and then the uncertainty of
bilateral teleoperation robotic system is estimated by using RBFNN. Using Lyapunov stability theorem, the stability of teleoperation robotic system under our
proposed control is proved. Finally, the effectiveness of the proposed control
algorithm is verified by Matlab Simulink, and position tracking and force estimation performance of the teleoperation robotic system are guaranteed.

Keywords: Adaptive RBFNN Control · Force Observer · Time Delay ·
Teleoperation Robotic System

This work was supported in part by the National Natural Science Foundation of China under
Grant 62225304, 62061160371 and Grant 62003032; in part by the Beijing Natural Science Foundation under Grant JQ20026; in part by the China Postdoctoral Science Foundation under Grant
2020TQ0031 and Grant 2021M690358; in part by the Beijing Top Discipline for Artificial Intelligent Science and Engineering, University of Science and Technology Beijing; and in part by
Macao Science and Technology, Development Fund under Grant FDCT/0031/2020/AFJ.

F. Sun et al. (Eds.): ICCCS 2022, CCIS 1732, pp. 315–330, 2023.
https://doi.org/10.1007/978-981-99-2789-0_27

1 Introduction

The teleoperation robotic systems have been widely employed in many fields, such as space operation, outer-space and underwater explorations, remote surgery, medical rehabilitation, handling hazardous chemicals [1]. Teleoperation robotic system is a system in which the human operator manipulates the master robot to generate desired trajectories, which are transmitted to the slave robots through communication channels. Then, the controller of the slave robot is designed to ensure that the slave robot can track the desired trajectory of the master robot within a certain error range. When the slave robot comes into contact with the environment, the position or force signal is transmitted to the master robot controller through the communication channel.

A major obstacle of teleoperation control is the dynamic uncertainties of robotic model, because of the complex nonlinear robot system. Another problem is the inevitable delay due to the presence of communication channels.

RBFNN has the ability to approximate the nonlinear function with arbitrary precision, so it has been widely applied in teleoperation robotic system [2–8]. In [9], an adaptive fuzzy control was proposed and output constraint was solved. Aiming at the steady-state performance of the system, the finite-time convergence method was studied [10]–[12]. For teleoperation robotic system with time delay, a lot of research has been discussed [13–15]. In [16], a nonlinear force observer was proposed and the transparency of the system was improved. In [17], a human motion intention prediction method based on an autoregressive (AR) model for teleoperation was proposed. Seen from [18–22], some intelligent methods were proposed for dealing with unknown dynamics and unknown states of robotic systems.

2 Related Knowledge

2.1 Master-Slave Teleoperation Robotic System

The dynamic model of the master and slave robots can be described as follows:

$$\begin{cases} M_m(q_m)\ddot{q}_m + C_m(q_m,\dot{q}_m)\dot{q}_m + G_m(q_m) = \tau_m + J_m{}^T(q_m)f_h \\ M_s(q_s)\ddot{q}_s + C_s(q_s,\dot{q}_s)\dot{q}_s + G_s(q_s) = \tau_s + J_s{}^T(q_s)f_e \end{cases} \tag{1}$$

where $q_i, \dot{q}_i$ and $\ddot{q}_i \in \mathbb{R}^n$ are position, velocity, and acceleration vectors respectively; $M_i(q_i) \in \mathbb{R}^{n \times n}$ is the symmetric positive-define inertia matrix; $C_i(q_i, \dot{q}_i) \in \mathbb{R}^{n \times n}$ is the Coriolis and centripetal vector; $G_i(q_i) \in \mathbb{R}^n$ is the gravity vector; and the subscript $i \in m, s$ refers to the master and slave robots, respectively. $\tau_i \in \mathbb{R}^n$ is the input control torque. J_i represents the Jacobian matrices. Moreover, f_h is the interaction force with humans exerted on the master robot and f_e is the interaction force with environment exerted on the slave robot. And we assumed f_h to be bounded, so we have

$$|f_h| \leq \eta_h \tag{2}$$

where η_h is known positive constants.

The robot dynamic models has some well-know useful properties as following.

Property 1. $M_i(q_i)$ is a symmetric positive-definite matrix, and it can be written as $\lambda_{\min}I \leq M_i(q_i) \leq \lambda_{\max}I$. I is identity matrix, $\lambda_{\min}$ and $\lambda_{\max}$ represent the minimum and maximum eigenvalues of $M_i(q_i)$.

Property 2. The equation $p^{\mathrm{T}}(\dot{M}_i(q_i) - 2C_i(q_i, \dot{q}_i))p = 0$ is always true for any vector $p \in \mathbb{R}^n$.

In reality, it is impossible to obtain accurate parameters dynamical model in bilateral teleoperation robotic system. Therefore, the nominal parts M_{oi}, C_{oi}, and G_{oi} are introduced to teleoperation dynamical model. The ΔM_i, ΔC_i, and ΔG_i are used to represent the uncertain parts in the teleoperation robotic system model. So, we have

$$\begin{cases} M_i(q_i) = M_{oi}(q_i) + \Delta M_i(q_i) \\ C_i(q_i, \dot{q}_i) = C_{oi}(q_i, \dot{q}_i) + \Delta C_i(q_i, \dot{q}_i) \\ G_i(q_i) = G_{oi}(q_i) + \Delta G_i(q_i) \end{cases} \tag{3}$$

then the teleoperation robotic dynamic models can be rewritten as

$$\begin{cases} M_{om}(q_m)\ddot{q}_m + C_{om}(q_m, \dot{q}_m)\dot{q}_m + G_{om}(q_m) = \tau_m + J_m^{\mathrm{T}}f_h - D_m(q_m, \dot{q}_m, \ddot{q}_m) \\ M_{os}(q_s)\ddot{q}_s + C_{os}(q_s, \dot{q}_s)\dot{q}_s + G_{os}(q_s) = \tau_s + J_s^{\mathrm{T}}f_e - D_s(q_s, \dot{q}_s, \ddot{q}_s) \end{cases} \tag{4}$$

where $D_m(q_m, \dot{q}_m, \ddot{q}_m)$ and $D_s(q_s, \dot{q}_s, \ddot{q}_s)$ can be defined as

$$\begin{cases} D_m(q_m, \dot{q}_m, \ddot{q}_m) = \Delta M_m(q_m)\ddot{q}_m + \Delta C_m(q_m, \dot{q}_m)\dot{q}_m + \Delta G_m(q_m) \\ D_s(q_s, \dot{q}_s, \ddot{q}_s) = \Delta M_s(q_s)\ddot{q}_s + \Delta C_s(q_s, \dot{q}_s)\dot{q}_s + \Delta G_s(q_s). \end{cases} \tag{5}$$

2.2 Nonlinear Force Observer

A force observer is proposed to estimates the interaction force on the end of the slave robot in Cartesian space. Let $\hat{f}_i$ be the estimation force and $\tilde{f}_i = f_i - \hat{f}_i$ be the force observer error. Based on the dynamic model of teleoperation robotic system, the force observer dynamic can be described as $\dot{\hat{f}}_i = -L_i J_i J_i^T \tilde{f}_i$, where L_i is the observer gain. An auxiliary variable is defined as $Z_i = \hat{f}_i - U_i$, which design to avoid introducing any acceleration measurement. Finally, the estimation force is described as

$$\begin{cases} U_i = L_i J_i M_i(q_i)\dot{q}_i \\ \dot{Z}_i = -L_i J_i J_i^T Z_i + J_i \dot{q}_i - L_i J_i \left(\tau_i + C_i(q_i, \dot{q}_i)\dot{q}_i - G_i(q_i) + J_i^T U_i \right) \\ \hat{f}_i = Z_i + U_i \end{cases} \tag{6}$$

For $i \in m, s$. Therefore, the derivative of force observer error can be defined as

$$\dot{\tilde{f}}_i = -L_i J_i J_i^T \tilde{f}_i - J_i \dot{S}_i \tag{7}$$

2.3   Useful Technical Lemmas

Lemma 1. *[13] For $x, y \in \mathbb{R}$, we have*

$$xy \le \frac{\varepsilon^m}{m}|x|^m + \frac{1}{n\varepsilon^n}|y|^n \tag{8}$$

where $\varepsilon > 0, m > 1, n > 1, (m-1)(n-1) = 1$. This is the famous Young's inequality.

Lemma 2. *[15] For Lyapunov function $V(x) \ge 0$, if we have $\dot{V}(x) \le -\alpha V(x) + C$, α and C are two positive scalars, then the signal $x(t)$ is uniformly bounded.*

3   Control Design and Stability Analysis

3.1   Control Design for Slave Robotic System

This section mainly introduces adaptive control design and stability analysis of the master robot and slave robot. Firstly, for slave robot, we design the adaptive RBFNN controller to realize trajectory tracking performance. We define the switching variable S_s as

$$S_s = \dot{q}_s + \Lambda_s e_s \tag{9}$$

where $\Lambda_s = \Lambda_s^T > 0$.
Then we have

$$\dot{q}_s = S_s - \Lambda_s e_s \tag{10}$$

Then we can get

$$
\begin{aligned}
M_{os}\dot{S}_s &= M_{os}(\ddot{q}_s + \Lambda_s \dot{e}_s) \\
&= M_{os}\Lambda_s \dot{e}_s + M_s(\ddot{q}_s) \\
&= M_{os}\Lambda_s \dot{e}_s - C_{os}\dot{q}_s - G_{os} + \tau_s + J_s^T f_e - D_s(q_s, \dot{q}_s, \ddot{q}_s) \\
&= M_{os}\Lambda_s \dot{e}_s - C_{os}S_s + C_{os}\Lambda_s e_s - G_{os} + \tau_s + J_s^T f_e - D_s(q_s, \dot{q}_s, \ddot{q}_s) \\
&= -C_{os}S_s + \tau_s + J_s^T f_e + o_s - G_{os}
\end{aligned}
\tag{11}
$$

where $o_s = M_{os}\Lambda_s \dot{e}_s + C_{os}\Lambda_s e_s - D_s(q_s, \dot{q}_s, \ddot{q}_s)$, it includes all the dynamical unknown parts in the system.

Then, the adaptive RBFNN controller is proposed as follows:

$$\tau_s = -J_s^T \hat{f}_e - \hat{W}_s^T \Phi_s(x_s) - K_1 S_s + G_{os} - \frac{1}{2}S_s \tag{12}$$

where $K_1 > 0$, and $\hat{f}_e$ is the estimation value of the environment interaction force exerted on the slave robot.

In practical applications, it is difficult to build an accurate and complete teleoperation robotic system model, the RBFNN can be used to estimate the uncertainty parts of the model:

$$Q_i = W_s^T \Phi_s(x_s) + \epsilon_i \tag{13}$$

where W_s^T represents the ideal weight vector, $\Phi_s(x_s)$ is a Gaussian functions, $x_s = [e_s^T \ \dot{e}_s^T \ q_{1s}^T \ \dot{q}_{1s}^T]$. ϵ_i is approximation error and has upper bound that $\epsilon_i \leq \bar{\epsilon}_i$, $\bar{\epsilon}_i$ is a positive constant. And we define,

$$\tilde{W}_s = W_s - \hat{W}_s \tag{14}$$

where $\hat{W}_s$ is the estimation of ideal weight W_s and $\tilde{W}_s$ represents the weight error. Substituting (15) into (14), we have

$$M_{os}\dot{S}_s = -C_{os}S_s + \tau_s + J_s^T f_e + o_s - G_{os}$$

$$= -(C_{os} + K_1)S_s + \tilde{W}_s^T \Phi_s(x_s) + \epsilon_s + J_s^T \tilde{f}_e + G_{os} - \frac{1}{2}S_s \tag{15}$$

where $\tilde{f}_e = f_e - \hat{f}_e$, $\tilde{f}_e$ represents the estimation error of the force observer.

Then adaption laws are given as follows:

$$\dot{\hat{W}}_s = -\Gamma_s \left[\phi_s(x_s)S_s + \sigma_s \hat{W}_s \right] \tag{16}$$

where Γ_s is constant gain matrix, σ_s is small positive constants.

3.2 Stability Analysis of Slave Robotic System

Considering the Lyapunov function candidate V_s as

$$V_s = V_1 + V_2 + V_3 \tag{17}$$

where V_1, V_2 and V_3 can be expressed as

$$V_1 = \frac{1}{2}S_s^T M_{os}S_s$$
$$V_2 = \frac{1}{2}\tilde{W}_s^T \Gamma_s^{-1}\tilde{W}_s \tag{18}$$
$$V_3 = \frac{1}{2}\tilde{f}_e^T \tilde{f}_e$$

The derivative of V_1 can be expressed as

$$\dot{V}_1 = S_s^T M_{os}\dot{S}_s + \frac{1}{2}S_s^T \dot{M}_{os}S_s \tag{19}$$

From Property 2, we have $S_s^T(\dot{M}_{os} - 2C_{os})S_s = 0$, and substituting (18), then we can obtain

$$\dot{V}_1 = -S_s^T K_1 S_s + \frac{1}{2}S_s^T(\dot{M}_{os} - 2C_{os})S_s + S_s^T \left[\tilde{W}_s^T \Phi_s(x_s) + \epsilon_s + J_s^T \tilde{f}_e - \frac{1}{2}S_s \right] \tag{20}$$

According to Lemma 1, we have following inequalities:

$$S_s^\mathrm{T} \epsilon_s \le \frac{1}{2} S_s^2 + \frac{1}{2} \|\bar{\epsilon}_s\|^2 \tag{21}$$

Substituting (21) into (20), we have

$$\dot{V}_1 = -S_s^T K_1 S_s + \frac{1}{2} \|\bar{\epsilon}_s\|^2 + S_s^T \left[\tilde{W}_s^\mathrm{T} \Phi_s(x_s) + J_s^T \tilde{f}_e \right] \tag{22}$$

V_2 taking the derivative of time, we can get

$$\begin{aligned}
\dot{V}_2 &= \tilde{W}_s^T \Gamma_s^{-1} \dot{\tilde{W}}_s \\
&= \tilde{W}_s^T \Gamma_s^{-1} \left[-\Gamma(\phi_s S_s + \sigma_s \hat{W}_s) \right] \\
&= \tilde{W}_s^T (-\phi_s S_s - \sigma_s \hat{W}_s)
\end{aligned} \tag{23}$$

According to Lemma 1, we have following inequalities:

$$-\sigma_s \tilde{W}_s^T \hat{W}_s \le -\frac{\sigma_s}{2} \left\| \tilde{W}_s \right\|^2 + \frac{\sigma_s}{2} \|W_s\|^2 \tag{24}$$

Substituting (24) into (23), we have

$$\dot{V}_2 \le -\tilde{W}_s^T \phi_s S_s - \frac{\sigma_s}{2} \left\| \tilde{W}_s \right\|^2 + \frac{\sigma_s}{2} \|W_s\|^2 \tag{25}$$

Taking the derivative of V_3 respect to time, we can get

$$\dot{V}_3 = \tilde{f}_e^T \dot{\tilde{f}}_e \tag{26}$$

Substituting (7) into (26), we have

$$\dot{V}_3 = -\tilde{f}_e^T L_s J_s J_s^T \tilde{f}_e - \tilde{f}_e^T J_s \dot{S}_s \tag{27}$$

In above all, substituting (22), (25) and (27), we have

$$\begin{aligned}
\dot{V}_s \le &-S_s^T K_1 S_s + S_s^T \left[\tilde{W}_s^\mathrm{T} \Phi_s(x_s) + \epsilon_s + J_s^T \tilde{f}_e - \frac{1}{2} S_s \right] \\
&- \tilde{W}_s^T \phi_s S_s - \frac{\sigma_s}{2} \left\| \tilde{W}_s \right\|^2 - \tilde{f}_e^T L_s J_s J_s^T \tilde{f}_e + \frac{\sigma_s}{2} \|W_s\|^2 - \tilde{f}_e^T J_s \dot{S}_s
\end{aligned} \tag{28}$$

After simplification, we have

$$\begin{aligned}
\dot{V}_s \le &-S_s^T K_1 S_s + \frac{1}{2} \|\bar{\epsilon}_s\|^2 - \frac{\sigma_s}{2} \left\| \tilde{W}_s \right\|^2 + \frac{\sigma_s}{2} \|W_s\|^2 - \tilde{f}_e^T L_s J_s J_s^T \tilde{f}_e \\
\le &-S_s^T K_1 S_s - \frac{\sigma_s}{2} \left\| \tilde{W}_s \right\|^2 - \tilde{f}_e^T L_s J_s J_s^T \tilde{f}_e + C
\end{aligned} \tag{29}$$

$$C = \frac{1}{2} \|\bar{\epsilon}_s\|^2 + \frac{\sigma_s}{2} \|W_s\|^2 \tag{30}$$

Hence we have

$$\dot{V}_s \le -\beta V + C \tag{31}$$

where

$$\beta = \min\left(\frac{\lambda_{\min}(K_1)}{\lambda_{\max}(M_{os})}, \min(\frac{\sigma}{\lambda_{\max}(\Gamma^{-1})}, \lambda_{\min}(L_s))\right) \tag{32}$$

Applying Lemma 2, it is obviously that all signal are uniformly bounded.

3.3   Control Design for Master Robotic System

It is similar to the control design of slave robot. We first define the tracking error as

$$e_m = q_m - q_s(t - T_s) \tag{33}$$

Differentiating e_m, we have

$$\dot{e}_m = \dot{q}_m - \dot{q}_s(t - T_s) \tag{34}$$

To avoid the introduction of acceleration signal, defining a switching variable as

$$S_m = \dot{q}_m + \Lambda_m e_m \tag{35}$$

where $\Lambda_s = \Lambda_s^T > 0$.

Then, we have

$$\dot{q}_m = S_m - \Lambda_m e_m \tag{36}$$

Then, we can obtain

$$
\begin{aligned}
M_{om}\dot{S}_m &= M_{om}(\ddot{q}_m + \Lambda_m \dot{e}_m) \\
&= M_{om}\Lambda_m \dot{e}_m + M_m(\ddot{q}_m) \\
&= M_{om}\Lambda_m \dot{e}_m - C_{om}\dot{q}_m - G_{om} + \tau_m + J_m^T f_h - D_m(q_m, \dot{q}_m, \ddot{q}_m) \\
&= M_{om}\Lambda_m \dot{e}_m - C_{om}S_m + C_{om}\Lambda_m e_m - G_{om} \\
&\quad + \tau_m + J_m^T f_h - D_m(q_m, \dot{q}_m, \ddot{q}_m) \\
&= -C_{om}S_m + \tau_m + J_m^T f_h + o_m - G_{om}
\end{aligned}
\tag{37}
$$

where $o_m = M_{om}\Lambda_m \dot{e}_m + C_{om}\Lambda_m e_m - D_m(q_m, \dot{q}_m, \ddot{q}_m)$, it includes all the dynamical unknown parts in the system.

Then, the adaptive RBFNN controllers of master robot is proposed as follows:

$$\tau_m = -\hat{W}_m^T \Phi_m(x_m) - K_2 S_m + G_{om} - \frac{1}{2}S_m + K_3 \dot{e}_m - \text{sign}(S_m^T) \odot J_m^T \eta_h \tag{38}$$

where $K_2 > 0$, $K_3 > 0$ and f_h is the human force exerted on the master robot.

Remark 1. The signal sgn returns the sign value of the vector, then we have

$$S_m^T \left(-\text{sign}(S_m^T) \odot J_m^T \eta_h + J_m^T f_h \right) \le 0 \tag{39}$$

And we define,

$$\tilde{W}_m = W_m - \hat{W}_m \tag{40}$$

where $\hat{W}_m$ is the estimation of ideal weight W_m and $\tilde{W}_m$ represents the weight error.
Substituting (44) into (43), we have

$$
\begin{aligned}
M_{om}\dot{S}_m = & -C_{om}S_m + \tau_m + J_m^T f_h + o_m - G_{om} \\
= & -(C_{om} + K_2)S_m + \tilde{W}_m^T \Phi_m(x_m) + \epsilon_m - \frac{1}{2}S_m \\
& + J_m^T f_h + K_3 \dot{e}_m - \text{sign}(S_m^T) \odot J_m^T \eta_h
\end{aligned}
\tag{41}
$$

Then adaption laws are given as follows:

$$\dot{\hat{W}}_m = -\Gamma_m \phi_m(x_m)S_m \tag{42}$$

where Γ_m is a constant gain matrix.

3.4   Stability Analysis of Master Robotic System

Considering the Lyapunov function candidate V_m as

$$V_m = V_1 + V_2 + V_3 \tag{43}$$

where V_1, V_2 and V_3 can be expressed as

$$
\begin{aligned}
V_1 &= \frac{1}{2}S_m^T M_m S_m \\
V_2 &= \frac{1}{2}\tilde{W}_m^T \Gamma_m^{-1} \tilde{W}_m \\
V_3 &= \frac{1}{2}e_m^T \Lambda_m K_3 e_m + \frac{1}{2}\int_{t-T_m}^{t} \dot{q}_m^T K_3 \dot{q}_m d\sigma
\end{aligned}
\tag{44}
$$

Taking the derivative of V_1 we can get

$$
\begin{aligned}
\dot{V}_1 = & S_m^T M_m \dot{S}_m + \frac{1}{2}S_m^T \dot{M}_m S_m \\
= & -S_m^T K_2 S_m + \frac{1}{2}S_m^T(\dot{M}_m - 2C_m)S_m + S_m^T[\tilde{W}_m^T \Phi_m(x_m) + \epsilon_m + \\
& J_m^T f_h + K_3 \dot{e}_m - sgn(S_m^T) \odot J_m^T \eta_h - \frac{1}{2}S_m]
\end{aligned}
\tag{45}
$$

Using Property 2 and substituting (45) into (45), we have

$$\dot{V}_1 \le -S_m^T K_2 S_m + S_m^T \tilde{W}_m^T \Phi_m(x_m) + \frac{1}{2}\|\bar{\epsilon}_m\|^2 + S_m^T K_3 \dot{e}_m \tag{46}$$

V_2 taking the derivative of time, we can get

$$
\begin{aligned}
\dot{V}_2 &= \tilde{W}_m^T \Gamma_m^{-1} \dot{\tilde{W}}_m \\
&= \tilde{W}_m^T \Gamma_m^{-1}(-\Gamma_m \phi_m S_m) \\
&= -\tilde{W}_m^T \phi_m S_m
\end{aligned}
\tag{47}
$$

Taking the derivative of V_3 respect to time, we have

$$
\dot{V}_3 = e_m^{\mathrm{T}} \Lambda_m K_3 \dot{e}_m + \frac{1}{2} \dot{q}_m^{\mathrm{T}} K_3 \dot{q}_m - \frac{1}{2} \dot{q}_m^{\mathrm{T}}(t - T_m) K_3 \dot{q}_m(t - T_m)
\tag{48}
$$

In above all, substituting (46), (47) and (48), we have

$$
\begin{aligned}
\dot{V}_m \leq &- S_m^T K_2 S_m + S_m^T \tilde{W}_m^T \Phi_m(x_m) + \frac{1}{2}\|\bar{\epsilon}_m\|^2 + S_m^T K_3 \dot{e}_m - \tilde{W}_m^T \phi_m S_m \\
&+ e_m^{\mathrm{T}} \Lambda_m K_3 \dot{e}_m + \frac{1}{2} \dot{q}_m^{\mathrm{T}} K_3 \dot{q}_m - \frac{1}{2} \dot{q}_m^{\mathrm{T}}(t - T_m) K_3 \dot{q}_m(t - T_m)
\end{aligned}
\tag{49}
$$

Through simplification, we can obtain

$$
\dot{V}_m \leq -S_m^T K_2 S_m + \frac{1}{2}\|\bar{\epsilon}_m\|^2 - \frac{1}{2} \dot{e}_m^{\mathrm{T}} K_3 e_m^{\mathrm{T}}
\tag{50}
$$

Since $V_m = V_1 + V_2 + V_3 \geq 0$ and $\dot{V}_m \leq 0$, $S_m, \dot{e}_m \in \mathcal{L}_2$ and $S_m, e_m \in \mathcal{L}_\infty$. It is obvious that $\dot{q}_m \in \mathcal{L}_\infty$. Then we can obtain $\dot{S}_m \in \mathcal{L}_\infty$, then we can get $|S_m| \to 0$ as $t \to \infty$. Moreover, it imply that $\ddot{q}_m \in \mathcal{L}_\infty$. Then we have $|\dot{e}_m| \to 0$ as $t \to \infty$. At the same time, we can also get $|\dot{q}_m| \to 0$ as $t \to \infty$.

4   Simulation

In this simulation, we use MATLAB Simulink to verify the feasibility of proposed method in teleoperation robotic systems. Both master and slave robot is considered as a two degrees of freedom (DOF) robot. And we assume that time-varing delay $T_m = T_s = 0.1 + 0.02\sin(2t) + 0.03\sin(3t) + 0.01\sin(5t)$s.

We assume the following initial positions and velocities: $q_{m1} = q_{m2} = -\pi/6$ rad, $\dot{q}_{m1} = \dot{q}_{m2} = 0$ rad/s, $q_{s1} = q_{s2} = -\pi/4$ rad, $\dot{q}_{s1} = \dot{q}_{s2} = 0$ rad/s. $t \in 0$. And we assume that $K_1 = K_2 = \mathrm{diag}[15, 15]$, $K_3 = \mathrm{diag}[3, 3]$, $\Gamma_m = \Gamma_s = 1$, $\Lambda_m = \Lambda_s = 1$.

4.1   Case 1

The simulation results are shown in Fig. 1–Fig. 5. The communication delays are shown in Fig. 1. The tracking performances in Case 1 are shown in Fig. 2, we can see that under the adaptive controller we designed, slave robot can follow the trajectory of the master robot. However, due to the existence of communication channel delay, there is a certain lag in position tracking. In Fig. 3 and Fig. 4, it shows the position errors and second errors respectively, we can see that the error signals are bounded as we have proof. And the tracking error are acceptable. We set f_e $3\sin(0.25\pi t)$ N from 8 s to 12 s and $2\sin(0.25\pi t)$ N from 16 s to 20 s.

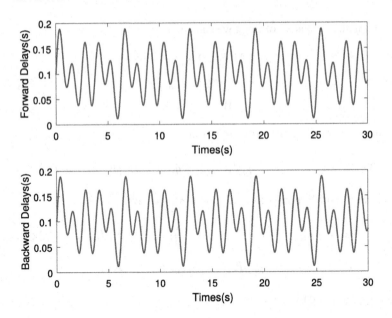

Fig. 1. Communication delays in forward and backward channel

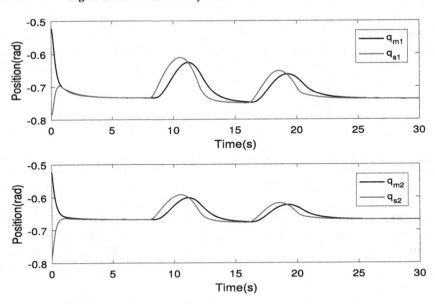

Fig. 2. Tracking performance in Case 1

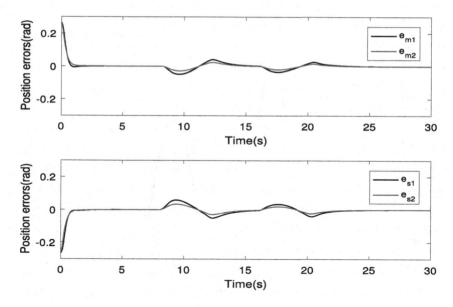

Fig. 3. Position errors in Case 1

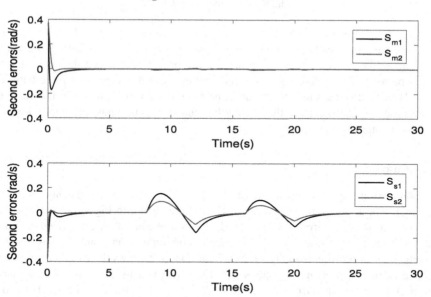

Fig. 4. Second errors in Case 1

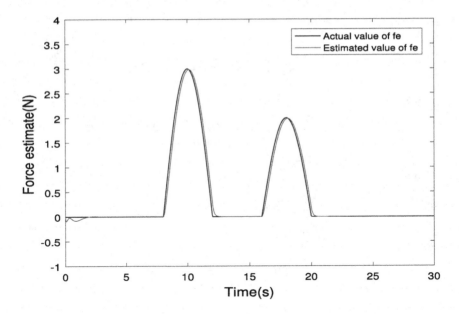

Fig. 5. Force observer in Case 1

The estimation results of force observer is shown in Fig. 5. In Case 1, we use it to simulate the general contact between the end of slave robot and the environment. It can be seen that the force observer can well estimate the interaction force between the robot and the environment. At the same time, there is a slight lag in the value of force observer. We speculate that it may be related to neural network approximation and communication channel delay.

4.2   Case 2

The simulation results are shown in Fig. 6–Fig. 9. The communication delays is shown in Fig. 1. The tracking performances in Case 2 are shown in Fig. 6, we can see that under the adaptive controller we designed, slave robot can follow the trajectory of the master robot. However, due to the existence of communication channel delay, there is a certain lag in position tracking. In Fig. 7 and Fig. 8, it shows the position errors and second er5rors respectively, we can see that the error signals are bounded as we have proof. And the tracking error are acceptable. We set f_e $3\sin(0.25\pi t)$ N from 0 s to 20 s.

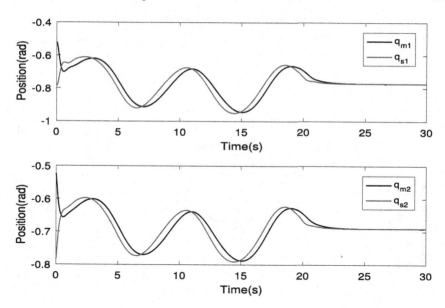

Fig. 6. Tracking performance in Case 2

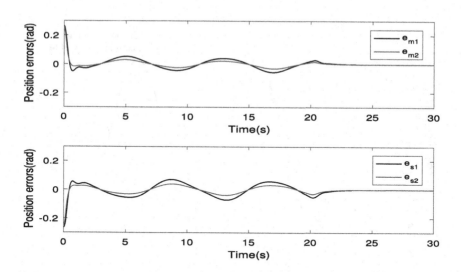

Fig. 7. Position errors in Case 2

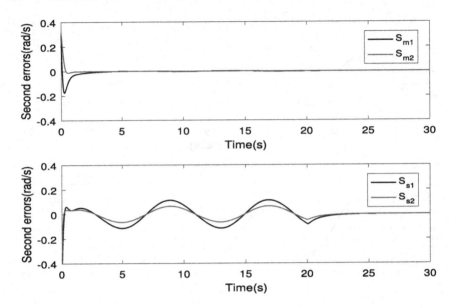

Fig. 8. Second errors in Case 2

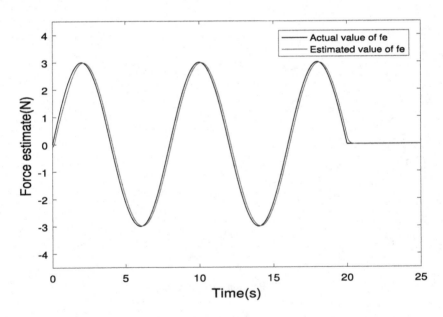

Fig. 9. Force observer in Case 2

Compared with Case 1, the tracking error is slightly increased due to the difficulty of tracking the signal. The estimation result of force observer is shown in Fig. 7. In Case 2, f_e is designed as sine function from 0s to 20s, we use it to simulate a repetitive and periodic contact force, which represents a kind of planned repetitive interaction tasks. It can be seen that the force observer can estimate the sinusoidal signal well. Feedback of this estimation to the master robot enables the operator to obtain force feedback, which ensures system security and improves performance.

5   Conclusion

In this paper, an adaptive RBFNN control architecture was proposed considering time delay and force feedback in bilateral teleoperation robotic system. RBFNN was used to estimate the uncertain parameters of bilateral teleoperation robotic system. At the same time, because the interaction force between the end of slave robot and environment may be difficult to measure, a force observer was introduced to estimate the interaction force. Lyapunov function was designed to ensure the stability of the closed-loop system. The simulation part verifies the stability of the teleoperation robotic system and the effectiveness of the designed controller. We will consider establishing an experimental platform to verify our proposed method in the future.

References

1. Chen, Z., Huang, F., Sun, W.: RBF-neural-network-based adaptive robust control for nonlinear bilateral teleoperation manipulators with uncertainty and time delay. IEEE/ASME Trans. Mechatron. **25**, 906–918 (2020)
2. Yuan, M., Chen, Z., Yao, B., Liu, X.: Fast and accurate motion tracking of a linear motor system under kinematic and dynamic constraints: an integrated planning and control approach. IEEE Trans. Control Syst. Technol. **29**, 804–811 (2021)
3. Liu, Z., Han, Z., Zhao, Z., He, W.: Modeling and adaptive control for a spatial flexible spacecraft with unknown actuator failures. Sci. China Inf. Sci. **64**, 152208:1–152208:16 (2021)
4. Yu, X., He, W., Li, Q., Li, Y., Li, B.: Human-robot co-carrying using visual and force sensing. IEEE Trans. Ind. Electron. **68**, 8657–8666 (2020)
5. He, W., Mu, X., Zhang, L., Zou, Y.: Modeling and trajectory tracking control for flapping-wing micro aerial vehicles. IEEE/CAA J. Automatica Sinica **8**, 148–156 (2021)
6. He, W., Wang, T., He, X., Yang, L., Kaynak, O.: Dynamical modeling and boundary vibration control of a rigid-flexible wing system. IEEE/ASME Trans. Mech. **25**, 2711–2721 (2020)
7. He, W., Meng, T., He, X., Sun, C.: Iterative learning control for a flapping wing micro aerial vehicle under distributed disturbances. IEEE Trans. Cybern. **49**, 1524–1535 (2019)
8. Yang, Y., Hua, C., Ding, H., Guan, X.: Finite-time coordination control for networked bilateral teleoperation. Robotica **33**, 451–462 (2015)
9. Yu, X., He, W., Li, H., Sun, J.: Adaptive fuzzy full-state and output-feedback control for uncertain robots with output constraint. IEEE Trans. Syst. Man Cybern. Syst. **51**, 6994–7007 (2021)
10. Yang, Y., Hua, C., Guan, X.: Finite time control design for bilateral teleoperation system with position synchronization error constrained. IEEE Trans. Cybern. **46**, 609–619 (2016)
11. Gao, H., He, W., Zhang, Y., Sun, C.: Adaptive finite-time fault-tolerant control for uncertain flexible flapping wings based on rigid finite element method. IEEE Trans. Cybern. **52**, 9036–9047 (2021)

12. Fu, Q., Mu, X., Wang, Y.: Minimal solution for estimating fundamental matrix under planar motion. Sci. China Inf. Sci. **64**, 209203 (2021)
13. He, W., Yan, Z., Sun, Y., Ou, Y., Sun, C.: Neural-learning-based control for a constrained robotic manipulator with flexible joints. IEEE Trans. Neural Netw. Learn. Syst. **29**, 5993–6003 (2018)
14. Hua, C., Yang, Y., Guan, X.: Neural network-based adaptive position tracking control for bilateral teleoperation under constant time delay. Neurocomputing **113**, 204–212 (2013)
15. Yang, L., Chen, K.R., Xu, Z.F.: Adaptive tracking control of a nonlinear teleoperation system with uncertainties in kinematics and dynamics. Adv. Mech. Eng. **11**, 1–10 (2019)
16. Dehghan, S.A.M., Koofigar, H.R.: Observer-based adaptive force-position control for nonlinear bilateral teleoperation with time delay. Control. Eng. Pract. **107**, 104679 (2021)
17. Luo, J., Huang, D., Li, Y., Yang, C.: Trajectory online adaption based on human motion prediction for teleoperation. IEEE Trans. Autom. Sci. Eng. **19**, 3184–3191 (2022)
18. Ding, L., Li, S., Gao, H., et al.: Adaptive partial reinforcement learning neural network-based tracking control for wheeled mobile robotic systems. IEEE Trans. Syst. Man Cybern. Syst. **50**, 2512–2523 (2020)
19. Yang, C., Huang, D., He, W., et al.: Neural control of robot manipulators with trajectory tracking constraints and input saturation. IEEE Trans. Neural Netw. Learn. Syst. **32**, 4231–4242 (2020)
20. Feng, Y., Dai, L., Gao, J., et al.: Uncertain pursuit-evasion game. Soft Comput. **24**, 2425–2429 (2020)
21. Liu, Y., Li, Z., Su, H., et al.: Whole-body control of an autonomous mobile manipulator using series elastic actuators. IEEE/ASME Trans. Mechatron. **26**, 657–667 (2021)
22. Liu, Y., Fu, Y., He, W., Hui, Q.: Modeling and observer-based vibration control of a flexible spacecraft with external disturbances. IEEE Trans. Ind. Electron. **66**, 8648–8658 (2018)

Evaluating Visual and Auditory Substitution of Tactile Feedback During Mixed Reality Teleoperation

Yannick Jonetzko[1]([⊠]), Judith Hartfill[2], Niklas Fiedler[1], Fangwei Zhong[3],
Frank Steinicke[2], and Jianwei Zhang[1]

[1] Technical Aspects of Multimodal Systems (TAMS), Universität Hamburg,
Vogt-Kölln-Straße 30, 22527 Hamburg, Germany
{yannick.jonetzko,niklas.fiedler,jianwei.zhang}@uni-hamburg.de
[2] Human-Computer Interaction (HCI), Universität Hamburg, Vogt-Kölln-Straße 30,
22527 Hamburg, Germany
{judith.hartfill,frank.steinicke}@uni-hamburg.de
[3] School of Artificial Intelligence, Peking University and Beijing Institute for General
Artificial Intelligence (BIGAI), Yiheyuan Road 2 and 5, Haidian District,
Beijing 100871, People's Republic of China
zfw@pku.edu.cn

Abstract. Mixed reality (MR) technology has shown enormous potential for real-time human-robot teleoperation. To provide the user with sensor feedback, typically extensive instrumentation is required, such as wearable devices. In this paper, we introduce an MR human-robot teleoperation system based on the Microsoft HoloLens 2 (HL2). The user can directly control the end-effector pose and the gripper of an arbitrary robot arm via hand tracking. Additionally, tactile information can be perceived without wearing data gloves. Therefore, the tactile information gathered by the robot during interaction with objects is substituted and presented to the user visually and auditory. We conducted a pilot user study to analyze the system's usability and the effects of substituted tactile feedback during typical teleoperation tasks. The results show that the system is applicable for teleoperation as the users reached an 87.1% success rate in the performed manipulation task. The users were satisfied with the easy-to-use teleoperation interface and reported a preference for multimodal tactile substitution modes. Regarding the performance, an improvement in the average applied force could be observed when providing tactile feedback.

Keywords: Teleoperation · Tactile Sensing · Multimodal Feedback · Mixed Reality

This research was funded by the German Research Foundation (DFG) and the National Science Foundation of China in the project Crossmodal Learning, TRR-169.

1   Introduction

Teleoperation allows the user to control a system or machine over distance. For instance, humans can teleoperate robots in the same laboratory, in another room, in another country, or even in space. However, it requires a lot of expertise and practice to precisely control such robots. In particular, in surgical tasks for medical purposes or handling hazardous materials, precision is an inevitable requirement. To improve the interaction with the robot and provide humans with feedback about the robot's actions, researchers develop easy-to-use systems [26,29].

One of the most challenging modalities during teleoperation is tactile perception. Realistic haptic or force feedback is often implemented through wearable devices such as data gloves or fingertip devices, which exert vibrations or small deformations to the skin [23]. However, this user instrumentation is often cumbersome and prevents a natural movement of the user's hands. On the other hand, there are approaches without any wearable devices. For instance, Carter et al. [6] present the principle of inducing contactless haptic cues using ultrasound feedback. An acoustically transparent display applies small vibrations to multiple points on the user's hand in an area above the device.

In recent years, the fields of augmented reality (AR) and virtual reality (VR) have been used for robot teleoperation to provide feedback, environment augmentation, and interaction interfaces [7,12,17]. Milgram and Kishino [19] summarize the combination of real and virtual environments as Reality-Virtuality continuum, which describes mixed reality as the area where both the real and virtual worlds are mixed. However, MR human-robot teleoperation (e.g. by using the Microsoft HoloLens) often cannot present the above-mentioned tactile information to the user.

This paper presents a novel setup for human-robot interaction with MR-based real-time teleoperation for one- and two-handed manipulation tasks. Additionally, to improve the performance while teleoperating a robot manipulator, we introduce different methods to substitute tactile information gathered by the robot's gripper. The tactile information is presented visually and auditory by the head-mounted display (HMD). To evaluate the new system, we conducted a pilot study with five participants. They performed a task with the robot while receiving different types of uni- and multimodal feedback (see Fig. 1).

Hence, the contributions of this work can be summarized as follows:

- Development of an MR-based real-time teleoperation system combining Microsoft HoloLens 2 with arbitrary robot arms
- Introduction of visual and auditory substitution of tactile feedback during MR teleoperation
- Pilot study to evaluate the usability of the MR teleoperation system with the mobile robotic PR2 platform

The remainder of this paper is structured as follows. Section 2 summarizes related work. In Sect. 3 we introduce the MR system and the different concepts of visual and auditory substitution of tactile feedback. Section 4 describes the pilot study, which we conducted to evaluate the usability of our approach. Afterward, we discuss the performance and the usability of the setup in Sect. 5. Section 6 concludes the paper and gives an outlook on future work.

Fig. 1. The MR teleoperation scenario. A user wears the HL2 and teleoperates the robot arm's end-effector pose (indicated by the sphere). Visually substituted tactile feedback is provided with arrows (condition S).

2 Related Work

In recent years, VR and AR have become more and more popular, in particular, in the entertainment and gaming industry, but also for training and simulation as well as for improving human-robot collaboration [9]. Different approaches show the usability of interactive AR robot interfaces and teleoperation [17,30,31].

Chan et al. [7] present a multimodal system to teleoperate an industrial robot arm on a predefined trajectory using augmented reality. In their experiment, the users program a trajectory on a table and move the robot afterward along this trajectory while controlling the force applied to the surface by the robot's end-effector. The researchers compare two different force feedback modalities, one of them as a visual arrow indicating the normal force, the other one as haptic vibrations at the user's forearm. As their results show that providing only haptic feedback performs best, they hypothesize that the additional visual feedback leads to cognitive overload.

Contrary to this hypothesis, other researchers determined performance improvements when providing multimodal feedback [5,13,20,25]. Herbst and Stark [13] propose a desktop VR setup. They substitute force magnitudes with visual, auditory, and haptic feedback. In two experiments, participants were asked to sort virtual blocks by their weight and push or pull the block with the least friction out of a stack. They tested each modality individually and in combination. The authors found that a combination of two modalities performs better in terms of execution time and number of transitions compared to the single modalities. The combination of all three modalities did not increase the performance any further.

In the field of medical robotics, different visual presentations of force feedback have been analyzed. Aviles-Rivero et al. [1] compare four different color-based visualizations on a 2D display of a robotic surgical system and found a strong preference for a system with the visual cue as close as possible to the tool, compared to systems with the visual cue in the upper right corner of the display.

Cooper et al. [8] evaluate the effects of substituted feedback in a VR environment with a user study. In their experiment, the participants were asked to perform a wheel change on a car. During the task, they received vibration feedback through tactile gloves, visual cues, and audio information with headphones. They tested each modality individually and in combination and measured the execution time to compare the performance. The subjects performed best when all modalities were used.

Compared to the approaches above, our setup allows teleoperating a 7 Degrees of Freedom (DoF) robotic arm without any wearable devices in an MR environment. To provide tactile information without haptic devices, we use visual and auditory substitutions.

3   System

In this section, we provide an overview of the hard- and software used for this MR-based teleoperation setup. The system enables us to teleoperate the end-effector pose of different robot arms by tracking the user's hand with an HMD. It is not necessary to wear any data gloves or tracking devices on the hands, allowing free hand movements.

3.1   Hardware

Our system consists of two parts: (i) the robotic platform and (ii) the MR head-mounted display HoloLens 2 developed by Microsoft [18]. A robot arm with at least 6-DoF is required for a reliable teleoperation and needs to be controlled using the robot operating system (ROS). To show the usability of the system, we describe it at the example of the mobile platform PR2 and conducted a pilot study as a proof of concept. The platform we used for the pilot study in this work is equipped with one 7-DoF arm with a parallel gripper at the end. This gripper has tactile pressure sensor arrays, one array at each finger [22].

3.2   Communication

Two different environments were integrated in this setup: (i) the robot framework ROS [21] and (ii) the Unity3D game engine [27], which is used on the HoloLens. In recent years, different approaches were developed to combine these environments. The most common one is ROS# [3], which is used in our setup. This framework provides the integration with the Universal Windows Platform (UWP) and handles the communication between ROS and Unity3D.

3.3   Registration

A major challenge in AR setups is the alignment of the virtual environment with the real world. The HoloLens 2 uses its depth sensors to localize itself and place virtual objects at fixed locations in the environment in real-time. Nevertheless, if the virtual objects are to be aligned with real ones, either a known fixed transformation or an object pose recognition algorithm is required. In our setup, we use the fiducial marker tracking library AprilTag [16]. The PR2 detects the marker with its head camera and provides the resulting transform to the Unity MR environment. On the HMD side, the marker is tracked by the front-facing camera of the HL2. Both transforms are transferred into the HMD's coordinate system and align the virtual robot with the real one. Kalaitzakis et al. [14] determine the accuracy of the AprilTag detection algorithm with an average position error of 2 cm, tested with different cameras. As we detect the tag with both the robot and the HMD, the error can add up to 4 cm. Furthermore, when calculating the fiducial marker's pose on the HL2 continuously, the frame rate drops from ~60 fps to ~10 fps and results in an unusable setup. In combination with the marker pose inaccuracy, we decided to register the transform once when starting the system. To minimize the position error afterward, the user can manually adjust position and orientation with a button panel in the virtual environment. This small inaccuracy is no issue in this setup, as the user sees the actual position of the robot's real end-effector during teleoperation.

3.4   Teleoperation

We directly teleoperate the arm via its 6-DoF end-effector pose in Cartesian space using a custom version of the jog_control[1] ROS package. The customization is presented in [28], allowing the teleoperation by an absolute pose, instead of relative position and orientation deltas, with an almost unnoticeable average delay of 0.35 s. The author shows that the framework is independent of the robot platform. As this is the only interface to the robot in the system presented in this paper, our setup can be integrated with different robots as long as it runs ROS. To calculate the end-effector pose, we use the hand tracking algorithm provided by Microsoft in their toolkit[2] for their MR Displays. This algorithm extracts all finger, palm, and wrist joints from the inbuilt depth camera's measurements and provides 25 joint poses overall. In this setup, the first finger and the thumb are used to calculate the goal position and rotation for the teleoperation. We use the knuckle and metacarpal joints for the pose, as their position is more stable than those of the fingertips. To control the gripper's opening state, the distance between the tips of index finger and thumb is directly mapped to the distance between the gripper's fingers. To give some feedback throughout the teleoperation, a small semi-transparent sphere is visualized at the goal pose (see Fig. 1).

---

[1] https://github.com/tork-a/jog_control [Accessed Aug. 2, 2022].
[2] https://github.com/microsoft/MixedRealityToolkit-Unity [Accessed Aug. 2, 2022].

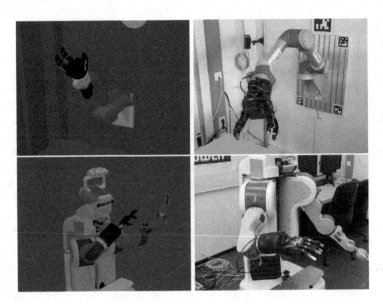

Fig. 2. The teleoperation system integrated on different robotic platforms (top: UR5 + robotic gripper, bottom: PR2 + Shadow Hand) in simulation (left) and on the real hardware (right).

3.5 Tactile Readings

Romano et al. [22] present a method to estimate the applied normal force with the PR2 and the pressure arrays at hand. Each tactile sensor consists of 22 taxels, 15 at the front, two at each side, two at the top, and one at the back. With the 15 front taxels' readings, the applied force is calculated by the sum of the cells' forces. To provide the user with this information, the average force value of both fingers is sent to the MR Display and is substituted by visual and auditory feedback.

3.6 Arbitrary Robots

As mentioned earlier, the teleoperation system we describe and evaluate can easily be integrated on arbitrary robots. We tested it successfully on three different platforms: On a PR2 system using the original left arm, on the right arm of the PR2 with an attached Shadow Dexterous Hand as forearm, and on a UR5 with an attached robotiq 3 finger gripper. We used the original PR2 left arm setup for the pilot study. In addition to Fig. 1, Fig. 2 shows the other two robots both in simulation and on the real hardware.

We designed the interfaces to be independent of the robot arm and its Degrees of Freedom as well as independent of the gripper so that they can be exchanged easily. In Fig. 3, the robot independent parts are marked in blue.

Accordingly, the orange parts need to be updated for the individual robot. We already implemented an interface for full five-fingered dexterous manipulation, which will be part of future work as we do not have a hand controller yet (see the white node in Fig. 3). Both the robotiq gripper and the Shadow Hand are currently abstracted as parallel grippers during the teleoperation. The HL2 can track both hands simultaneously, allowing us to teleoperate two arms in parallel. We implemented this on our two-arm PR2 platform. As jog control is taking care of collisions, all movements were collision-free.

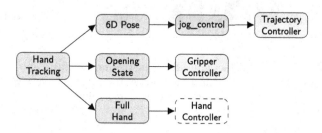

Fig. 3. Abstract teleoperation process for robotic arms and parallel gripper control. Dexterous teleoperation of multifinger hands is not yet supported.

4   User Study

We conducted a pilot study using the mobile robotic PR2 platform to test the user performance and usability of the proposed system to answer the following two research questions:

1. *Does providing tactile information in the form of visual and auditory feedback improve the teleoperation performance regarding execution time, average applied force, and success rate?*
2. *Is the new MR robot setup applicable for teleoperation tasks regarding the usability of non-expert users?*

We designed an object manipulation task to evaluate different levels of tactile substitution methods, including no feedback as well as uni- and multimodal conditions. Furthermore, we used different hardness levels of the manipulated objects to analyze whether this affects the applied force.

4.1   Tactile Substitution

To test the influence of substituted tactile information by visual and audio feedback, the user study was conducted with (i) no feedback, (ii) visual-only, (iii) audio-only, and (iv) multimodal visual + audio feedback.

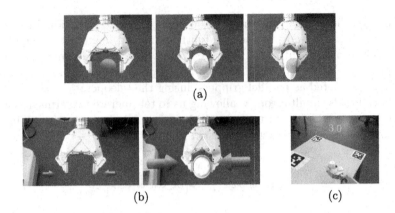

Fig. 4. Visually substituted tactile feedback in the form of (a) a sphere, (b) arrows, and (c) text. Proportional to the measured force, the sphere changes color, the arrow changes size, and the text shows the actual measured value.

Visualization. We developed three visualizations that encode the tactile information with different visual properties:

Colored Sphere (C). A small sphere is visualized as a position reference for the teleoperation under all conditions. The sphere is placed on top of the gripper to have it as close as possible to the region of interest without occluding the manipulated object. In condition C, the color of the sphere will change proportionally to the applied force from white (0 N) over green (5 N) and yellow (7.5 N) to red (10 N max) (see Fig. 4 a).

Arrows (S). Two arrows are visualized next to each finger of the gripper. The arrows' size will change proportionally to the applied force, with the maximum size at 10 N. To stay comparable with the other substitutions, both arrows are visualized the same size and not individually for each finger (see Fig. 4 b).

Text (T). The force is visualized as a number in the top center of the user's field of view to avoid occlusion (see Fig. 4 c).

Audio (A). When the gripper is in contact with both fingers, a constant sine wave is played through the HMD's internal speakers. The frequency varies proportionally to the average measured forces of both fingers between 200 Hz (0 N) and 600 Hz (10 N). Similar to a variometer's feedback, the constant tone starts to beep if the maximum frequency of 600 Hz is exceeded.

4.2 Manipulation Task

In this experiment, the subjects were asked to stack four plastic cups onto a fifth by teleoperating the robot. The participants were instructed to be as fast and precise as possible and apply a minimal amount of pressure to the cup. As it is necessary to maintain the applied pressure to prevent damaging the objects, this task seems reasonable to evaluate the tactile feedback substitution. To simplify the task, all cups were placed in a row, the region where the objects were to be grasped was marked with red to improve comparability, as the bottom of the cups is harder (see Fig. 5 a). The task was considered completed when each cup was either stacked or fallen over. To test if the hardness of the cups influences the average applied force or the execution time, each condition was performed with three different cups shown in Fig. 5 b. The hardness level is decreasing from cup 1 to cup 3.

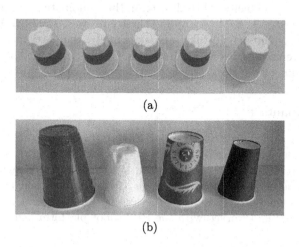

(a)

(b)

Fig. 5. The manipulation task. (a) Cup stacking task: The four cups on the left are supposed to be stacked on the cup on the right. The red area indicates the region where the object should be grasped. (b) Different cups were used to test the influence of the object's hardness. The left cup is used in the training phase, the other ones are cups 1 to 3 from left to right with decreasing hardness levels. (Color figure online)

4.3 Measures

To evaluate the system and the tactile substitutions, we measured multiple factors. Regarding the performance, execution time, average applied force, and success rate are the factors of interest. The execution time was stopped from the first teleoperated movement until the last cup was stacked or fallen over. When both fingers detected a contact, the force was recorded and averaged afterward. For the success rate, we counted the stacked cups (maximum 4 per trial).

To evaluate the overall user experience of the proposed system, the following standard questionnaires were used: NASA Task Load Index (NASA TLX) [10], System Usability Scale (SUS) [4], Simulator Sickness Questionnaire (SSQ) [15], and AttrakDiff2 [11].

Furthermore, we developed a custom questionnaire to collect feedback on the different substitutions.

4.4   Procedure

At the beginning of each session, the participants were informed about the overall procedure and were introduced to the setup and task. They were also asked to fill out a demographics questionnaire and the first part of the SSQ. Each participant got some warm-up time to get familiar with the robot and grasping a cup and was introduced to the different tactile feedback substitution modes. In this phase, a different kind of cup than the ones in the experiment was used (see Fig. 5 b). Each participant completed 24 trials overall, the combination of eight conditions (*No, C, S, T, A, C+A, S+A, T+A*) and three cups were shuffled. After 8 and 16 trials, the participants could take a break. Finally, the participants were asked to fill out the remaining questionnaire forms. One session took about 60 min and the wearing time of the HMD was approximately 45 min.

4.5   Participants

For the pilot study, we recruited 5 participants from the staff of our working groups (1 female, 4 male) aged 23 to 57 ($M = 34.4$, $SD = 13.18$). All of them had normal or corrected-to-normal vision, and none reported a known eye disorder or displacement of equilibrium. Three participants had used an AR headset before, and three reported having experience with hand tracking technology. Four participants had worked with the PR2 robot platform before.

4.6   Results

Performance. In Table 1, the average execution times and success rate of all eight conditions are listed. *No* and *T+A* needed the least average execution times with 104 and 102 s, respectively. With *A* and *T*, the participants needed ~111 s. The conditions *S* and *C*, with and without *A*, resulted in the longest average execution times with over 120 s. The success rate is stable over all conditions and ranges from 83.3% to 91.6% resulting in an average rate of 87.1%. Figure 6 shows the average applied force over all cup types for all eight conditions. It is noticeable that the most force was exerted when no feedback was given.

We also measured the difference of the average applied forces between no feedback and any feedback, separated for the three cup types (see Fig. 5 b). On cup one, an average force of 5.1 N is exerted, on cup two 4.5 N, and on cup three 3.0 N. The average saved force is 1.9 N for cup one, 0.44 N for cup two, and 0.82 N for cup three.

Table 1. Average execution time & Success rate

	No	C	S	T	A	C+A	S+A	T+A
time	104.6 s	120.2 s	127.2 s	112.2 s	111.2 s	120.8 s	125.7 s	102.7 s
rate	88.3%	83.3%	85%	88.3%	86.6%	85%	91.6%	88.3%

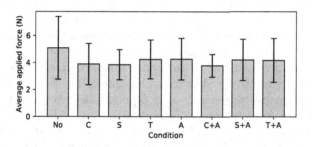

Fig. 6. Average applied force for the individual conditions. *No* is no feedback, *C* is the colored sphere, *S* are the size changing arrows, *T* is the text, and *A* is the audio feedback.

Usability. Simulator sickness increased in an expected way from 153.38 ($SD = 85.8$) before the experiment to 228.17 ($SD = 171.62$) after the experiment. We found a high SUS mean score of 77 ($SD = 13.04$), which can be interpreted to be above average (69.5) [2]. Therefore, the system usability is considered as good.

The results for the NASA TLX are shown in Table 2. Values range from 0–100; higher values refer to a higher task load. The final score consists of six subscales that address different aspects of task complexity. We found especially low values in the overall performance, showing a good user satisfaction and feeling of success.

The results of the AttrakDiff2 questionnaire are summarized in Fig. 7. The underlying model of this questionnaire divides attractiveness into a pragmatic quality (PQ), referring to the amount a product enables a person to fulfill a task, and a hedonic quality (HQ) [11]. The HQ describes the degree to which a product stimulates a user or communicates a certain identity. The participants mostly rated the attractiveness of the system positively on both dimensions.

The collected data shows a preference for multimodal tactile substitutions over substitutions using only one modality. Three of the 5 participants liked *S+A* most, while the other two preferred *C+A* over all other conditions. Similarly, *S+A* and *C+A* were stated the most helpful (2 participants each), while only one person found *S* most helpful. The distraction of each tactile substitution mode was relatively low, with mean values of 1.2 (*S*), 1 (*C*), 1 (*T*), and 1.2 (*A*) (range 0–4). However, each mode was rated distracting to some degree by at least one participant, which suggests personal preference to be an essential factor.

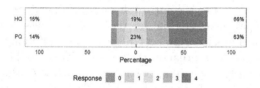

Fig. 7. Percentages for Hedonic Quality (HQ) and Pragmatic Quality (PQ) dimensions of the AttrakDiff2 questionnaire. Responses on 5-point Likert scale.

The participants were asked to rate each visualization against the others on a 5-point Likert scale. We converted the answers into a scoring system, where 0 points were given when no tendency was reported and one and two points were given respectively for each level towards one visualization. C, S, and T scored 14, 10, and 1 point, respectively, showing a clear preference for size and color over text. The participants' agreement on a 5-point Likert scale to the statement "The information from the tactile sensors helped me a lot" was quite mixed, resulting in a mean value of 2.4. Finally, we collected feedback on the difficulty of the different cup types. Three participants perceived cup 1 to be the easiest, while cup 2 was the easiest for the other two.

5   Discussion

5.1   Performance

To answer the first research question (see Sect. 4), the provided feedback improves the performance, as the average applied force decreases when providing feedback. In contrast to that, we do not observe any difference within the individual feedback conditions. We assume that the participants were already fully occupied by concentrating on the stacking tasks, which means that the differences between the individual modalities and conditions need to be tested with easier tasks. The same applies to the success rate, as it seems that the addition of feedback does not influence it.

Table 2. NASA TLX scores.

	M	SD
Mental Demand	56.7	25.28
Physical Demand	53.4	13.94
Temporal Demand	53.4	18.26
Overall Performance	26.67	9.13
Effort	60.0	19.00
Frustration Level	40.0	22.36
Overall Workload	56.1	11.35

With the three cups, we wanted to test the influence of the object's hardness. As the average applied force is comparable and the force difference for cup 1 (the most rigid cup type) is larger than for the others, we assume that more force is saved when grasping harder objects. Sigrist et al. [24] indicate that too much information can cause perceptual overload in AR. This could explain the comparably long execution time for conditions including S or C. Both augmentations are visualized at the end-effector of the robot. This may attract too much of the user's attention. In contrast to the execution time, none of the two visualizations affect the average applied force compared to the other feedback conditions.

5.2   Usability

The pilot study showed that the proposed system is applicable for teleoperation tasks with regard to the usability of the system and acceptance by non-expert users (see Sect. 4). The above-average SUS, as well as the high attractiveness ratings, indicate user satisfaction. Although we only recruited participants from our working groups and most of them had a technical background and some experiences with the PR2 platform and AR, none of them was familiar with tele-operating a robotic arm using hand tracking. The different tactile substitution methods were evaluated as helpful, and we found a preference for (redundant) multi-over unimodal methods.

6   Conclusion and Future Work

In this paper, we present an MR human-robot interaction setup, allowing the user to intuitively teleoperate an arbitrary robot arm in real-time. With the pilot user study, we show with the example of a 7-DoF robot arm of a PR2 platform that the system can be used by non-expert users to perform precise manipulation tasks, which is indicated by the success rate of 87.1%. With substituted tactile feedback, the average applied force on grasped objects can be reduced. We assumed more deviation in the performance between the different feedback modalities than we could find with the study. There seems to be a user preference for multimodal substitutions for tactile feedback over unimodal ones. On the other hand, multimodal feedback seems to be less efficient. This contradiction needs to be further investigated. In future studies, we will choose easier tasks to take the users' attention away from the task, allowing for a higher concentration on the provided feedback.

A weak point of the system is the inaccuracy in the registration. This problem does not occur for teleoperation, as the user gets direct position feedback from the real robot and does not rely on any precisely positioned augmentations. Before we can perform collaboration tasks with autonomously moving robots, we need to focus on this problem. Compared to the first version, the new capability of the HoloLens 2 to track full-hand postures opens the potential for more dexterous teleoperated manipulation tasks. This possibility can also be explored in autonomous side-by-side cooperation and interactive, collaborative tasks.

References

1. Aviles-Rivero, A.I., et al.: Sensory substitution for force feedback recovery: a perception experimental study. ACM Trans. Appl. Percept. **15**(3), 1–19 (2018). ISSN 1544–3558, 1544–3965. https://doi.org/10.1145/3176642
2. Bangor, A., Kortum, P., Miller, J.: Determining what individual SUS scores mean: adding an adjective rating scale. J. Usability Stud. **4**(3), 114–123 (2009)
3. Bischoff, M.: ROS#, June 2019. https://github.com/siemens/ros-sharp/releases/tag/1.7.0. Accessed 2 Aug 2022

4. Brooke, J.: SUS-a quick and dirty usability scale. Usability Eval. Ind. **189**(194), 4–7 (1996)
5. Burke, J.L., et al.: Comparing the effects of visual-auditory and visual-tactile feedback on user performance: a meta-analysis. In: ICMI 2006 Proceedings of the 8th International Conference on Multimodal Interfaces, pp. 108–117. Association for Computing Machinery, Banff (2006)
6. Carter, T., Seah, S.A., Long, B., Drinkwater, B., Subramanian, S.: UltraHaptics: multi-point mid-air haptic feedback for touch surfaces. In: Proceedings of the 26th Annual ACM Symposium on User Interface Software and Technology, pp. 505–514 (2013)
7. Chan, W.P., Quintero, C.P., Pan, M.K., Sakr, M., Van der Loos, H.M., Croft, E.: A multimodal system using augmented reality, gestures, and tactile feedback for robot trajectory programming and execution. In: Proceedings of the ICRA Workshop on Robotics in Virtual Reality, pp. 21–25. IEEE, Brisbane (2018)
8. Cooper, N., Milella, F., Pinto, C., Cant, I., White, M., Meyer, G.: The effects of substitute multisensory feedback on task performance and the sense of presence in a virtual reality environment. PLoS ONE **13**(2), e0191846 (2018)
9. Green, S.A., Billinghurst, M., Chen, X., Chase, J.G.: Human-robot collaboration: a literature review and augmented reality approach in design. Int. J. Adv. Rob. Syst. **5**(1), 1 (2008)
10. Hart, S.G., Staveland, L.E.: Development of NASA-TLX (Task Load Index): results of empirical and theoretical research, vol. 52, pp. 139–183. Elsevier (1988). ISBN 978-0-444-70388-0. https://doi.org/10.1016/S0166-4115(08)62386-9
11. Hassenzahl, M.: The interplay of beauty, goodness, and usability in interactive products. Hum. Comput. Interact. **19**(4), 319–349 (2004)
12. Hedayati, H., Walker, M., Szafir, D.: Improving collocated robot teleoperation with augmented reality. In: HRI 2018: Proceedings of the 2018 ACM/IEEE International Conference on Human-Robot Interaction, pp. 78–86. Association for Computing Machinery, Chicago (2018)
13. Herbst, I., Stark, J.: Comparing force magnitudes by means of vibro-tactile, auditory, and visual feedback. In: IEEE International Workshop on Haptic Audio Visual Environments and their Applications, p. 5. IEEE, Ottawa (2005)
14. Kalaitzakis, M., Carroll, S., Ambrosi, A., Whitehead, C., Vitzilaios, N.: Experimental comparison of fiducial markers for pose estimation. In: 2020 International Conference on Unmanned Aircraft Systems (ICUAS), pp. 781–789. IEEE (2020)
15. Kennedy, R.S., Lane, N.E., Berbaum, K.S., Lilienthal, M.G.: Simulator sickness questionnaire: an enhanced method for quantifying simulator sickness. Int. J. Aviat. Psychol. **3**(3), 203–220 (1993)
16. Krogius, M., Haggenmiller, A., Olson, E.: Flexible layouts for fiducial tags. In: 2019 IEEE/RSJ International Conference on Intelligent Robots and Systems (IROS), pp. 1898–1903. IEEE, Macau (2019)
17. Krupke, D., Steinicke, F., Lubos, P., Jonetzko, Y., Görner, M., Zhang, J.: Comparison of multimodal heading and pointing gestures for co-located mixed reality human-robot interaction. In: IEEE/RSJ International Conference on Intelligent Robots and Systems (IROS), pp. 5003–5009. IEEE, Madrid (2018)
18. Microsoft: HoloLens 2-Overview, Features, and Specs — Microsoft HoloLens, February 2021. https://www.microsoft.com/en-us/hololens/hardware. Accessed 2 Aug 2022
19. Milgram, P., Kishino, F.: A taxonomy of mixed reality visual displays. IEICE Trans. Inf. Syst. **77**(12), 1321–1329 (1994)

20. Prewett, M.S., et al.: The benefits of multimodal information: a meta-analysis comparing visual and visual-tactile feedback. In: ICMI 2006: Proceedings of the 8th International Conference on Multimodal Interfaces, pp. 333–338. Association for Computing Machinery, Banff (2006)

21. Quigley, M., et al.: ROS: an open-source robot operating system. In: ICRA Workshop on Open Source Software, vol. 3, p. 5. IEEE, Kobe (2009)

22. Romano, J.M., Hsiao, K., Niemeyer, G., Chitta, S., Kuchenbecker, K.J.: Human-inspired robotic grasp control with tactile sensing. IEEE Trans. Rob. **27**(6), 1067–1079 (2011)

23. Schorr, S.B., Okamura, A.M.: Fingertip tactile devices for virtual object manipulation and exploration. In: CHI 2017: Proceedings of the 2017 CHI Conference on Human Factors in Computing Systems, pp. 3115–3119. Association for Computing Machinery, Denver (2017)

24. Sigrist, R., Rauter, G., Riener, R., Wolf, P.: Augmented visual, auditory, haptic, and multimodal feedback in motor learning: a review. Psychon. Bull. Rev. **20**(1), 21–53 (2013)

25. Sun, M., Ren, X., Cao, X.: Effects of multimodal error feedback on human performance in steering tasks. J. Inf. Process. **18**, 284–292 (2010). ISSN 1882–6652. https://doi.org/10.2197/ipsjjip.18.284

26. Tavakoli, M., Patel, R.V., Moallen, M., Aziminejad, A.: Haptics for Teleoperated Surgical Robotic Systems, vol. 1. World Scientific, Singapur (2008)

27. Unity Technologies: Unity Real-Time Development Platform — 3D, 2D VR & AR Engine, February 2021. https://unity.com/. Accessed 2 Aug 2022

28. Wieczorek, F.H.: Universal teleoperation ROS interface for robotic manipulators. Bachelor's thesis, Universität Hamburg (2020)

29. Yang, C., Luo, J., Pan, Y., Liu, Z., Su, C.Y.: Personalized variable gain control with tremor attenuation for robot teleoperation. IEEE Trans. Syst. Man Cybern. Syst. **48**(10), 1759–1770 (2017)

30. Yew, A., Ong, S., Nee, A.: Immersive augmented reality environment for the teleoperation of maintenance robots. Procedia CIRP **61**, 305–310 (2017)

31. Zaeh, M.F., Vogl, W.: Interactive laser-projection for programming industrial robots. In: 2006 IEEE/ACM International Symposium on Mixed and Augmented Reality, pp. 125–128. IEEE, Santa Barbara (2006)

Multi-target Detection and Classification for Intelligent Vehicle Based on Deep Learning

Hongbo Gao[1], Huiping Su[1], Xi He[1], Yanzhen Liao[2], Yulin Wu[1], Juping Zhu[1], and Fei Zhang[3(✉)]

[1] School of Information Science and Technology,
University of Science and Technology of China, Hefei 230026, China
[2] School of Data Science, University of Science and Technology of China,
Hefei 230026, China
[3] School of Mathematical Sciences, Anhui University, Hefei 230039, China
fzhang@ahu.edu.cn

Abstract. A multi-target detection and classification method was presented for intelligent vehicle. This method is based on deep learning. Multi-target detection and classification under the traffic road scene is hard to realize since the complexity of road traffic conditions and the diversity of demand. By finding a suitable deep learning algorithm, each detection and recognition task is completed. Based on You Only Look Once (YOLO) v5 algorithm, the task of road multi-target detection is completed. Based on Deep Simple Online and Real Time Tracking (Deep SORT) algorithm, the target tracking of YOLOv5 detection results is realized. Based on Multi-Task Convolutional neural network (MTCNN), the license plate recognition is realized. The built system is adopted to guarantee target detection and tracking of input video. The experiment results are presented and the results show the effectiveness and high accuracy of the target detection and tracking.

Keywords: Multi-target detection · Multi-target classification · Deep learning · Intelligent vehicle

1 Introduction

Target detection-classification are one of the key application fields of intelligent vehicle, which solves traffic congestion and greatly improve production efficiency

This work was supported in part by the National Natural Science Foundation of China (Grant Nos. U20A20225 and U2013601), the Natural Science Foundation of Hefei, China (Grant No. 2021032), the Key Research and Development Plan of Anhui Province (Grant No. 202004a05020058), the Fundamental Research Funds for the Central Universities, the Science and Technology Innovation Planning Project of Ministry of Education of China, NVIDIA NVAIL program, the CAAI-Huawei Mind Spore Open Fund (Grant No. CAAIXSJLJJ-2022-011A). And experiments are conducted on NVIDIA DGX-2.

F. Sun et al. (Eds.): ICCCS 2022, CCIS 1732, pp. 346–353, 2023.
https://doi.org/10.1007/978-981-99-2789-0_29

and traffic efficiency [1]. For intelligent driving, it is necessary to accurately detect road targets in real time, such as pedestrians, cars, bicycles, lane lines, traffic signs, obstacles and so on. Moreover, it is necessary to track and predict the movement of the detected target to predict whether it will affect the current normal driving.

Traditional target detection firstly lists all the features of the target, stores them in the knowledge base with structured symbols, designs corresponding logical rules, and determines whether the object meets the features of the target category by means of question and answer [2]. However, it not only consumes a lot of computing resources, but also has poor real-time performance. In addition, it is impossible to enumerate all the features of each type of detection target, and it is impossible to really sum up the feature symbols that belong to each category, and to construct a general logical structure and algorithm rules.

In recent years, with the development of deep learning, deep learning has become an effective method to solve the problem of target detection, and a large number of inductive learning work is completed by computers [3–5]. Among them, the Deep Convolutional Neural Network (CNN) is the most prominent one [6], which is invariant to geometric transformation, deformation, illumination and so on. It effectively overcomes the difficulties caused by the changeable appearance of non-motor vehicles, and can adaptively construct feature description driven by training data, which has the characteristics of high flexibility and strong generalization ability. In 2013, R-CNN [7], as a pioneer in the application field of deep learning target detection, innovatively combined traditional machine learning with deep learning, and then SPP-Net [8] was proposed to solve some defects of RCNN in detection. Without affecting the detection accuracy, the detection speed is greatly improved. It has good robustness under complex background and some strong interference, and has good detection effect at all scales. In 2015, Liu Wei, et al., proposed SSD algorithm [9]. SSD is a multi-target detection algorithm that directly predicts the target category. Firstly, it extracts the feature map through CNN network model, and then regresses and classifies the feature map. SSD algorithm adds multi-scale feature map function, which can regress candidate frames of different sizes on different levels of feature maps, detect targets of different sizes, and improve recognition accuracy. Similar to SSD, YOLO algorithm also introduces regression algorithm, and generates prediction frame by regression algorithm, which achieves real-time detection and promotes the research of target detection to a new level [10]. At present, YOLO series algorithms have been updated to the 5th generation version, and the 1st to 4th generations have been open source in the market. Many deep learning frameworks can well implement YOLO series algorithms.

Based on YOLOv5 algorithm, and using BDD100K dataset training model, the task of road multi-target detection is completed, and the published Deep SORT algorithm is studied, which is used to realize target tracking based on YOLOv5 detection results. The network models such as MTCNN are studied to realize license plate recognition. Based on the algorithms used, a system is built to complete the functions of target detection and tracking for input video.

The main contributions of this study are summarized as follows:

1) YOLOv5 and Deep SORT algorithms are adopted to guarantee multi-target detection and target tracking.
2) MTCNN, Spatial Transformer Networks (STNet) and License Plate Recognition via Deep Neural Networks (LPRNet) models are utilized to realize license plate recognition.
3) A system is built, which guarantees the target detection, tracking effectiveness and accuracy for input video.

The structure distribution of other section is as follows: Section 2 briefly introduces YOLOv5, Deep SORT, MTCNN, LPRNet and STNet. Experiment and experiment results are shown in Sect. 3. Finally, Sect. 4 concludes the whole paper and proposes the future work.

2   Basic Algorithm Principle

In this section, We first introduce YOLOv5 algorithm for target detection task, next Deep SORT algorithm for target tracking task, then MTCNN, LPRNet and STNet models for license plate recognition task.

2.1   YOLOv5 Algorithm and Deep SORT Algorithm

YOLO algorithm directly inputs the whole picture information to predict, and adopts convolution neural network structure. Image features are extracted by convolution layer, and output probability is predicted by fully connected layer. The specific network structure of YOLOv5 is shown in Fig. 1(a).

Deep SORT algorithm [11] is a commonly used multi-target tracking algorithm at present. The flow of Deep SORT core algorithm is shown in the Fig. 1(b). It can be known that the basic idea of Deep SORT algorithm is prediction and matching, and the two basic operations are Kalman filter and Hungarian algorithm.

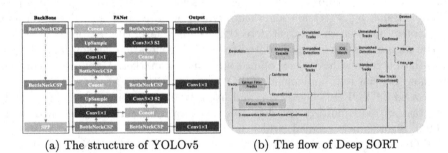

(a) The structure of YOLOv5   (b) The flow of Deep SORT

Fig. 1. Data visualization results.

2.2 MTCNN, STNet and LPRNet Models

MTCNN is a famous real-time detection model for face recognition, which can learn "face recognition", "border regression" and "face key point recognition" at the same time. After modification, it can also be used to detect the position of license plate in a picture. STNet is used to adjust the position of the image, including rotation, translation, scaling and other functions, so that we can get the best input image of license plate recognition, so as to improve the detection accuracy. LPRNet is a license plate recognition model based on deep neural network. It is a lightweight network, which can carry out real-time and high-precision license plate recognition. At present, it has been applied to real traffic monitoring scenes. Using these three network models, the tasks of license plate detection, adjustment and recognition can be well realized.

3 Experiment and Experiment Results

Using multi-module design, the task is divided into four stages: detection, identification, tracking, data analysis and visualization. Each sub-module of the system is independent of each other, and the modification of the sub-module only involves local modification within the module, which will not affect other parts of the system, thus greatly improving the anti-repair ability of the whole system.

The identified target types include 10 common targets on roads, such as pedestrians, cars, trucks and bicycles. The target detection and tracking of input video can be realized. The system processing flow is shown in Fig. 2.

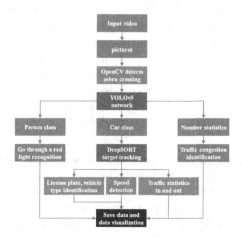

Fig. 2. Project management process.

Read a picture in the input video, process the image with Opencv, detect the position of sidewalk through sliding window, and classify and mark the ground signs artificially. Then, the pictures are sent to YOLO V5 network for target detection and classification. When the traffic light category is detected, the image area is sent to the CNN three-level classification network to judge the current traffic light type. When the car category is detected, the detection frame corresponding to the car category is added to the target tracking list. When a pedestrian is detected, comprehensively consider the output value of CNN network and the sidewalk detection value obtained in the first step, judge whether the pedestrian runs a red light, and then send it to the target tracking list. By tracking the target, the displacement of the first 8 frames and the current frame can be obtained, the moving speed and direction of the vehicle can be calculated, and whether the vehicle is speeding or not can be judged. Send the pictures in the vehicle target tracking frame to the vehicle type recognition network and the license plate detection network. Output the result and bind it to the vehicle ID. Judging the behavior of a vehicle by its driving direction at a red light. The obtained data information such as traffic flow, vehicle speed and other information results are stored in CSV table. The data in CSV table is displayed visually by pychart, and the visual interface is written by pyQT.

3.1   Experiment Setup

For each functional module, it is necessary to determine the most suitable algorithm or network structure.

Object Detection. For target detection, there are many available datasets on the network, such as KITTI, Cityscapes, Mapillary and so on. Among them, BDD100K dataset published by Berkeley is the largest and most diverse open driving dataset in the field of computer vision. This dataset contains 100,000 high-definition videos, which are video datasets obtained by driving vehicles in four regions of the United States for sampling. Therefore, this paper chooses the Image data of BDD100K dataset as the training sample.

Moreover, the object category is changed to 13 road targets to be identified. Secondly, this paper trains the largest X model and the smallest S model at the same time, and selects them according to different requirements.

License Plate Recognition. The module uses the triple network structure of MTCNN+STNet+LPRNet to realize the tasks of license plate detection, adjustment and recognition. In the input picture, MTCNN network structure is used to detect the position of the license plate and extract it. And choosing the method of non-maximum suppression to decide which regression box is worth inputting into R-Net to continue training. Using the idea of yolo algorithm for reference, the images in the dataset are counted once, and an aspect ratio which is convenient for target detection is obtained.

The license plate is adjusted into a standard rectangle through ST-Net network, and then input into LPR-Net network. The LPR-Net process is a classification process, and the numerical value corresponding to the license plate information can be output by inputting the adjusted license plate picture. Set the input image as an RGB image with 94×24 pixels. The important feature information of license plate number is obtained by feature extraction network. In which feature mapping is enhanced by embedding and linking global context. A 1×1 convolution is added in the convolution layer, and the depth of feature mapping is adjusted to 68 character classes with 26 letters of "Jing", "Chuan", "Wan", "A" to "Z" and ten numbers of "0" to "9". Because the output result of LPR-Net network is not the standard license plate number after the previous recognition, we need to translate the LPR-Net output with a preset dictionary and translate it into the standard license plate number output.

The dataset used in this part of the training is CCPD, an open source dataset of China University of Science and Technology. This is a large-scale domestic parking lot license plate dataset, which is used for license plate recognition. Besides ordinary conditions, the license plate photos also include various complicated weather conditions such as blur, tilt, rain and snow. The file name of each image in the CCPD dataset contains all the location information and text information of the license plate. We only use normal license plate pictures, but the number still reaches 130,000.

Vehicle Type Identification. For automobile classification, residual neural network can be used. And this paper adopts a 34-layer residual neural network, namely Resnet34. The core idea of Res-Net is to divide the output of feature extraction network into two parts, namely identity mapping and residual mapping, i.e., $y = x + F(x)$.

Residual network changes the learning goal, and only needs to learn the difference between input and output instead of learning the complete output, which reduces the difficulty of learning. At the same time, the input is directly transmitted to the output, which protects the integrity of information and solves the problem of information loss in the traditional CNN network. For vehicle type recognition, the vehicle dataset released by StanFord is adopted, which contains 16,185 images including 196 vehicle types. Finally, the pre-trained resnet34 classification network is used to replace the output category of the last layer, and the data of the previous layer is frozen for migration learning, and finally the model information of the vehicle is obtained.

Target Tracking. The module uses DeepSORT algorithm as the core algorithm, which can keep track of the detected objects in the video and prevent the ID jump caused by excessive occlusion. At the same time, the matching process is a cycle, and max age is set to 70, that is, tracks and Detections are matched from the first frame of mismatch to the next 70 frames, and tracks without mismatching are preferentially selected for matching with Detections, while tracks with longer mismatch time are matched with Detections later. Through this matching operation, the previously blocked target can be matched with Track again, and the number of ID hops can be reduced.

3.2   Experiment Results and Analysis

The system designed by pyQT is easy to operate and has clear functions. Through the operation interface, it is convenient to input video, view processing results and visualize traffic information. Select the video to be processed through the menu bar on the left. When the road situation is complicated, you can manually mark the road signs such as the front lane line and the left lane line. After the video is input, frame the video through opencv, and then send each frame of pictures into the system for detection. The detection result is shown in Fig. 3(a).

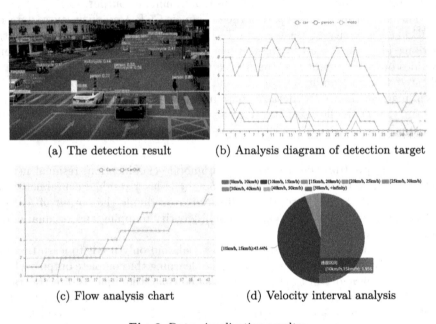

(a) The detection result   (b) Analysis diagram of detection target

(c) Flow analysis chart   (d) Velocity interval analysis

Fig. 3. Data visualization results.

It can be seen that, in the processing results, all road targets in the video are successfully detected, and for car targets, the identified license plate information, vehicle type information and speed information are marked beside the detection frame. For pedestrians in the video, it is judged whether to run a red light by comprehensively considering CNN network output value and sidewalk detection value obtained in the first step, and if illegal behavior is found, it will be saved. As shown in Fig. 3(b)(c)(d). The traffic flow and speed information in the video can be visualized by chart and displayed in the form of pictures.

4   Conclusion and Future Work

In this paper, the problem of road multi-targets detection, recognition and tracking has been investigated. YOLOv5 has been adopted for multi-target detection,

Deep SORT has been utilized to realize target tracking, MTCNN-based method solves the problem of the license plate recognition. The framework divides the system into corresponding modules according to their functions, which are independent of each other and can be improved separately for each module. The presented results have been presented to show the efficiency of the method.

The replaceability and extensibility of the proposed framework for road target identification, detection and tracking have been successfully realized. However, there are still some issues such as improve the detection speed, the influence of complex environment, such as rain and snow, the extracted features are easily affected, which will be researched in the future.

References

1. Li, D., Gao, H.: A hardware platform framework for an intelligent vehicle based on a driving brain. Engineering **4**(2018), 464–470 (2018)
2. Gao, H., Cheng, B., Wang, J., et al.: Object classification using CNN-based fusion of vision and LIDAR in autonomous vehicle environment. In: 2011 IEEE Conference on Prognostics and Health Management, pp. 1–6. IEEE, Denver, Colorado, USA (2011)
3. Gao, H., Cheng, B., Wang, J., et al.: Object classification using CNN-based fusion of vision and LIDAR in autonomous vehicle environment. IEEE Trans. Industr. Inf. **14**(9), 4224–4231 (2018)
4. Gao, H., Su, H., Cai, Y., et al.: Trajectory prediction of cyclist based on dynamic Bayesian network and long short-term memory model at unsignalized intersections. Sci. China Inf. Sci. **64**(7), 1–13 (2021)
5. Xie, G., Gao, H., Qian, L., et al.: Vehicle trajectory prediction by integrating physics-and maneuver-based approaches using interactive multiple models. IEEE Trans. Industr. Electron. **65**(7), 5999–6008 (2017)
6. Long, J., Shelhamer, E., Darrell, T.: Fully convolutional networks for semantic segmentation. IEEE Trans. Pattern Anal. Mach. Intell. **39**(4), 640 (2017)
7. Leegwater, A.J., Meuwly, D., Sjerps, M., et al.: Performance study of a score-based likelihood ratio system for forensic fingermark comparison. J. Forensic Sci. **62**(3), 626–640 (2017)
8. Wang, H., Zhang, C.: Forensic automatic speaker recognition based on likelihood ratio using acoustic-phonetic features measured automatically. J. Forensic Sci. Med. **1**(2), 119–123 (2015)
9. Liu, W., et al.: SSD: single shot multibox detector. In: Leibe, B., Matas, J., Sebe, N., Welling, M. (eds.) ECCV 2016. LNCS, vol. 9905, pp. 21–37. Springer, Cham (2016). https://doi.org/10.1007/978-3-319-46448-0_2
10. Redmon, J., Divvala, S., Girshick, R., et al.: You only look once: unified, real-time object detection. In: Proceedings of the IEEE Conference on Computer Vision and Pattern Recognition, pp. 779–788. IEEE, Las Vegas, Nevada, USA (2016)
11. Wojke, N., Bewley, A., Paulus, D.: Simple online and realtime tracking with a deep association metric. In: 2017 IEEE International Conference on Image Processing, pp. 3645–3649. IEEE, Beijing, China (2017)
12. Yin, X., Liu, X.: Multi-task convolutional neural network for pose-invariant face recognition. IEEE Trans. Image Process. **27**(2), 964–975 (2017)
13. Zherzdev, S., Gruzdev, A.: LPRNet: license plate recognition via deep neural networks. arXiv preprint. arXiv: 1806.10447 (2018)

Online Static Obstacle Avoidance and Offline Static Obstacle Avoidance Framework Based on Interaction Probabilistic Movement Primitives

Jian Fu$^{(\boxtimes)}$, Feng Yang , Yadong Zhong , and Zhu Yang

School of Automation, Wuhan University of Technology, Wuhan 430070, China
fujian@whut.edu.cn

Abstract. How to make robots self-adaptive to obstacle avoidance in the process of human-robot collaboration is one of the challenges in the community. In an actual environment, robots often encounter unanticipated obstacles that make it difficult to complete a task. So we in this paper proposed an obstacle avoidance framework based on Interaction Probabilistic Movement Primitives (iProMP), which combines online static obstacle avoidance with offline static obstacle avoidance. For unanticipated obstacles in human-robot collaboration, we find obstacle avoidance trajectories by solving the Lagrange equation, and then the product of Gaussian distribution is used to fuse the two iProMP trajectories to smoothly switch from the original trajectory to the obstacle avoidance trajectory to achieve fast online static obstacle avoidance. However, the obstacle avoidance trajectory is not optimal. When human-robot collaboration is over, the obstacle is usually not immediately cleared, and the unanticipated obstacles become the anticipated obstacles. In order to obtain a better obstacle avoidance trajectory, Path Integral Policy Improvement with Covariance Matrix Adaptation algorithm is used to train the demonstration trajectories to obtain new iProMP parameters, using the new parameters of human-robot cooperation to realize offline static obstacle avoidance. Experimental results based on two-dimensional trajectory obstacle avoidance and UR5 obstacle avoidance demonstrate the feasibility and effectiveness of the proposed framework.

Keywords: Interaction Probabilistic Movement Primitives ·
Human-robot collaboration · Obstacle avoidance · Path Integral Policy
Improvement

1 Introduction

At present, robots are widely used in factories, homes, hospitals, and other places. Most robots perform tasks according to specified commands, and some need to cooperate with humans to perform tasks. Moreover, the trend of robots interacting with humans is becoming more and more obvious. These phenomena suggest that we need to improve the level of rapid-learning skills of robots.

F. Sun et al. (Eds.): ICCCS 2022, CCIS 1732, pp. 354–367, 2023.
https://doi.org/10.1007/978-981-99-2789-0_30

Learning from demonstrations (LfD) is a mature solution, and the Movement Primitives (MP) in LfD are especially widely used. The Dynamic Movement Primitives (DMPs) [1] proposed by Schaal formalizes the trajectory planning and control theory of dynamic nonlinear systems without causing instability. However, DMPs also have many weaknesses including their inability to adapt to unforeseen situations and changing intermediate points.

A probabilistic formulation of the MP (Probabilistic Movement Primitives, ProMP) was proposed in [2], Compared to DMPs, ProMP is a more general framework that encodes the distribution of possible trajectories that can be tuned to the required start and target states to achieve a motion plan. Maeda et al. [3] used Interaction ProMP (iProMP) to build the joint distributions between the human and the robot. Recently adaptive human-robot interaction was proposed in [4]. However, iProMP does not achieve obstacle avoidance. Taking this into account, Fu et al. [5] proposed a new method to endow a robot with the ability of human-robot collaboration and online obstacle avoidance simultaneously.

Many researchers also have studied ProMP and obstacle avoidance from many aspects. Koert et al. [6] utilized offline planning to avoid obstacles not present during original demonstrations. However, the computationally expensive sample based Kullback-Leibler Divergence prevents online replanning. Colomé et al. [7] proposed a simple yet effective quadratic optimization-based obstacle avoidance method for ProMPs, but he can only perform anticipated obstacle, and unanticipated obstacle is crucial in human-robot collaboration. Qian et al. [8] proposed an environment-adaptive probabilistic interaction primitive method using LfD. However, it is training for obstacles that already exist during original demonstrations. Ashith Shyam et al. [9] addressed obstacle avoidance problems by combining local trajectory optimisation with learning from demonstrations, but he can also only perform anticipated obstacle.

In this paper, we proposed an online static obstacle avoidance algorithm and an offline static obstacle avoidance algorithm based on iProMP. In the process of human-robot collaboration, the robot using iProMP plans out its path by observing the human trajectory, if midway appears obstacles, then through an online obstacle avoidance algorithm to plan a new path, and then using the property that the product of Gaussian distribution is still Gaussian distribution, obtaining a trajectory can be smoothly switched from the trajectory before obstacle avoidance to the trajectory after obstacle avoidance. When the obstacle is observed, the offline obstacle avoidance algorithm is used to update the parameters of iProMP in idle time, so that an optimal obstacle avoidance trajectory can be planned using iProMP after the human trajectory is observed again.

2   The ProMP Algorithm

In this section, we briey introduce the structure of ProMP and iProMP and their parameter estimation methods.

2.1   Probabilistic Movement Primitives

The ProMP model is a movement primitive that uses a probability distribution to describe the trajectory. A robot joint is usually called a degree of freedom(DoF), each DoF in the ProMP is represented by position q_t at time t. Therefore, N Gaussian basis functions can be used to parameterize a smooth trajectory by linear regression, expressed as

$$q_t = \phi_t^T \mathbf{w} + \epsilon \qquad (1)$$

where $\phi_t \in \mathbb{R}^{N \times 1}$ is the Gaussian basis function matrix at time t. $\mathbf{w} \in \mathbb{R}^{N \times 1}$ is the parameter of ProMP, and it represents the weight of each basis function. $\epsilon \sim \mathcal{N}(0, \Sigma_q)$ is gaussian noise with zero mean. Considering the problem with Q DoFs, the form of ProMP can be expressed as

$$\mathbf{y}_t = \begin{bmatrix} q_{1,t} \\ \vdots \\ q_{Q,t} \end{bmatrix} = \begin{bmatrix} \phi_{1,t}^T & & \\ & \ddots & \\ & & \phi_{Q,t}^T \end{bmatrix} \begin{bmatrix} \mathbf{w}_1 \\ \vdots \\ \mathbf{w}_Q \end{bmatrix} + \begin{bmatrix} \epsilon_1 \\ \vdots \\ \epsilon_Q \end{bmatrix}$$
$$= \boldsymbol{\Phi}_t^T \boldsymbol{w} + \boldsymbol{\Sigma}_y \qquad (1)$$

where $\mathbf{y}_t \in \mathbb{R}^{Q \times 1}$ is the position of each DoF at time t, $q_{i,t}$ is the position of ith DoF, $\phi_{i,t}$ is the Gaussian basis function matrix for the ith DoF at time t, $\boldsymbol{\Phi}_t^T = diag(\phi_{1,t}^T, \ldots, \phi_{Q,t}^T)$ is an observation matrix, and $\boldsymbol{w} = [\mathbf{w}_1, \ldots, \mathbf{w}_Q]^T \in \mathbb{R}^{NQ}$ is the parameter of multivariate ProMP. Let the trajectory of time length T be denoted by $\mathbf{y}_{1:T}$, Then, the observation probability of the whole trajectory is expressed as

$$p(\mathbf{y}_{1:T}|\boldsymbol{w}) = \Pi_1^T \mathcal{N}(\mathbf{y}_t|\boldsymbol{\Phi}_t^T \boldsymbol{w}, \boldsymbol{\Sigma}_y) \qquad (2)$$

For ProMP, parameters are generally calculated from multiple demonstration trajectories through the least squares algorithm, and can be expressed as

$$\mathbf{w}_i = (\phi_{i,*} \phi_{i,*}^T + \lambda)^{-1} \phi_{i,*} \boldsymbol{q}_{i,*} \qquad (3)$$

where $\phi_{i,*} = [\phi_{i,1}^T, \ldots, \phi_{i,T}^T], \boldsymbol{q}_{i,*} = [q_{i,1}, \ldots, q_{i,T}]^T$ is the value of the ith demonstration, and λ is the regularization term. If there are M demonstration trajectories, we can calculate the parameter value $\boldsymbol{w}$ of each demonstration trajectory according to Eq. 3. We assume that the parameter $\boldsymbol{w}$ satisfies gaussian distribution $p(\boldsymbol{w}) = \mathcal{N}(\boldsymbol{w}|\boldsymbol{\mu}_w, \boldsymbol{\Sigma}_w)$, so the mean and covariance of $\boldsymbol{w}$ is

$$\begin{cases} \boldsymbol{\mu}_w = \frac{1}{M} \sum_{i=1}^M \boldsymbol{w}_i \\ \boldsymbol{\Sigma}_w = \frac{1}{M-1} \sum_{i=1}^M (\boldsymbol{w}_i - \boldsymbol{\mu}_w)(\boldsymbol{w}_i - \boldsymbol{\mu}_w)^T \end{cases} \qquad (4)$$

where $\boldsymbol{w}_i$ is the parameter of ProMP corresponding to the ith demonstration. then we can get the trajectory probability distribution at time t by integrating with respect to $\boldsymbol{w}$

$$
\begin{aligned}
p\left(\mathbf{y}_t\right) &= \int p(\mathbf{y}_t \mid \boldsymbol{w})p(\boldsymbol{w})d\boldsymbol{w} \\
&= \int \mathcal{N}\left(\mathbf{y}_t \mid \boldsymbol{\Phi}_t^{\mathrm{T}}\boldsymbol{w}, \boldsymbol{\Sigma}_y\right)\mathcal{N}\left(\boldsymbol{w} \mid \boldsymbol{\mu}_w, \boldsymbol{\Sigma}_w\right)d\boldsymbol{w} \qquad (5) \\
&= \mathcal{N}\left(\mathbf{y}_t \mid \boldsymbol{\Phi}_t^{\mathrm{T}}\boldsymbol{\mu}_w, \boldsymbol{\Phi}_t^{\mathrm{T}}\boldsymbol{\Sigma}_w\boldsymbol{\Phi}_t + \boldsymbol{\Sigma}_y\right)
\end{aligned}
$$

Usually the ProMP needs to pass the expectation point $\mathbf{y}_t^* = [q_{1,t}^*, \ldots, q_{Q,t}^*]^{\mathrm{T}}$ at time t, whose variance is $\boldsymbol{\Sigma}_y^*$, and then the parameters of ProMP need to be reestimated to calculate the posterior probability $p(\boldsymbol{w}|\mathbf{y}_t^*) = \mathcal{N}(\boldsymbol{w}|\boldsymbol{\mu}_w^*, \boldsymbol{\Sigma}_w^*)$. According to Bayesian Theory $p(\boldsymbol{w}|\mathbf{y}_t^*) = p(\mathbf{y}_t^*|\boldsymbol{w})p(\boldsymbol{w})/p(\mathbf{y}_t^*)$, the estimation equation of $\boldsymbol{\mu}_w^*$ and $\boldsymbol{\Sigma}_w^*$ is shown below

$$
\begin{cases}
\boldsymbol{K} = \boldsymbol{\Sigma}_w\boldsymbol{\Phi}_t\left(\boldsymbol{\Sigma}_y^* + \boldsymbol{\Phi}_t^{\mathrm{T}}\boldsymbol{\Sigma}_w\boldsymbol{\Phi}_t\right)^{-1} \\
\boldsymbol{\mu}_w^* = \boldsymbol{\mu}_w + \boldsymbol{K}\left(\mathbf{y}_t^* - \boldsymbol{\Phi}_t^{\mathrm{T}}\boldsymbol{\mu}_w\right) \\
\boldsymbol{\Sigma}_w^* = \boldsymbol{\Sigma}_w - \boldsymbol{K}\boldsymbol{\Phi}_t^{\mathrm{T}}\boldsymbol{\Sigma}_w
\end{cases} \qquad (6)
$$

By substituting the updated parameters into Eq. 5, the new trajectory probability can be obtained to achieve the task of passing point.

2.2 Interaction Probabilistic Movement Primitives

ProMP's goal is to establish the relationship between the weights of the basis functions of a single agent, while iProMP can fuse multiple agents to establish the relationship between the weights of the basis functions of multiple agents, and utilize this relationship to achieve collaboration among multiple agents. During the human-robot collaboration, it assumes that the human has P DoFs, and the robot has Q DoFs. The form of iProMP can be described as

$$
\begin{aligned}
\bar{\mathbf{y}}_t &= [q_{1,t}^H, \ldots, q_{P,t}^H, q_{P+1,t}^R, \ldots, q_{P+Q,t}^R]^{\mathrm{T}} \\
&= \begin{bmatrix} \boldsymbol{\phi}_{1,t}^{\mathrm{T}} & & & & \\ & \ddots & & & \\ & & \boldsymbol{\phi}_{P,t}^{\mathrm{T}} & & \\ & & & \boldsymbol{\phi}_{P+1,t}^{\mathrm{T}} & \\ & & & & \ddots \\ & & & & & \boldsymbol{\phi}_{P+Q,t}^{\mathrm{T}} \end{bmatrix} \begin{bmatrix} \mathbf{w}_1^H \\ \vdots \\ \mathbf{w}_P^H \\ \mathbf{w}_{P+1}^R \\ \vdots \\ \mathbf{w}_{P+Q}^R \end{bmatrix} + \begin{bmatrix} \epsilon_1 \\ \vdots \\ \epsilon_P \\ \epsilon_{P+1} \\ \vdots \\ \epsilon_{P+Q} \end{bmatrix} \qquad (7) \\
&= \bar{\boldsymbol{\Phi}}_t^{\mathrm{T}}\bar{\boldsymbol{w}} + \bar{\boldsymbol{\Sigma}}_y
\end{aligned}
$$

where the variables with superscript H represent the data of the human and those with superscript R represent the data of the robot. $\bar{\boldsymbol{\Phi}}_t^{\mathrm{T}} = diag(\boldsymbol{\phi}_{1,t}^{\mathrm{T}}, \ldots, \boldsymbol{\phi}_{P+Q,t}^{\mathrm{T}})$ is an observation matrix. $\bar{\boldsymbol{w}} = [(\mathbf{w}_1^H)^{\mathrm{T}}, \ldots, (\mathbf{w}_P^H)^{\mathrm{T}}, (\mathbf{w}_{P+1}^R)^{\mathrm{T}}, \ldots, (\mathbf{w}_{P+Q}^R)^{\mathrm{T}}]^{\mathrm{T}}$ is the parameter of iProMP, it includes the parameters of human ProMP and the parameters of robot ProMP.

Similar to ProMP, The parameters are obtained from the demo trajectory

$$\begin{cases} \mathbf{w}_i^H = \left(\phi_{i,*}^H \left(\phi_{i,*}^H \right)^{\mathrm{T}} + \lambda \right)^{-1} \phi_{i,*}^H q_{i,*}^H \\ \mathbf{w}_j^R = \left(\phi_{j,*}^R \left(\phi_{j,*}^R \right)^{\mathrm{T}} + \lambda \right)^{-1} \phi_{j,*}^R q_{j,*}^R \end{cases} \tag{8}$$

where i and j represent the ith DoF of human and the jth DoF of the robot, respectively. $\phi_{i,*}^H$ and $\phi_{j,*}^R$ represent the vector of Gaussian basis functions of the human and the robot, respectively.

Since only human trajectory is observed, but not robot trajectory, the observation vector is

$$\overline{y}_t^* = [q_{1,t}^H, \ldots, q_{P,t}^H, 0_{P+1,t}^R, \ldots, 0_{P+Q,t}^R]^{\mathrm{T}} \tag{9}$$

The new parameters $p(\overline{w}|\overline{y}_t^*) = \mathcal{N}(\overline{w}|\mu_{\overline{w}}^*, \Sigma_{\overline{w}}^*)$ are still obtained by calculating posterior probabilities using Bayesian Theory

$$\begin{cases} \mathbf{K} = \Sigma_{\overline{w}} \overline{\Phi}_t^{obs} \left(\Sigma_y^* + (\overline{\Phi}_t^{obs})^{\mathrm{T}} \Sigma_{\overline{w}} \overline{\Phi}_t^{obs} \right)^{-1} \\ \mu_{\overline{w}}^* = \mu_{\overline{w}} + \mathbf{K} \left(y_t^* - (\overline{\Phi}_t^{obs})^{\mathrm{T}} \mu_{\overline{w}} \right) \\ \Sigma_{\overline{w}}^* = \Sigma_{\overline{w}} - \mathbf{K}(\overline{\Phi}_t^{obs})^{\mathrm{T}} \Sigma_{\overline{w}} \end{cases} \tag{10}$$

where $\overline{\Phi}_t^{obs}$ is distinguished between the observable part of a human and the unobservable part of a robot, represented as

$$(\overline{\Phi}_t^{obs})^{\mathrm{T}} = \begin{bmatrix} \left(\phi_{1,t}^H \right)^{\mathrm{T}} & \cdots & 0 & 0 & \cdots & 0 \\ \vdots & \ddots & \vdots & \vdots & \ddots & \vdots \\ 0 & \cdots & \left(\phi_{P,t}^H \right)^{\mathrm{T}} & 0 & \cdots & 0 \\ 0 & \cdots & 0 & 0_{P+1,t}^R & \cdots & 0 \\ \vdots & \ddots & \vdots & \vdots & \ddots & \vdots \\ 0 & \cdots & 0 & 0 & \cdots & 0_{P+Q,t}^R \end{bmatrix} \tag{11}$$

In the process of collaboration, when the human state is observed, we can estimate the expectation of parameter $\mu_{\overline{w}}^*$, and then calculate the state of the robot, which is utilized to drive the robot.

3   Path Integral Policy Improvement

The Path Integral Policy Improvement (PI^2) [10] method is to first find a random Hamilton-Jacobo-Bellman (HJB) equation, the dynamic equation and the optimal control law can be brought into the HJB equation to transform the HJB equation into a nonlinear second-order partial differential equation, and then further transformed into a linear partial differential equation, and then combined with Feynman-Kac formula to transform the linear partial differential equation into path integral problem, to derive the PI^2 algorithm. In the original PI^2 algorithm, the perturbation ε obeys the zero mean normal distribution $\mathcal{N}(0, \boldsymbol{\Sigma})$, when generating a trajectory, each time step adds different perturbations to the parameter θ, and the covariance of the perturbations is fixed. Path Integral Policy Improvement with Covariance Matrix Adaptation ($PI^2 - CMA$) is proposed by Freek Stulp [11], their improvement on PI^2 is to add the same perturbation to the parameters θ of all time steps, but the perturbed covariance matrix is self-adaptive, which greatly improves the performance of PI^2.

θ is defined as the parameter to be optimized, τ is the trajectory, K as the number of trajectories, $\mathcal{J}$ as the loss, and $\mathcal{S}_{k,t}$ as the cumulative cost of the kth trajectory from time t to the end of time T, i.e.

$$\mathcal{S}_{k,t} = \sum_{j=t}^{T} \mathcal{J}(\tau_{j,k}) \tag{12}$$

where $\mathcal{J}(\tau_{j,k})$ is the cost of the kth trajectory at time j. And then you can calculate the probability of K trajectories

$$P_{k,t} = \frac{\exp\{-\frac{1}{\eta}\mathcal{S}_{k,t}\}}{\sum_{k=1}^{K} \exp\{-\frac{1}{\eta}\mathcal{S}_{k,t}\}} \tag{13}$$

where $P_{k,t}$ is the probability of the kth trajectory at time t, η is a constant, scaling the cumulative cost. Then, when the cumulative cost at time t is larger, the probability of the corresponding trajectory at time t is smaller. Thus, when updating parameters θ and covariance $\boldsymbol{\Sigma}$, the trajectory with less cumulative cost has a larger contribution to the corresponding parameters. The update formula of the new parameter is

$$\theta_t^{new} = \sum_{k=1}^{K} P_{k,t}\theta_{k,t} \tag{14}$$

$$\boldsymbol{\Sigma}_t^{new} = \sum_{k=1}^{K} P_{k,t}(\theta_{k,t} - \theta_t^{new})(\theta_{k,t} - \theta_t^{new})^T \tag{15}$$

Then, the time-independent parameters are updated

$$\theta^{new} = \frac{\sum_{t=0}^{T}(T-t)\theta_t^{new}}{\sum_{i=0}^{T}(T-i)} \tag{16}$$

$$\boldsymbol{\Sigma}^{new} = \frac{\sum_{t=0}^{T}(T-t)\boldsymbol{\Sigma}_t^{new}}{\sum_{i=0}^{T}(T-i)} \tag{17}$$

The algorithm pseudo-code of $PI^2 - CMA$ is as

Algorithm 1. PI^2 – CMA

1: Given the number of trajectories sampled per time is K, parameter θ and covariance matrix Σ
2: **repeat**
3:     **for** $k = 1; k <= K; k + +$ **do**
4:        Generate policy parameters with added perturbation $\theta_k = \theta + \varepsilon$
5:        The corresponding trajectory τ_k is obtained by the parameter θ_k
6:     **end for**
7:     **for** $t = 0; t < T; t + +$ **do**
8:        **for** $k = 1; k <= K; k + +$ **do**
9:           Calculate the cumulative cost of trajectory k^{th} according to Formula 12
10:           Calculate the probability of trajectory k^{th} according to Formula 13
11:        **end for**
12:        Update parameter θ_t^{new} according to Formula 14
13:        Update parameter Σ_t^{new} according to Formula 15
14:     **end for**
15:     Update parameter θ^{new} according to Formula 16
16:     Update parameter Σ^{new} according to Formula 17
17: **until** algorithm convergence

---

4   iProMP Obstacle Avoidance

4.1   Online Static Obstacle Avoidance

In the process of human-robot cooperation, obstacle avoidance is a function that the robot should have. In general, we consider the obstacle is a sphere. So when a robot is trying to avoid a sudden obstacle, a simple approach can be to condition the iProMP so as to go through a point which is at a certain threshold distance D from the obstacle. However, there are infinite many solutions of points at a distance D of an obstacle. Therefore, we define the following optimization goals to decide which one to take is needed.

$$\text{argmin}_w \ (\boldsymbol{w} - \boldsymbol{w}_u)^{\mathrm{T}}(\boldsymbol{w} - \boldsymbol{w}_u)$$
$$\text{s.t.} \ (\mathbf{y}_0 - \boldsymbol{\Phi}_t^{\mathrm{T}} \boldsymbol{w})^{\mathrm{T}}(\mathbf{y}_0 - \boldsymbol{\Phi}_t^{\mathrm{T}} \boldsymbol{w}) = D^2 \tag{18}$$

where $\boldsymbol{w}_u$ is the parameter predicted from the trajectory of the human, $\mathbf{y}_0$ is the Cartesian coordinate of the center of the obstacle. We form the Lagrangian equation $L(\boldsymbol{w})$ with the Lagrangian multiplier λ

$$L(\boldsymbol{w}) = (\boldsymbol{w} - \boldsymbol{w}_u)^{\mathrm{T}}(\boldsymbol{w} - \boldsymbol{w}_u) + \lambda(D^2 - (\mathbf{y}_0 - \boldsymbol{\Phi}_t^{\mathrm{T}} \boldsymbol{w})^{\mathrm{T}}(\mathbf{y}_0 - \boldsymbol{\Phi}_t^{\mathrm{T}} \boldsymbol{w})) \tag{19}$$

Our goal is to find a $\boldsymbol{w}^* = \text{argmin}_w L(\boldsymbol{w})$, according to the formula 19, let $\partial L(\boldsymbol{w})/\partial \boldsymbol{w} = 0$, we get $\boldsymbol{w}^*$ as

$$\boldsymbol{w}^* = (I - \lambda \boldsymbol{\Phi}_t \boldsymbol{\Phi}_t^{\mathrm{T}})^{-1}(\boldsymbol{w}_u - \lambda \boldsymbol{\Phi}_t \mathbf{y}_0) \tag{20}$$

By using the latest parameter $\boldsymbol{w}^*$, we get the iProMP trajectory that can avoid obstacle. For the obstacle existing before the trajectory planning, we can directly update the trajectory in this way. However, for the robot that has passed through part of the trajectory, it needs to smoothly transition from the original iProMP trajectory to the iProMP trajectory that can avoid the obstacle.

For the iProMP p_{old} before obstacle avoidance and the iProMP p_{new} after obstacle avoidance, the transition from the iProMP before obstacle avoidance to the iProMP after obstacle avoidance can be obtained by multiplying the Gaussian distribution of different primitives by adding different weights

$$p^* \propto p_{old}^{\omega^{old}} * p_{new}^{\omega^{new}} \tag{21}$$

where ω^{old} and ω^{new} are the weights of the corresponding iProMP, and $\omega^{old} + \omega^{new} = 1$. We take the hyperbolic tangent function to achieve smooth transition of the iProMP, that is $\omega_t^{new} = (tanh(t - t_o) + 1)/2$. Since the hyperbolic tangent function has a range of (-1,1), we add 1 to the hyperbolic tangent function and divide it by 2 to make sure it's in the range of (0,1). The t_o value can be determined after the obstacle is found at a certain time to realize the smooth transition of the iProMP.

For the Gaussian distribution after fusion $p^*(\mathbf{y}_t) = \mathcal{N}(\mathbf{y}_t | \boldsymbol{\mu}_t^*, \boldsymbol{\Sigma}_t^*)$, so according to the product rule of the Gaussian distribution, its mean and covariance are

$$\begin{aligned} \boldsymbol{\Sigma}_t^* &= ((\boldsymbol{\Sigma}_t^{old}/\omega_t^{old})^{-1} + (\boldsymbol{\Sigma}_t^{new}/\omega_t^{new})^{-1})^{-1} \\ \boldsymbol{\mu}_t^* &= \boldsymbol{\Sigma}_t^*((\boldsymbol{\Sigma}_t^{old}/\omega_t^{old})^{-1}\boldsymbol{\mu}_t^{old} + (\boldsymbol{\Sigma}_t^{new}/\omega_t^{new})^{-1}\boldsymbol{\mu}_t^{new}) \end{aligned} \tag{22}$$

Since the product of Gaussian distribution is still Gaussian distribution, this method can be used to fuse iProMP after such fusion in subsequent updates.

4.2   Offline Static Obstacle Avoidance

After obstacles appeared in the process of movement, we could use the above method to plan an obstacle avoidance trajectory in real-time. However, such trajectory might not be an optimal one. Therefore, we explored according to the demonstration trajectories combined with $PI^2 - CMA$ algorithm in the offline period and updated iProMP parameters.

Firstly, the loss of obstacles is defined as

$$c = \begin{cases} 0, & d \geq \epsilon, \\ \epsilon - d, & d < \epsilon. \end{cases} \tag{23}$$

where d represents the Euclidean distance from the end of the robot to the center of the obstacle, the obstacle is approximated as a sphere, with ϵ representing the square of the radius of the sphere.

Secondly, the parameter value after perturbation should not deviate from the parameter value of the demonstration trajectory as much as possible, which is defined as

$$s = (\boldsymbol{w} - \boldsymbol{w}_d)^{\mathrm{T}}(\boldsymbol{w} - \boldsymbol{w}_d) \tag{24}$$

where w is the parameter value after perturbation, w_d is the parameter value of the demonstration trajectories.

Finally, the starting and ending positions of the robot should be the same as the demonstration trajectory, which is defined as

$$f = ||\mathbf{x}_1^d - \mathbf{x}_1||^2 + ||\mathbf{x}_f^d - \mathbf{x}_f||^2 \tag{25}$$

where $\mathbf{x}_1^d$ and $\mathbf{x}_f^d$ are the initial position and end position of the demonstration trajectory respectively, $\mathbf{x}_1$ and $\mathbf{x}_f$ are the initial position and end position of the trajectory generated by the parameter value after perturbation respectively, $||\cdot||$ is the Euclidean norm.

So the total loss can be expressed as

$$S = \alpha_1 s + \alpha_2 c + \alpha_3 f \tag{26}$$

where $\alpha_1, \alpha_2, \alpha_3$ is the corresponding weight.

The offline obstacle avoidance process using $\text{PI}^2 - \text{CMA}$ algorithm is as follows

---

Algorithm 2. Offline Obstacle Avoidance Algorithm

1: Gets the parameters of each demonstration trajectory w_d
2: The $\text{PI}^2 - \text{CMA}$ algorithm is used to obtain the new parameters of each trajectory w
3: The new mean and covariance parameters are calculated according to w
4: iProMP is subsequently executed using the new mean parameter

---

The framework summary of the whole obstacle avoidance algorithm is shown in the Fig. 1. This framework not only judges obstacle once, when the robot has already avoided one obstacle and found a new obstacle, the obstacle avoidance algorithm can also be planning out a new obstacle avoidance path, due to the product of Gaussian distribution is Gaussian distribution, which can continue to incorporate two iProMP, get new smooth obstacle avoidance path. After updating parameters with $\text{PI}^2 - \text{CMA}$ algorithm, the online obstacle avoidance algorithm can still be used to avoid new obstacles, and then $\text{PI}^2 - \text{CMA}$ algorithm can be used for training in an offline period.

5   Experiments

5.1   Two Dimensional Trajectory Obstacle Avoidance Experiment

This experiment is a two-dimensional human-robot collaboration experiment. Forty-five demonstration data were used to train iProMP, and then the mean value of the human demonstration trajectories was used as the trajectory observed by the robot. Here, when the robot observed 20% of the human trajectory, the iProMP was used to plan the trajectory of the robot. In order to verify

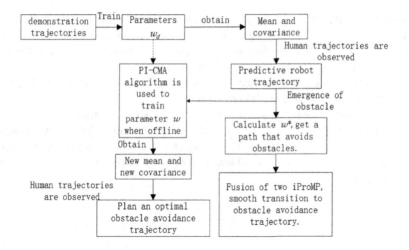

Fig. 1. Obstacle avoidance framework.

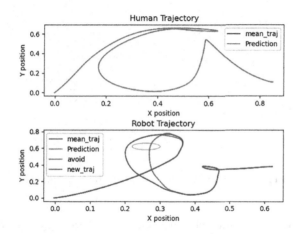

Fig. 2. Two-dimensional online obstacle avoidance.

the obstacle avoidance effect of our proposed algorithm, we added obstacles in the middle of robot movement, and its online obstacle avoidance effect is shown in Fig. 2.

In the Fig. 2, the blue line represents the mean trajectory, and the orange line represents the predicted trajectory before obstacle avoidance. Since the human mean trajectory was used as the observation value of the robot, the predicted trajectory and the mean trajectory almost coincide. When the robot had covered some distance, it found an obstacle halfway through, which was the yellow circle in the picture. At this point, the robot planned a green line obstacle avoidance trajectory in real-time according to the online obstacle avoidance algorithm, and obtained a new red line trajectory which smoothly transitions from the original predicted trajectory to the trajectory after obstacle avoidance according to the fusion algorithm of two iProMP trajectories.

After the robot found this obstacle, it started $PI^2 - CMA$ algorithm to train demonstration trajectories in idle time. When the human trajectory was observed again, an optimal obstacle avoidance trajectory could be obtained directly by using the new parameters. The obstacle avoidance effect after offline training is shown in Fig. 3.

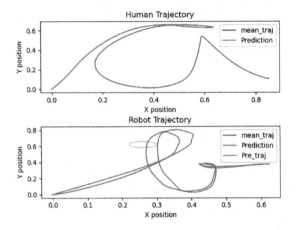

Fig. 3. Two-dimensional offline obstacle avoidance.

In the Fig. 3, the blue line represents the mean trajectory, and the orange line is the predicted trajectory using previous parameters. In order to better compare with the online obstacle avoidance method, we still let the robot observe the same human's mean trajectory, so the orange line and the blue line almost coincide. The green line is the prediction trajectory updated by the new parameters obtained by $PI^2 - CMA$ algorithm. Compared with the online obstacle avoidance algorithm, we can be found that the online obstacle avoidance algorithm pursues to avoid obstacles as fast as possible, and does not care about the shape of the original trajectory, resulting in a great change from the original trajectory. The offline $PI^2 - CMA$ algorithm will not only consider avoiding obstacles, but also keep the shape of the original trajectory as far as possible, which is more in line with the requirements.

5.2   UR5 Obstacle Avoidance Experiment

This experiment is UR5 human-robot collaboration experiment. Ninety-five demonstration data were used to train iProMP, and then the mean value of human demonstration trajectories was used as the trajectory observed by the robot. Here, when the robot observed 20% of the human trajectory, the iProMP was used to plan the trajectory of the robot. Similar to two-dimensional obstacle avoidance, its online obstacle avoidance effect is shown in the Fig. 4. In the Fig. 4, the blue line represents the predicted trajectory before obstacle avoidance. When the robot had covered some distance, it found an obstacle halfway through, which was the green sphere in the picture. At this point, the robot planned a green line obstacle avoidance trajectory in real-time according to the online obstacle avoidance algorithm, and obtained a new red line trajectory which smoothly transitions from the original predicted trajectory to the trajectory after obstacle avoidance according to the fusion algorithm of two iProMP trajectories.

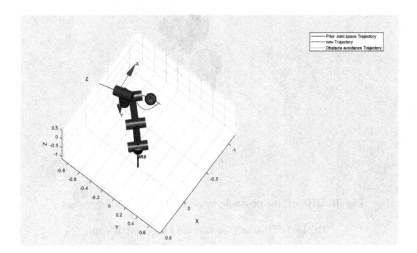

Fig. 4. UR5 online obstacle avoidance.

The obstacle avoidance effect of UR5 after offline training is shown in Fig. 5, and the red line in the figure represents the predicted trajectory of previous parameters. In order to better compare with the online obstacle avoidance method, we still let the robot observe the same human mean trajectory. The blue line is the predicted trajectory updated by using the new parameters obtained by $PI^2 - CMA$ algorithm. Compared with the online obstacle avoidance algorithm, the offline algorithm is more consistent with the original trajectory shape.

Table 1 compares the losses of the two experiments. It can be seen that the loss of offline obstacle avoidance algorithm is smaller than that of online obstacle avoidance algorithm, indicating that the obstacle avoidance trajectory planned by offline obstacle avoidance algorithm is better than that of online obstacle avoidance algorithm.

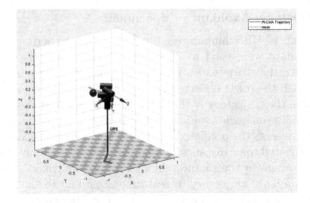

Fig. 5. UR5 offline obstacle avoidance.

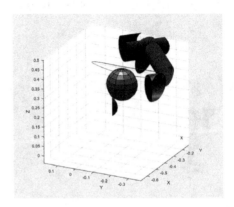

Fig. 6. UR5 offline obstacle avoidance partial enlargement.

Table 1. Obstacle avoidance loss comparison.

Cost	Online obstacle avoidance	Offline obstacle avoidance
Two Dimensional Trajectory	101.33	2.81
UR5	55103.00	513.80

6   Conclusion

We proposed an offline obstacle avoidance algorithm based on $PI^2 - CMA$ algorithm, and an online obstacle avoidance algorithm which can smooth transition to obstacle avoidance trajectory by fusing two iProMP trajectories. This framework can better realize obstacle avoidance. Two-dimensional trajectory obstacle avoidance experiment and UR5 obstacle avoidance experiment show that this framework can better realize online static obstacle avoidance and offline static obstacle avoidance, which proves the feasibility of our framework.

References

1. Schaal, S.: Dynamic movement primitives -A framework for motor control in humans and humanoid robotics. https://doi.org/10.1007/4-431-31381-8_23
2. Paraschos, A., Daniel, C., Peters, J., Neumann, G.: Probabilistic movement primitives. In: NIPS (2013)
3. Maeda, G., Ewerton, M., Lioutikov, R., Ben Amor, H., Peters, J., Neumann, G.: Learning interaction for collaborative tasks with probabilistic movement primitives. In: 2014 IEEE-RAS International Conference on Humanoid Robots, pp. 527–534 (2014). https://doi.org/10.1109/HUMANOIDS.2014.7041413
4. Fu, J., Du, J., Teng, X., Fu, Y., Wu, L.: Adaptive multi-task human-robot interaction based on human behavioral intention. IEEE Access 9, 133762–133773 (2021). https://doi.org/10.1109/ACCESS.2021.3115756
5. Fu, J., Wang, C.Q., Du, J.Y., Luo, F.: Concurrent probabilistic motion primitives for obstacle avoidance and human-robot collaboration. In: Yu, H., Liu, J., Liu, L., Ju, Z., Liu, Y., Zhou, D. (eds.) ICIRA 2019. LNCS (LNAI), vol. 11745, pp. 701–714. Springer, Cham (2019). https://doi.org/10.1007/978-3-030-27529-7_59
6. Koert, D., Maeda, G., Lioutikov, R., Neumann, G., Peters, J.: Demonstration based trajectory optimization for generalizable robot motions. In: 2016 IEEE-RAS 16th International Conference on Humanoid Robots (Humanoids), pp. 515–522 (2016). https://doi.org/10.1109/HUMANOIDS.2016.7803324
7. Colomé, A., Torras, C.: Demonstration-free contextualized probabilistic movement primitives, further enhanced with obstacle avoidance. In: 2017 IEEE/RSJ International Conference on Intelligent Robots and Systems (IROS), pp. 3190–3195 (2017). https://doi.org/10.1109/IROS.2017.8206151
8. Qian, K., Xu, X., Liu, H., Bai, J., Luo, S.: Environment-adaptive learning from demonstration for proactive assistance in human-robot collaborative tasks. Robot. Auton. Syst. 151, 104046 (2022). https://doi.org/10.1016/j.robot.2022.104046.https://www.sciencedirect.com/science/article/pii/S0921889022000185
9. Shyam, R.A., Lightbody, P., Das, G., Liu, P., Gomez-Gonzalez, S., Neumann, G.: Improving local trajectory optimisation using probabilistic movement primitives. In: 2019 IEEE/RSJ International Conference on Intelligent Robots and Systems (IROS), pp. 2666–2671 (2019). https://doi.org/10.1109/IROS40897.2019.8967980
10. A. Theodorou, E., Buchli, J., Schaal, S.: A generalized path integral control approach to reinforcement learning. J. Mach. Learn. Res. 11(11), 3137–3181 (2010)
11. Stulp, F., Sigaud, O.: Path integral policy improvement with covariance matrix adaptation. ArXiv abs/1206.4621 (2012)

An All-terrain Mobile Platform for Multi-Modal Perception and Traversability Estimation

Cong Sun[1], Jingyu Xie[1], Yeqiang Qian[2(✉)], and Ming Yang[1]

[1] Department of Automation, Shanghai Jiao Tong University,
Shanghai 200240, China
[2] University of Michigan-Shanghai Jiao Tong University Joint Institute,
Shanghai Jiao Tong University, Shanghai 200240, China
qianyeqiang@sjtu.edu.cn

Abstract. In this paper, we present a fast, multi-modal perception solution for vehicles or robots traversing in a wide range of terrains, specially designed for off-road environments. We combine images from fisheye cameras and an infra-red camera to generate a dead-zone-free view of the surroundings. We also propose an approach that utilizes LiDAR and GNSS/IMU data to analyze the elevation and traversability of the scene. The integrated perception system runs at a frequency of 12 Hz and immune to adverse weather conditions. We fitted the system onto an all-terrain robot platform and evaluated the overall performance in various off-road and urban-road terrains.

Keywords: Multi-modal perception · Traversability estimation · Semantic segmentation

1 Introduction

The recent advancement in the Advanced Driver Assistance System (ADAS) and autonomous vehicle technology have proved their significance in highly standardized and structured environments [1–4]. However, the adaptation of such intelligent driving frameworks from urban structured road to off-road environments is a relatively unworked area, where the system is desired to accommodate to more complex terrains, a wider range of in-path obstacles, and adverse weather conditions [5]. Hence, in a more natural terrain, a rapid and dead-zone-free perception of incoming terrain changes and obstacles is crucial to either the autonomous driving system or the drivers themselves. Moreover, the desired perception system should remain operational in low-light and severe weather conditions to guarantee the

This work was supported by the National Natural Science Foundation of China (62103261/62173228/61873165) and Postdoctoral Science Foundation of China (2020M681301).

full-time path planning and maneuverability of the vehicle. Such traversing power is advantageous in areas including defense, search and rescue, and forestry.

The segmentation of terrains can be formulated by Simultaneous Localization and Mapping (SLAM) approaches using LiDAR and Global Navigation Satellite System (GNSS)/Inertial Measurement Unit (IMU). Overbye et al. estimated terrain gradients using a 32-channel surround LiDAR and GNSS/IMU [6]. The information contained in a 2D occupancy map works well in a structured road [7] but becomes insufficient in off-road terrains with much complex features. Extending the maps into 3D greatly increases the amount of information and poses a heavy burden to the storage and the processor [8].

Deep-learning-based approaches rely mostly on vision information and fine-grained labels to segment the scene and discriminate the regions with different traversability [9,10]. However, the datasets for off-road scenes [11] are rare and the performance of off-road vehicles or robots can be rather diverse, making it difficult to define traversable areas or even obstacles. Hence, the performance is limited by transferring models from a different robot platform.

Multi-modal perception fuses information from both cameras and LiDAR and processes a much wider range of features from the environment - roughness of the ground, color, probabilities of obstacles, and elevation [12,13]. The main concerns become fully utilizing computational resources for the processing of each type of data and the data fusion strategies.

Considering the requirements of off-road autonomous vehicles and relating tasks, we made a bold exploration in adapting state-of-the-art perception techniques to off-road driving conditions. An all-weather perception system based on a mobile robot platform was designed and built, comprising vision cameras, LiDAR, and GNSS/IMU. A set of experiments were conducted on the platform, optimizing its stability, efficiency, and range of perception. The latest version of the platform is capable of collecting multi-modal data in a wide range of environments. We also propose a fast-running procedure for the generation of the round view of ego vehicle and a method for terrain traversability estimation.

2   System Design and Algorithm Framework

(a) Front view   (b) Back view

Fig. 1. Hardware components of the mobile platform.

The design of the perception system took full consideration of the detection range of the sensors comprised, the resolution and sampling frequency of the data, and data fusion scheme. Based on the foreseeable tasks, the results of the perception were both visualized on a monitor screen as the driving assistance to the driver and transmitted to the path planning module.

Figure 1 shows the components of the perception system assembled based on an all-terrain robot platform with a top speed of 2 m/s. To minimize the blind area of vision, four fish-eye cameras with a horizontal field of view (FOV) of 192° were installed with a 90° offset along the horizontal plane. The coverage also yields a relatively large intersecting space between two adjacent cameras, which is friendly to the image stitching algorithm and also more robust to occlusion occasions.

The system relies on a passive infra-red camera fitted with a linear lens for night works. In properly illuminated cases, the processing of infra-red data can be fused with the fish-eye camera module. While in low-light situations, the infra-red camera works individually and provides a vision with an FOV of 90° to ensure the normal running of the vehicle.

The perception of depth information is achieved using a 32-line blind spot LiDAR and a MEMS LiDAR. The MEMS LiDAR was installed on top of the robot for mid-to-far-range depth perception while the blind spot LiDAR was fitted on the front face of the robot for pre-warning of close-range collision. The combined range of perception reached 200 meters in depth with a close-range FOV of 180°. Unlimited by the illumination of surroundings, the LiDAR module works all-time and functions as a proximity sensor of in-path obstacles and an elevation estimator. The perception system is also equipped with an integrated inertial navigation module, which comprises an IMU and a GNSS that output unified position, pose, velocity and acceleration information. The LiDAR work with the integrated navigation module and leverage motion compensation to generate the local point cloud maps for runtime path planning.

Figure 2 shows the functionality and the hardware architecture of the platform. The computing unit (i.e., an Industrial Personal Computer or IPC) collects raw data from all sensors and creates individual nodes for real-time processing. Then the outputs are fused based on a certain scheme to generate the final multi-modal perception of the surrounding. The visualization of the results can also be checked by the driver on a monitor.

3   Methods

3.1   Generation of the Round-View Image

The general processing scheme from the raw images of the fish-eye cameras to a view that is compatible with the observation habits of a human is as follows.

- Transformation from the four fisheye views to a single rectified equirectangular view.
- Merging of adjacent views based on a weighted stitching algorithm.
- Transformation from the equirectangular view back to the 3D unit sphere for panorama roaming.

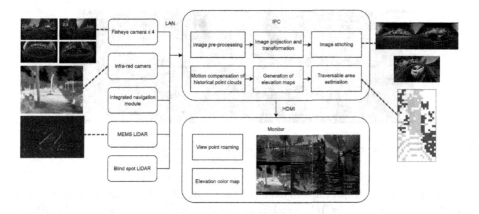

Fig. 2. System architecture of the mobile platform.

Stage 1. For cameras with distinct optic centers, the stitching of images requires the transformation of different 2D views planes to a shared unit sphere based on the spherical projection model and then to the equirectangular plane.

Given a desired resolution of the equirectangular view, pixels from each fisheye view can be inversely projected to avoid hollow areas.

Stage 2. The overlapping section brought by the wide FOV of the cameras can be utilized in image fusion and stitching. However, since the cameras have distinct optic centers, objects at different depths appear differently due to various disparity, making the merging of adjacent views unable to apply common template matching methods. The proposed method in this paper follows an optimized APAP [14] method and divides the overlapping section into multiple sub-grids, and apply SURF feature detection and Toet's grayscale difference analysis [15] to each grid and derives the corresponding homogeneous matrix. Eventually, the transformation between the two adjacent views is expressed by a set of homogeneous matrices.

The merging of the paired image sections is based on ROIs with weighted interests. Areas with low interests (such as the sky) are merged in a way that pixel intensities are correlated with pixel coordinates as follows,

$$I(i,j) = \beta I_1(i,j) + (1 - \beta)I_2(i,j) \tag{1}$$

where i, j indicate the row and column indices of the image respectively and $\beta \in [0, 1]$ represents the weight of merging.

Areas with more features are carefully merged where more weight is invested to the side with more obvious features, which ensures clear and sharp edges of the objects and mitigates motion blur and ghosting.

Stage 3. There are two common ways of visualizing the final stitched image. One of them is equirectangular view shown in Fig. 3. And in a digitalized image

Fig. 3. The equirectangular view of the stitched image by four cameras.

where the upper limits for longitude, W and latitude, H are defined as integers with a ratio of 2:1, the relationship between a point in latitude and longitude $P(\lambda_p, \varphi_p)$ and its corresponding coordinates on a 2D image plane, (u_p, v_p), is given by

$$\lambda_p = \frac{2\pi u_p}{W} \tag{2}$$

$$\varphi_p = \frac{\pi v_p}{H} \tag{3}$$

The other way of visualization is panorama roaming, which takes the driver's attention of objects into consideration. Viewing from a point outside the sphere with a FOV that is compatible with human habits makes it possible to display areas with more interest (such as road objects and terrain surfaces) on the monitor. Specifically, the distances from obstacles to ego vehicle can be easily determined in a top-down perspective, which is meaningful in driving safety. The establishment of such view is demonstrated in Fig. 4. Given a point $Q(u_q, v_q)$ on the 2D view plane with a resolution of w by h, d the distance between the center of the sphere O and the center of the view plane $(u_0, v_0) = (w/2, h/2)$, the conversion from P on the equirectangular plane to point Q on the target view plane is given by

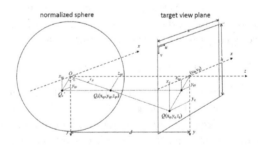

Fig. 4. Panorama roaming schematic from a spherical view to a 2D view plane.

$$\frac{u_q - u_0}{\sqrt{(u_q - u_0)^2 + (v_q - v_0)^2 + d^2}} = \sin\frac{\pi v_p}{H} \cdot \cos\frac{2\pi u_p}{W} \qquad (4)$$

$$\frac{v_q - v_0}{\sqrt{(u_q - u_0)^2 + (v_q - v_0)^2 + d^2}} = \sin\frac{\pi v_p}{H} \cdot \sin\frac{2\pi u_p}{W} \qquad (5)$$

$$\frac{d}{\sqrt{(u_q - u_0)^2 + (v_q - v_0)^2 + d^2}} = \cos\frac{\pi v_p}{H} \qquad (6)$$

where the horizontal FOV is determined by d, and the roaming is achieved through inverse projection with a rotational matrix $\mathbf{R}$ applied.

3.2 Ground Segmentation and Traversable Area Analysis

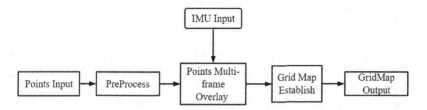

Fig. 5. Procedure of the generation of elevation maps.

Generation of Ground Elevation Map. The density of the points in a single scan of the LiDAR is in many cases inadequate for object clustering and terrain reconstruction. The quality of the point cloud map can be improved by overlaying multiple frames of point cloud based on motion compensation. It was achieved in this system by mapping 4 historical point cloud frames to the current frame applying extrinsic matrices derived from IMU data. It is given by

$$\mathbf{R}T_k = \mathbf{R}T_{k-1} \cdot \mathbf{R}T_{k-2} \cdot \mathbf{R}T_{k-3} \cdot \mathbf{R}T_{k-4} \qquad (7)$$

The resulting point cloud has denser points on targets and also highlights small objects that are undetectable in original point cloud.

Points after the motion compensation are then projected on to a grid map. Figure 6 shows the resulting elevation map in the FOV of the MEMS LiDAR with each grid cell color coded by its height, ranging from -0.5 m to 2 m and above.

Calculation of Frontal Slope Gradient. In addition to height information, the grid map also provides normal vector for each grid cell, determining its degree of terrain irregularities or slope. The complexity in an off-road environment requires a quicker response to incoming ramps and pits, especially at close proximity. This part of design involved the division of depth from 0 to 50 m into three zones, 0–15 m, 15–30 m, and 30–50 m. The overall degree of slope was computed individually for each zone and abrupt elevation changes were marked to highlight possible obstacles in the field of view, as shown in Fig. 6(c).

(a)   (b)   (c)

Fig. 6. (a) Photo of the platform traversing in an off-road terrain. (b) Elevation map color coded by height data from -0.5 to 2 m and above. (c) Visualization of frontal slope gradient. The overall slopes in different zones are displayed at the bottom (unit: degree). Grid cells with abrupt elevation changes are marked with their corresponding heights (unit: meter).

Traversable Area Analysis. Traversable area analysis is an important task for the robot to explore in off-road environment. It contributes greatly to efficient and reliable path planning.

Firstly, each point is labelled with its ring value. The angle of each point relative to the laser coordinate X-Y plane is computed. Then the points are classified based on different ring values. The formulation is shown below:

$$\theta = \arctan(z, \sqrt{x^2 + y^2}) \tag{8}$$

where, θ is the angle, (x, y, z) is the coordinates of the point.

Secondly, given that sharp corners, smooth corners and flat points are characteristic points in point clouds, the curvature of five points before and after the current point and current point are calculated. Actually, the curvature is the variance. The formulation is shown below:

$$c = \frac{1}{|S| \cdot \|X_{k,i}^L\|} \| \sum_{j \in S, j \neq i} (X_{k,i}^L - X_{k,j}^L)\| \tag{9}$$

In application, we calculate the curvature of the i-th point by

$$c = \frac{1}{2n} \sum_{j}^{(-n,n)} (X_{k,i}^L - X_{k,i-j}^L) \tag{10}$$

where c is the calculated curvature; n is the neighborhood of the point, it means n points are selected forward and n points backward along the array; k is the k-th points cloud; X is the coordinates of the points.

By setting certain threshold, the feature points can be obtained. The feature points in the raw point cloud were first extracted based on the information of historical frames and their normal vectors were computed. Then the points were colored based on the angle between the normal vectors and the horizontal plane to indicate the traversable areas and areas with a higher traversing risk, shown in Fig. 7. The pink points represent the traversable areas, the yellow ones represent the nearly traversable areas and red ones represent the impassable areas.

(a)   (b)

Fig. 7. Color coded point clouds based on normal vectors, showing frames from (a) self-collected data in a rural area and (b) the RELLIS-3D [16] dataset.

(a)   (b)   (c)

Fig. 8. (a)Photo of an off-road scene. (b)Misclassified areas (marked with white circles) in individual normal vector analysis. (c) Visualization of traversable area analysis.

For normal vector analysis on individual points, it is by nature not possible to eliminate all misclassified points on a roughly flat surface, as the areas marked with white circle shown in Fig. 8(b). They are generated due to measurement noise, uneven reflectivity of the object surface or being tiny objects (such as pebbles and leaves). This situation was properly mitigated through a specially adapted version of plane fitting method proposed in [17]. Misclassification was corrected by determining whether they were in the planes and whether they were too far from the fitted plane. Then a grid map was generated based on the analysis on a post-compensation point cloud, showing red areas, the non-passable areas, and yellow areas, the passable areas, in Fig. 8(c).

4   Experimental Results and Analysis

4.1   Integrated Visualization of Perception Results

Fig. 9. The integrated visualization of the real-time perception results on the monitor.

Figure 9 shows a screenshot of the real-time processing of the mobile platform moving on the sidewalk. It visualizes the infra-red camera view (bottom left), the color-coded elevation map (bottom middle), the status of the robot showing instantaneous IMU and GNSS data (bottom right), the view point roaming of the surrounding (top right, currently showing the top-down view) and front view (top left), which can be switched to any of the four views from the fisheye cameras or the equirectangular view as in Fig. 3.

4.2   Qualitative Analysis of Traversable Area Segmentation

The proposed traversable area estimation method was tested on the open-source dataset RELLIS-3D. This dataset was built using a 32-channel Velodyne LiDAR with a sampling frequency 10 Hz. The dataset classifies the points into 18 classes. The original categorization was remapped for our purpose into passable and non-passable categories based on their traversability.

Table 1 shows the quantitative results of our traversability estimation algorithm compared with several state-of-the-art deep-learning-based methods. The benchmark of deep-learning-based methods were set up on Nvidia GTX 1080Ti. Our method achieves the best performance in precision and is also compatible in the rest metrics. The processing time of our method is also the shortest, which is essential in robot platforms and requires no use of GPU.

Table 1. Evaluation of the traversable area estimation on RELLIS-3D

	Accuracy	Precision	Recall	IoU	Time consumed
Our method	78.9%	**92.2%**	74.6%	71.8%	**21 ms**
SalsaNext [18]	**87.0%**	86.3%	**88.0%**	**77.2%**	31 ms
KPConv [19]	84.0%	82.0%	87.0%	73.1%	911 ms
BEVNet [20]	82.4%	85.0%	74.7%	66.1%	58 ms

4.3 Time Efficiency

The IPC running on the platform was loaded with an Intel® i7-10700 processor and a 16 GB RAM with no graphic card to enhance the reliability in adverse environment and weather conditions. Table 2 shows the time consumed by each module in the statistical form. The processing pipeline includes the pre-processing of fisheye images, image projection and stitching, generation of elevation maps and traversable area estimation. The projection of the images from different coordinate system leveraged hash tables on regions with relatively fixed homogenous matrix as well as the asynchronized topic subscription in ROS to speed up the whole process.

Table 2. Time consumption of each module (unit: millisecond)

Thread	Module	Time
Perception with vision	Pre-processing of camera images	18.73
	Image projection and stitching	32.77
Perception with LiDAR	Generation of elevation maps	5.32
	Traversable area estimation	21.0
Total time consumed		77.82

5 Conclusion

We have designed and constructed a multi-modal perception flatform for unmanned driving in off-road environments. We stitched images from fisheye cameras and infra-red cameras, reprojected the pixels on a normalized plane to generate a dead-zone free view of the surroundings. We also proposed an approach that performed traversable area estimation and generated elevation maps using LiDAR and GNSS/IMU data. The system can detect abrupt terrain gradient changes and obstacles precisely even in the dark. The integrated perception system has been tested in various off-road and urban-road terrains and showed great potential in the processing of multi-modal data.

References

1. Badue, C. et al.:Self-driving cars: a survey. Expert Syst. Appl. **165**, 113816 (2021)
2. Sadat, A., Casas, S., Ren, M., Wu, X., Dhawan, P., Urtasun, R.: Perceive, predict, and plan: safe motion planning through interpretable semantic representations. In: Vedaldi, A., Bischof, H., Brox, T., Frahm, J.-M. (eds.) ECCV 2020. LNCS, vol. 12368, pp. 414–430. Springer, Cham (2020). https://doi.org/10.1007/978-3-030-58592-1_25
3. Wu, P., Chen, S., Metaxas, D.N.: MotionNet: joint perception and motion prediction for autonomous driving based on bird's eye view maps. In: 2020 IEEE/CVF Conference on Computer Vision and Pattern Recognition, pp. 11382–11392. (2020)
4. Zhuang, H., Zhou, X., Wang, C., Qian, Y.: Wavelet transform-based high-definition map construction from a panoramic camera. J. Shanghai Jiaotong Univ. (Sci.) **26**(5), 569–576 (2021). https://doi.org/10.1007/s12204-021-2346-9
5. Ososinski, M., et al.: Automatic driving on ill-defined roads: An adaptive, shape-constrained, color-based method. Journal of Field Robotics **32**(4), 504–533 (2015)
6. Overbye, T., Saripalli, S.: Fast local planning and mapping in unknown off-road terrain. In: 2020 IEEE International Conference on Robotics and Automation, pp. 5912–5918. IEEE (2020)
7. Mentasti, S., Matteucci, M.: Multi-layer occupancy grid mapping for autonomous vehicles navigation. In: 2019 AEIT International Conference of Electrical and Electronic Technologies for Automotive, pp. 1–6 (2019)
8. Payeur, P., Hebert, P., Laurendeau, D., Gosselin, C.: Probabilistic octree modeling of a 3D dynamic environment. In: Proceedings of International Conference on Robotics and Automation, 2, pp. 1289–1296 (1997)
9. Gao, B. et al.:Fine-grained off-road semantic segmentation and mapping via contrastive learning. In: 2021 IEEE/RSJ International Conference on Intelligent Robots and Systems, pp. 5950–5957 (2021)
10. Sharma, S., et al.: Semantic segmentation with transfer learning for off-road autonomous driving. Sensors **19**(11), 2577 (2019)
11. Valada, A., Oliveira, G.L., Brox, T., Burgard, W.: Deep multispectral semantic scene understanding of forested environments using multimodal fusion. In: Kulić, D., Nakamura, Y., Khatib, O., Venture, G. (eds.) ISER 2016. SPAR, vol. 1, pp. 465–477. Springer, Cham (2017). https://doi.org/10.1007/978-3-319-50115-4_41
12. Douillard, B. et al.: Hybrid elevation maps: 3D surface models for segmentation. In: 2010 IEEE/RSJ International Conference on Intelligent Robots and Systems, pp. 1532–1538. IEEE (2010)
13. Jaspers, H., Himmelsbach, M., Wuensche, H.: Multi-modal local terrain maps from vision and LiDAR. In: 2017 IEEE Intelligent Vehicles Symposium, pp. 1119–1125. IEEE (2017)
14. Zaragoza, J. et al.:As-projective-as-possible image stitching with moving DLT. In: Proceedings of the IEEE conference on computer vision and pattern recognition, pp. 2339–2346. (2013)
15. Toet, A., Walraven, J.: New false color mapping for image fusion. Opt. Eng. **35**(3), 650–658 (1996)
16. Peng, J. et al.: Rellis-3D dataset: Data, benchmarks and analysis. In: 2021 IEEE International Conference on Robotics And Automation, pp. 1110–1116. IEEE (2021)
17. Himmelsbach, M., Hundelshausen, F., Wuensche, H.: Fast segmentation of 3D point clouds for ground vehicles. In: 2010 IEEE Intelligent Vehicles Symposium, pp. 560–565. IEEE (2010)

18. Cortinhal, T., Tzelepis, G., Erdal Aksoy, E.: SalsaNext: fast, uncertainty-aware semantic segmentation of lidar point clouds. In: Bebis, G., et al. (eds.) ISVC 2020. LNCS, vol. 12510, pp. 207–222. Springer, Cham (2020). https://doi.org/10.1007/978-3-030-64559-5_16

19. Thomas, H. et al.:KPConv: flexible and deformable convolution for point clouds. In: Proceedings of the IEEE/CVF International Conference on Computer Vision, pp. 6411–6420 (2019)

20. Shaban, A. et al.: Semantic terrain classification for off-road autonomous driving. In: Conference on Robot Learning, pp. 619–629. PMLR (2022)

Author Index

F. Sun et al. (Eds.): ICCCS 2022, CCIS 1732, pp. 381–382, 2023.
https://doi.org/10.1007/978-981-99-2789-0

Printed in the United States
by Baker & Taylor Publisher Services